T0344676

INTERACTIVE DISPLAYS

Wiley-SID Series in Display Technology

Series Editor:
Anthony C. Lowe & Ian Sage

INTERACTIVE DISPLAYS
NATURAL HUMAN-INTERFACE TECHNOLOGIES

Edited by

Achintya K. Bhowmik
Intel Corporation, USA

This edition first published 2015
© 2015 by John Wiley & Sons, Ltd

Registered office: John Wiley & Sons, Ltd, The Atrium, Southern Gate, Chichester, West Sussex, PO19 8SQ, UK

Editorial offices: 9600 Garsington Road, Oxford, OX4 2DQ, UK
The Atrium, Southern Gate, Chichester, West Sussex, PO19 8SQ, UK
111 River Street, Hoboken, NJ 07030-5774, USA

For details of our global editorial offices, for customer services and for information about how to apply for permission to reuse the copyright material in this book please see our website at www.wiley.com/wiley-blackwell.

The right of the author to be identified as the author of this work has been asserted in accordance with the UK Copyright, Designs and Patents Act 1988.

All rights reserved. No part of this publication may be reproduced, stored in a retrieval system, or transmitted, in any form or by any means, electronic, mechanical, photocopying, recording or otherwise, except as permitted by the UK Copyright, Designs and Patents Act 1988, without the prior permission of the publisher.

Designations used by companies to distinguish their products are often claimed as trademarks. All brand names and product names used in this book are trade names, service marks, trademarks or registered trademarks of their respective owners. The publisher is not associated with any product or vendor mentioned in this book.

Limit of Liability/Disclaimer of Warranty: While the publisher and author(s) have used their best efforts in preparing this book, they make no representations or warranties with respect to the accuracy or completeness of the contents of this book and specifically disclaim any implied warranties of merchantability or fitness for a particular purpose. It is sold on the understanding that the publisher is not engaged in rendering professional services and neither the publisher nor the author shall be liable for damages arising herefrom. If professional advice or other expert assistance is required, the services of a competent professional should be sought.

Library of Congress Cataloging-in-Publication Data

Interactive displays / edited by Achintya K. Bhowmik.
        1 online resource.
    Includes bibliographical references and index.
    Description based on print version record and CIP data provided by publisher; resource not viewed.
    ISBN 978-1-118-70620-6 (ePub) – ISBN 978-1-118-70622-0 (Adobe PDF) – ISBN 978-1-118-63137-9
    (cloth) 1. Information display systems. 2. Interactive multimedia. 3. Human-computer interaction. I.
Bhowmik, Achintya K., editor of compilation.
    TK7882.I6
    621.3815′422–dc23

                                                                                        2014016292

A catalogue record for this book is available from the British Library.

ISBN: 978-1-118-63137-9

Set in 10/12pt, TimeLTStd by Laserwords Private Limited, Chennai, India
Printed and bound in Singapore by Markono Print Media Pte Ltd

1   2015

Contents

| **3** | **Voice in the User Interface** | **107** |

Andrew Breen, Hung H. Bui, Richard Crouch, Kevin Farrell, Friedrich Faubel,
Roberto Gemello, William F. Ganong III, Tim Haulick, Ronald M. Kaplan,
Charles L. Ortiz, Peter F. Patel-Schneider, Holger Quast, Adwait Ratnaparkhi,
Vlad Sejnoha, Jiaying Shen, Peter Stubley and Paul van Mulbregt

About the Author

Dr. Achintya K. Bhowmik is the general manager and chief technology officer of the perceptual computing group at Intel Corporation, where he leads the research, development, and productization of advanced computing solutions based on natural human-interaction technologies, including visual sensing, speech recognition, biosensing, immersive displays, multimodal user interfaces and applications.

Previously, he served as the chief of staff of the personal computer group, Intel's largest business unit. Prior to that, he led the advanced video and display technology group, responsible for developing power-performance optimized multimedia processing architecture and display technologies for Intel's computing products. His previous work included high-definition display systems based on all-digital liquid-crystal-on-silicon microdisplay technology, fast electro-optic modulation in organic molecular crystals, and integrated optoelectronic circuits for high-speed communication networks.

He has been an adjunct professor at the Kyung Hee University in Seoul, Korea, teaching graduate-level courses on display technologies and image processing. He has also taught mobile sensing and computer vision at the University of California, Santa Cruz Extension. He has had 150 publications, including two books and 27 issued patents. He is a winner of the SID Distinguished Paper award.

He is the regional vice president of SID Americas, chair of the SID display electronics program committee, and a senior member of the IEEE. He is an associate editor for the *Journal of the Society for Information Display*, and editor for two special volumes on *Advances in OLED Displays* and *Interactive Displays*. He is on the board of directors for OpenCV, the organization behind the open source computer vision library.

List of Contributors

Achintya K. Bhowmik, Intel Corporation, USA

Adwait Ratnaparkhi, Nuance Communications, Inc.

Andrew Breen, Nuance Communications, Inc.

Charles L. Ortiz, Nuance Communications, Inc.

Corey Pittman, University of Central Florida, Orlando, Florida

Daniël Van Nieuwenhove, SoftKinetic Sensors, Belgium

Friedrich Faubel, Nuance Communications, Inc.

Geoff Walker, Intel Corporation, Santa Clara, California

Heiko Drewes, LFE Medieninformatik, Ludwig-Maximilians-Universität, München, Germany

Holger Quast, Nuance Communications, Inc.

Hung H. Bui, Nuance Communications, Inc.

Jiaying Shen, Nuance Communications, Inc.

Jim Larimer, ImageMetrics, Half Moon Bay, California

Joseph J. LaViola Jr., University of Central Florida, Orlando, Florida

Kevin Farrell, Nuance Communications, Inc.

Lazaros Nalpantidis, Robotics, Vision and Machine Intelligence lab., Department of Mechanical and Manufacturing Engineering, Aalborg University Copenhagen, Denmark

Nikolaus Karpinsky, Department of Mechanical Engineering, Iowa State University, Ames, Iowa

Norman Poh, Department of Computing, University of Surrey, UK

Paul van Mulbregt, Nuance Communications, Inc.

Peter F. Patel-Schneider, Nuance Communications, Inc.

Peter Stubley, Nuance Communications, Inc.

Philip J. Bos, Kent State University, Kent, Ohio

Phillip A. Tresadern, University of Manchester, UK

Richard Crouch, Nuance Communications, Inc.

Rita Wong, University of Surrey, UK

Roberto Gemello, Nuance Communications, Inc.

Ronald M. Kaplan, Nuance Communications, Inc.

Sarah Buchanan, University of Central Florida, Orlando, Florida

Song Zhang, Department of Mechanical Engineering, Iowa State University, Ames, Iowa

Tim Haulick, Nuance Communications, Inc.

Tyler Bell, Department of Mechanical Engineering, Iowa State University, Ames, Iowa

Vlad Sejnoha, Nuance Communications, Inc.

William F. Ganong III, Nuance Communications, Inc.

Series Editor's Foreword

The interactive use of machines by humans goes back thousands of years. Possibly the first vending machine was invented by Philo of Byzantium in 220 BCE. By the insertion of a coin, it would deliver a measured quantity of soap to a washstand. This was a mechanical device with an escapement movement. It was an advanced machine, certainly state of the art, but whether it had a perceptible influence on the behavioural development of society is debatable.

Moving forward more than 2200 years, we find ourselves in a markedly different situation. Just 50 years ago the first capacitive touch screen was reported. It took a further 30 years to become sufficiently developed to enter the high-end commercial market in laptop computers, point-of-sale terminals and the suchlike and hand held consumer devices. After another decade the present explosion in touch enabled devices began, at least for hand held devices, its march to ubiquity.

That is the effective starting point for this latest addition to the Wiley-SID Series in Display Technology. Written by a highly qualified group of experts in their various fields, the book covers the major means of interaction: touch, voice and vision. The first two of these are discussed in a single chapter each and vision in the next five, which reflects on the disparate nature of available or soon to be available vision technologies. There follow two chapters which present different ways in which multiple methods can be used to develop multimodal interaction with displays. The book concludes by discussing methods by which 3D images presented on a display can be made more lifelike, more akin to the way in which we observe nature with our eyes, by retaining phase information which is lost with present display systems which preserve only intensity data.

From the above, the reader will conclude that this book offers a comprehensive review of current and emerging technologies. However, it does much more than this by discussing the social impact that advanced interaction will create. Much of this will be positive, but there are some potentially negative aspects. All of these are important issues for public awareness and are worthy of debate. On the positive side are ease of use, intuitive reasoning by the computer/telephone system to predict, discuss and manage complex outcomes from a simple voice command and the enablement of users with physical disabilities who are presently debarred from full use of today's products. On the negative side is intrusion by communication systems into many more aspects of a user's life than presently occur; the mobile phone has been skeptically described as a tracking device which enables its user to make phone calls, but future systems will probe more deeply into our behaviour patterns. Security systems that rely on biometric data to verify identity are in principle more secure than present chip and PIN technology, but if such systems are compromised, then the exposure will be more widespread.

The future will offer a richer human-machine interactive experience than many who will be its users can presently imagine. Past science fiction has already been overtaken by reality. There are issues which must be debated and resolved before some of the possibilities offered by technology will become acceptable by providers and users alike. This book provides the information to inform all aspects of such a debate. It will be an important book for scientists and technologists involved in the subject and for those who develop interactive products. It will also be important to a much wider readership which has an interest in or a need to know how interactive displays will influence future societal and interpersonal behaviour.

Anthony Lowe
Braishfield, UK, 2014

Preface

So, what are "interactive displays"? We define them to be the displays that not only show visual information on the screens, but also sense and understand human actions and receive direct user inputs. Interactive displays that can "feel" the touch of our fingers are already ubiquitous, especially on mobile devices and all-in-one computers. Now, the addition of human-like perceptual sensing and recognition technologies is allowing the development of a new class of interactive displays and systems that can also "see", "hear", and "understand" our actions in the three-dimensional space in front of and around them.

We use multisensory and multimodal interface schemes to comprehend the physical world surrounding us and to communicate with each other in our daily lives, seamlessly combining multiple interaction modalities such as touch, voice, gestures, facial expressions, and eye gaze. If we want human-device interactions to approach the richness of human-human interactions, then we must endow the devices with technologies to sense and understand such natural user inputs and activities. The addition of natural human-interfaces can thus bring lifelike experiences to human-device interactions.

The ways in which we interact with computers have already gone through a transformation in recent decades, with graphical user interfaces that use a mouse and keyboard as input devices replacing the old command-line interfaces that used text-based inputs. We are now witnessing the next revolution, with the advent of natural human interfaces where the user interacts with computing devices using touch, gesture, voice, etc. The ultimate goal of implementing a human-device interface scheme is to make the interaction experiences natural, intuitive and immersive for the user. While the limitations of the technologies at hand require designers and engineers to make compromises and to settle for a subset of these goals for specific product implementations, significant advances have been taking place in recent years towards realizing this objective.

This book presents an in-depth review of the technologies, applications, and trends in the rapidly emerging field of interactive displays, focusing on natural human interfaces. The first chapter of the book starts with a review of the basics of human sensing and perception processes, and an overview of human-device interactions utilizing natural interface technologies based on the sensing and inference of touch, voice, and vision. The subsequent chapters delve into the details of each of these input and interaction modalities, providing in-depth discussion on the fundamentals of the technologies and their applications in human interface schemes, as well as combinations of them, towards realizing multisensory and multimodal interactions. The book concludes with a chapter on the fundamental requirements, technology development status, and outlook towards realizing "true" 3D interactive displays that would provide lifelike immersive interaction experiences.

I would like to thank the series editor, Anthony Lowe, for recognizing the need for a book on interactive displays as part of the Wiley-SID Series on Display Technology. I am grateful to the experts from across the industry and academia who contributed to this book. I also appreciate the support of the production staff at John Wiley.

Finally, I dedicate this book to Shida, Rohan, and Ava, without whose encouragement and support I would not have been able to undertake and complete this project.

Achintya K. Bhowmik
Cupertino, California, USA
February 2014

List of Acronyms

Acronym	Meaning	Page (first appearance)
2D	Two Dimensional	10
3D	Three Dimensional	15
AAM	Active Appearance Model	317
AD	Absolute Differences	220
ADC	Analog-to-Digital Converter	40
AEC	Acoustic Echo Cancellation	127
AFE	Analog Front End	40
AG	Anti-glare	48
AI	Artificial Intelligence	109
AiO	All-in-One	28
AM	Acoustic Model	118
AMOLED	Active-Matrix OLED	88
AMR	Analog Multi-touch Resistive	31
ANN	Artificial Neural Networks	119
APR	Acoustic Pulse Recognition	32
ASIC	Application-Specific Integated Circuit	42
ASR	Automatic Speech Recognition	120
ASTM	American Society for Testing and Materials	246
ASW	Adaptive Support Weights	221
ATM	Automated Teller Machine	70
ATO	Antimony Tin Oxide	47
BCI	Brain-Computer Input	301
BE	Back-end	130
BOM	Bill of Materials	92
CAD	Computer-aided Design	287
CAT	Cluster Adaptive Training	134
CCD	Charge-coupled Device	37
CERN	The European Organization for Nuclear Research	35
CMOS	Complementary Metal−oxide−semiconductor	37

Acronym	Meaning	Page (first appearance)
COGAIN	Communication by Gaze Interaction	264
CPU	Central Processing Unit	75
CRF	Conditional Random Field	137
CRT	Cathode Ray Tube	28
CT	Computerized Tomography	298
DET	Detection Error Tradeoff	326
DFP	Digital Fringe Projection	182
DLP	Digital Light Processing	196
DMD	Digital Micro-mirror Device	196
DMR	Digital Multi-touch Resistive	31
DNN	Deep Neural Networks	119
DP	Dynamic Programming	225
DRS	Discourse Representation Structure	146
DRT	Discourse Representation Theory	146
DSI	Disparity Space Image	219
DSP	Digital Signal Processor	40
DST	Dispersive Signal Technology	32
DTW	Dynamic Time Warping	129
DWT	Digital Waveguide Touch	71
ECS	Eye Contact Sensors	257
EEG	Electroencephalography	301
EER	Equal Error Rate	318
EM	Electromagnetic	98
EMG	Electromyography	301
EMI	Electromagnetic Interference	48
EMMA	Extensible MultiModal Annotation	140
EOG	Electrooculography	256
EPD	Electronic Paper Display	98
ETRA	Eye-Tracking Research and Applications	264
FE	Front-end	130
FOV	Field of View	79
FPC	Flexible Printed Circuit	46
FPGA	Field-Programmable Gate Array	225
FSM	Finite State Machine	112
FST	Finite State Transducers	119
FTIR	Frustrated Total Internal Reflection	80
FTP	Fourier Transform Profilometry	188
G2P	Grapheme-to-Phoneme	132
GLMM	Generalized Linear Mixed Model	333
GMM	Gaussian Mixture Model	113
GPS	Global Positioning System	122
GPU	Graphics Processing Unit	95
GSI	Gesture and Speech Infrastructure	288
GUI	Graphical User Interface	263

Acronym	Meaning	Page (first appearance)
HCI	Human-Computer Interaction	251
HLDA	Heteroscedastic Linear Discriminant Analysis	118
HMI	Human-Machine Interface	98
HMM	Hidden Markov Model	112
HRI	Human-Robot Interaction	295
HSL	Hue-Saturation-Luminance	220
HTER	Half Total Error Rate	325
IEEE	Institute of Electrical and Electronics Engineers	168
IOB	Inside-Outside-Begins	136
IP	Intellectual Property	65
IPS	In-Plane Switching	89
IR	Infrared	32
ITO	Indium Tin Oxide	32
iVSM	Interpolated Voltage-Sensing Matrix	58
JFA	Joint Factor Analysis	129
LBP	Local Binary Patterns	323
LCD	Liquid Crystal Display	38
LCDM	Luminosity-Compensated Dissimilarity Measure	220
LDA	Linear Discriminant Analysis	316
LDPP	Learning Discriminative Projections and Prototypes	331
LED	Light Emitting Diode	69
LM	Language Model	111
LoG	Laplacian of Gaussian	227
LVCSR	Large Vocabulary Continuous Speech Recognition	120
MAGIC	Mouse and Gaze Input Cascaded	263
MAP	*maximum a posteriori*	117
MARS	Multi-touch Analog-resistive Sensor	59
MCE	Minimum Classification Error	118
ME	Modulation Efficiency	243
MFCC	Mel-Frequency Cepstral Coefficients	112
MLLR	Maximum Likelihood Linear Regression	117
MMIE	Maximum Mutual Information Estimation	118
MMSE	Minimum Mean-Square Error	124
MOBIO	Mobile Biometric	313
MPE	Minimum Phone Error	118
MRI	Magnetic Resonance Imaging	298
MTC	Member Team Committee	305
NAP	Nuisance Attribute Projection	129
NCC	Normalized Cross Correlation	220
NER	Named Entity Recognition	136
NFI	Near Field Imaging	39
NIR	Near Infrared	182
NIRS	Near-infrared Spectroscopy	302
NIST	National Institute of Standards and Technology	130

Acronym	Meaning	Page (first appearance)
NL	Natural Language	152
NLG	Natural Language Generation	108
NLMS	Normalized Least Mean Square	127
NLU	Natural Language Understanding	108
NRE	Non-recurring Engineering	92
OCA	Optically Clear Adhesive	43
ODM	Original Design Manufacturer	30
OEM	Original Equipment Manufacturer	30
OGS	One-Glass Solution	44
OLED	Organic Light Emitting Diode	28
OPWM	Optimal Pulse Width Modulation	207
OS	Operating System	29
OWL	Web Ontology Language	140
PC	Personal Computer	29
PCA	Principal Component Analysis	316
p-Cap	Projected Capacitive	28
PCB	Printed Circuit Board	71
PDA	Personal Digital Assistant	28
PDF	Probability Density Function	113
PET	Polyethylene Terephthalate	43
PET	Positron Emission Tomography	298
PIN	Positive-Intrinsic-Negative	79
PLP	Perceptual Linear Predictive analysis	118
POI	Point of Information	56
POMDP	Partially Observable Markov Decision Process	143
POS	Point of Sale	28
PPI	Pixels per Inch	93
PSD	Planar Scatter Detection	32
PSOLA	Pitch Synchronous Overlap and Add	132
PWM	Pulse Width Modulation	207
QA	Question Answering	151
QDA	Quadratic Discriminant Analysis	326
RAM	Random Access Memory	65
RASTA	Relative Spectral analysis	118
RBF	Radial Basis Function	332
RDF	Resource Description Framework	140
RDFS	RDF Schema	140
RDP	Reliability-based Dynamic Programming	225
RFI	Radio-Frequency Interference	95
RGB	Red-Green-Blue	220
RL	Reinforcement Learning	143
ROC	Receiver Operating Characteristic	326
RRFC	Reversing Ramped-Field Capacitive	36

Acronym	Meaning	Page (first appearance)
RSS	Rich Site Summary	131
s3D	Stereo-pair 3D	343
SAD	Sum of Absolute Differences	221
SAW	Surface Acoustic Wave	31
SAYS	Speak-As-You-Swipe	292
SBM	Squared Binary Method	207
SD	Squared Differences	220
SDK	Software Development Kit	294
SDRT	Segmented Discourse Representation Theory	151
SID	Society for Information Display	73
SLM	Statistical Language Model	156
SMS	Short Message Service	116
SNR	Signal-to-Noise Ratio	42
SPIE	The International Society for Optics and Photonics	191
SPWM	Sinusoidal Pulse Width Modulation	207
SRGS	Speech Recognition Grammar Specification	115
STFT	Short-term Fourier Transform	124
TCON	Timing Controller	92
TFA	Total Factor Analysis	129
TFT	Thin-Film Transistor	43
TIR	Total Internal Reflection	71
TOF	Time of Flight	233
TTS	Text-to-Speech Synthesis	130
UI	User Interface	107
UID	Unique Identity	315
US-VISIT	United States Visitor and Immigrant Status Indicator Technology	315
VA	Virtual Assistant	108
VR	Voice Recognition	108
VTLN	Vocal Tract Length Normalization	118
WFST	Weighted Finite State Transducers	119
WFT	Windowed Fourier Transform	188
WIMP	Windows, Icons, Menus, Pointer	287
WTA	Winner Takes All	219

1

Senses, Perception, and Natural Human-Interfaces for Interactive Displays

Achintya K. Bhowmik
Intel Corporation, USA

1.1 Introduction

Visual displays are now an integral part of a wide variety of electronic devices as the primary human interface to the computing, communications and entertainment systems which have become ubiquitous elements of our daily lives at home, work, or on the go. Whether it be the watches on our wrists, or the mobile phones that we are carrying everywhere with us in our pockets or purses, or the tablets that we are using for surfing the web and consuming multimedia content, or the laptop and desktop computers on which we are getting our work done, or the large-screen television sets at the center of our living rooms, or the presentation projectors in the business meetings, the visual display is the "face" of all these devices to us, the users.

The same applies to a plethora of vertical applications, such as the check-in kiosks at airports, check-out kiosks at retail stores, signages at shopping malls, public displays at museums – the list goes on and on. The wide array of applications and insatiable market demands have fuelled worldwide research and development to advance visual display technologies and products of all form factors in the past decades, ranging from mobile displays to large screens [1–5].

A quick glance at the market size helps us grasp just how pervasive visual displays have become in our lives. In the last five years, according to the display industry analysis firm IHS, the industry shipped nearly 17 billion flat-panel displays [6]. Also, to get a sense of the rate of adoption, the annual shipment of visual displays has grown more than 50% over this period.

In general, an electronic device performs three basic functions: receive instructions from the user, execute certain processing functions according to the instructions and information

Interactive Displays: Natural Human-Interface Technologies, First Edition. Edited by Achintya K. Bhowmik.
© 2015 John Wiley & Sons, Ltd. Published 2015 by John Wiley & Sons, Ltd.

received, and present output or results of the processing to the user. As an example, when the author was typing this chapter on his laptop computer, he used a keyboard and a mouse to input the information, the world-processing software application executed on the micro-processor translated the keystrokes and mouse taps into the desired text and format, and the liquid-crystal display of the laptop computer displayed the text on the screen as a real-time visual representation or output. Hence, the display subsystem in such devices already plays a critical role by presenting information to the user – and, until recently, barring some excep-tions, the majority of the electronic devices sported a display device whose sole function was just that – to display the visual information.

However, human-computer interaction and user interface paradigms have been undergoing a surge of innovation and rapid evolution in recent years. The ways we interact with com-puters had already gone through a transformation in the past decades, with the graphical user interfaces that use a mouse and keyboard as input devices replacing the old command-line interfaces that used text-based inputs. We are now witnessing the next revolution with the advent of more natural user interfaces, where the user interacts with the computing devices with touch, gesture, voice, facial expressions, eye gaze, and even thoughts!

Advanced sensors, systems, algorithms, and applications are being developed and demon-strated for natural and engaging interactions, where the computing devices understand the users' expressions and emotions in addition to the intent. These new interface technologies and the ensuing new class of applications present exciting opportunities for the display technology and the consumer electronics industry at large. With the integration of natural user interfaces, the display device morphs from a one-way interface device that merely shows visual con-tent, to a two-way interaction device that also directly receives user inputs and thus enables interactive applications and immersive experiences. The proliferation of touch-screens and touch-optimized interfaces and applications has already brought this transformation to mobile displays, and now the adoption of an extended array of natural interfaces promise to redefine the whole spectrum of displays and systems by making them more interactive.

This book presents a comprehensive treatment of the natural human interface technologies and applications that are enabling the emergence of highly interactive displays and systems. So, what are "interactive displays"? We define them to be the displays that not only show visual information on the screens, but also sense and understand human actions and receive direct user inputs. Equipped with human-like sensing and perception technologies, a "truly" interactive display will "feel" and detect our touch, "hear" and respond to our voice, "see" and recognize our faces and facial expressions, "understand" and interpret our gestural instructions conveyed by the movement of the hands and fingers or other body parts, and even "infer" our intent based on the context.

While these goals may seem rather ambitious, as the examples shown in Figure 1.1 illustrate, systems of various form factors and applications with natural user interaction technologies are already making a large impact in the market by offering easy and intuitive human interfaces. As reviewed throughout the book, significant advances are taking place in natural sensing and inference technologies as well as system integration and application developments, which are expected to bring about new frontiers in interactive displays.

The block diagram shown in Figure 1.2 depicts the generic functional modules and flow of an interactive display system. The interactions between the user and the display system are orchestrated by the interfaces, namely the input and the output blocks shown in the beginning and at the end. The input block consists of sensors that transform the physical stimuli resulting

Figure 1.1 Interactive displays and systems of a wide range of form factors and applications are already gaining a large foothold in the market, and some examples are shown above. The displays in many of these systems assume a new role of direct human-interface device, besides the traditional role of displaying visual information to the user. (See color figure in color plate section).

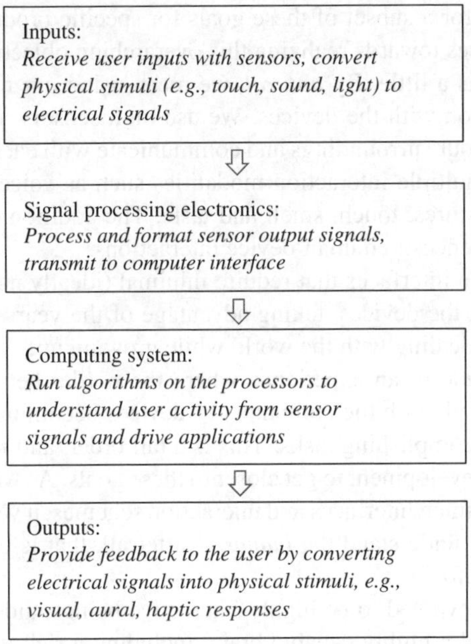

Inputs:
Receive user inputs with sensors, convert physical stimuli (e.g., touch, sound, light) to electrical signals

⇩

Signal processing electronics:
Process and format sensor output signals, transmit to computer interface

⇩

Computing system:
Run algorithms on the processors to understand user activity from sensor signals and drive applications

⇩

Outputs:
Provide feedback to the user by converting electrical signals into physical stimuli, e.g., visual, aural, haptic responses

Figure 1.2 Functional block diagram of an interactive display system. The input and the output blocks orchestrate the interactions between the user and the display, while the signal processing and computing functions facilitate these interactions.

from user inputs into electrical signals, while the output block performs the reverse function of providing system responses to user actions in the form of physical stimuli that the users can sense and perceive. The blocks in between perform the necessary signal processing and computing functions to facilitate these interactions.

In this chapter, we first review the basics of human sensing and perception, especially the mechanisms and processes that we deploy in our day-to-day interactions with the physical world. Building on this, we then provide an overview of human-computer interactions utilizing natural interface technologies based on touch-, voice-, and vision-based sensing and interactions, following a brief review of the most successful legacy interfaces. The subsequent chapters delve into the details of each of these input and interaction modalities, providing in-depth discussion on the fundamentals of the technologies and their applications in human interface schemes, as well as combinations of them towards realizing intuitive multisensory and multimodal interactions. The book concludes with a chapter on the fundamental requirements, technology development status, and outlook towards realizing "true" 3D interactive displays that would provide lifelike immersive interaction experiences.

1.2   Human Senses and Perception

We start with the assertion that the ultimate goal of implementing a human-device interface scheme is to make the interaction experience natural, intuitive, and immersive for the user. While the limitations of the technologies at hand require the designers and engineers to make compromises and settle for a subset of these goals for specific product implementations, we continue to make advances towards realizing this overarching objective.

Let us elaborate on this a little. By *natural*, we mean using our natural faculties for communication and interaction with the devices. We use multisensory and multimodal interface schemes to comprehend our surroundings and communicate with each other in our daily lives, seamlessly combining multiple interaction modalities such as voice, facial expressions, eye gaze, hand and body gestures, touch, smell and taste. The addition of natural interfaces can thus bring lifelike experiences to human-device interactions.

By *intuitive*, we refer to interfaces that require minimal (ideally no) training for the user to engage and interact with the devices, taking advantage of the years of training that we have already gone through in dealing with the world while growing up!

By *immersive*, we allude to an experience where the border between the real world and the virtual world is blurred, with the computers or devices becoming extensions of our body and brain to aid us in accomplishing tasks. This is a tall order, and it will require decades of continued research and development to get closer to these goals. As we endeavor to understand and implement lifelike human interfaces and interaction schemes, it would serve us well to take a look in the mirror, and understand the *human* – after all, that is the first word in the term *human-computer interaction*!

We, the humans, have evolved to be highly interactive beings, aided by a sophisticated set of perceptual sensors and a highly capable brain, including a rich visual perception system, aural and auditory capabilities, touch-sensitive skin and tactile perception, in addition to the chemical sensations of smell and taste via sensors embedded in the nose and the tongue. Well above half of the human brain is dedicated to processing perceptual signals which enable us to understand the space, beings, and objects around us, and interact in contextually-aware natural and intuitive ways.

Let us take a deeper look into three of our perceptual sensors and inference processes – specifically, the eyes and the visual perception process, the ears and the auditory perception process, the skin and the tactile perception process. One reason we limit our discussion to these three modalities of perception is that substantial parts of our interaction with the physical world utilize these mechanisms and, as we will see, imitations of these functions are implementable in electronics devices with the current state-of-the-art technologies in order to design and build highly interactive displays and systems. It would be nice also to implement smell and taste capabilities in human-computer interactions, but that will have to wait to be the subject matter of another book in the decades to come after further advances are made.

Let us consider the human interfaces with interactive display systems, such as those shown in Figure 1.1, from the neurophysiological perspective. The interaction process can be decomposed into three predominant phases: sensing, perception and recognition, and action. From the viewpoint of the human, the *sensing* process involves collecting the visual output of the display in the form of light waves entering the eyes, the audio output from the speakers in the form of sound waves entering the ears, and a feel of the display surface by touching with the fingertips. These perceptual sensors convert the physical stimuli into neural signals via transduction processes, which are relayed to the cerebral cortex of the brain, where the *perception* of "seeing", "hearing", and "touching" takes place, followed by recognition and understanding.

Based on the results of the perception and recognition processes, we then drive our body parts into *action*. For example, we converge and focus our eyes to the desired elements of the visual content on the display, guide our fingers to touch and activate specific content on the screen, tune our hearing attention to the audio output, sport an appropriate expression on our face, and even articulate a gesture with our fingers and hand.

First, let us review the visual perception process. We will only cover the salient aspects that are relevant to our subsequent discussion on implementing interactive displays, and refer the interested readers to more detailed accounts covered in other publications dedicated to human perception [7, 8]. The human eye is a marvel of evolution, in terms of the sheer complexity of its architecture, the efficacy of its function, and the central role that it plays in our perception of the world in conjunction with the visual cortex in the occipital lobe of the brain. As shown in Figure 1.3, the human eye resembles a camera in some key aspects of its construction, complete with a lens system to focus the incoming light from the scene onto the retina at the back of the eye, which contains the light sensor cells called the photoreceptors. There are two types of photoreceptors in the eye – the color-sensitive "cone" cells and the achromatic "rod" cells that convert the light into neural signals.

How about the *resolution* of this *camera* and the *bandwidth* for communicating with the *processor*? The retina contains a large number of photoreceptors – about eight million cones and about 120 million rods in each eye – and yet the visual system is cleverly designed to signal spatial and temporal changes in the scene, rather than the raw intensity levels detected by the photoreceptors to keep the bandwidth of communication between the eye and the brain down to practical levels.

The visual acuity is sharpest for central vision, when we point our eye to an object and the image is formed within a relatively small area around the optical axis of the eye. This is due to the largest concentration of the cone photoreceptors being in and around the fovea within a small region of the retina, which maps to a disproportionately large area in the visual cortex relative to other parts of the retina. Another important attribute of a camera is the dynamic range of light sensitivity; the human eye spans over ten orders of magnitude, well beyond the capability of our modern-day digital cameras.

While each of our eyes is an elegant *camera*, we have two of them. The human visual system consists of 3D and depth perception capability, with a binocular imaging scheme along with other visual cues such as motion parallax, occlusion, focus, etc., allowing us to navigate and interact with objects in the 3D space with apparent ease. Binocular vision has evolved to be prevalent among most biological systems; recent discovery of fossil records puts it as far back as more than 500 million years ago, to the arthropods of the Early Cambrian era [9]. The advent

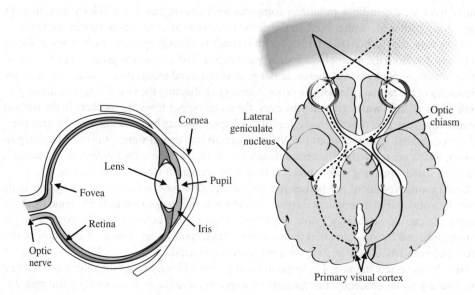

Figure 1.3 Left, anatomy of the human eye. Right, human visual system utilizes a binocular imaging scheme consisting of two eyes. The left visual field is sensed by the right sides of both eyes and is mapped to the right half of the primary receiving area in the visual cortex, whereas the right visual field follows a complementary pathway. The depth of the objects in scene are discerned from the binocular disparity, along with other visual cues such as motion parallax, occlusion, focus, etc.

of powerful visual systems is believed to be a trigger for the Cambrian Explosion of evolution [10]. The partially overlapping visual fields of the two laterally displaced eyes results in a "binocular disparity", where an object is observed to be laterally shifted by one of the eyes with respect to the other. As we will see later, the binocular disparity is inversely proportional to the distance of the object from the viewer.

With such a visual system aiding the perception of distance, the prey had a better chance of spotting an encroaching predator and escaping, and the predator had a better chance of triangulating the position of a prey and hunting. Binocular vision is thus believed to be a key enabler of evolutionary successes, and an attribute of the earliest mammals. In the modern era, we use our sophisticated binocular visual system to interact in the 3D world. Figure 1.3 also shows a simplified depiction of the sensory pathways connecting the eyes to the visual cortex.

Next, let us consider the key aspects of auditory perception, including the ears and the hearing process. Like the eye, the human ear has an elegant construction, with some astounding capabilities as the sound sensor. The ear, our natural *microphone*, is sensitive to over 12 orders of magnitude in sound intensity, and three orders of magnitude in sound frequency (20 Hz–20 kHz)!

As depicted in Figure 1.4, the ear flap, also called the *pinna*, directs the air waves carrying the sound signals into the auditory canal that includes the eardrum or the *tympanic membrane*. The pressure vibrations are amplified through the middle ear components, the *malleus*, *incus* and *stapes*, which are the smallest bones in the human body and are also referred to as the *hammer*, *anvil*, and *stirrup*, indicative of how they amplify and transmit the sound signals

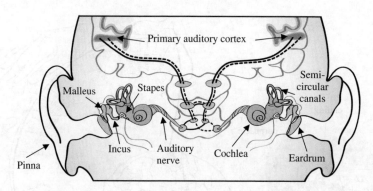

Figure 1.4 The anatomy of the human ear and the binaural construction, illustrating the simplified neural pathways between the cochlea in the inner ears and the auditory cortex in the temporal lobe of the brain. The binaural signals, along with frequency cues, are used for localization of the source for the sound signals.

to the inner ear section. Finally, the vibration waves of the sound are converted into neural signals via nerve impulses in the inner ear, more specifically by the hair cells at the converging spiral-shaped *cochlea*. These neural signals are subsequently dispatched to the auditory cortex of the brain, located in the temporal lobe, and are processed for perception.

As in the case of eyes, we also have two of these natural *microphones*, enabling a binaural perception scheme that, in addition to frequency cues, is capable of localizing the sources of sound accurately within the 3D space. While the binaural 3D perception, along with the extreme pressure sensitivity, was crucial to our evolutionary success, it is also crucial in helping us navigate and interact with the 3D physical world in our daily lives today. Figure 1.4 also shows the simplified neural pathway between the ears and the auditory cortex in the brain.

Finally, we turn to touch sensitivity and the tactile perception process. The perceptual process for our sensation to touch, also called the cutaneous sense, starts with the *mechanoreceptors* within the skin that pick up mechanical pressure in the corresponding skin area due to contact and activate neural responses. Figure 1.5 depicts the four principal types of mechanoreceptors. While the sensory organs for vision (the eyes) and for hearing (the ears) are located in the skull, with relatively short neurophysiological pathways to the cortex, the sensory organ for touch (the skin) covers the entire body. As a result, the signals from the touch receptors often have to travel long distances (e.g., from the fingertips to the head). The spinal cord serves as the "information highway" for the touch sensors, relaying the signals from the receptors to the *somatosensory* cortex in the parietal lobe, the part of the brain in the top region of the head that processes sensation to touch.

The seminal work of the neurosurgeon Wilder Penfield in the 1950s on tactile sensitivities has shown that adjacent parts of the human body map to adjacent areas of the cortex [11]. More interestingly, this mapping study has established the relative proportions of the somatosensory cortex that are dedicated to various body parts. This is presented by the concept of the *cortical homunculus*, as shown in Figure 1.5. Not to be mistaken as an arbitrary caricature, the homunculus depicts a scale model of the human body that represents the relative spaces that the corresponding body parts occupy on the somatosensory cortex. As the figure shows, the cortex dedicated to processing touch signals from the fingers far outweighs that dedicated to the entire arm and wrist, a vindication for the touch-screen based user-interface designers who

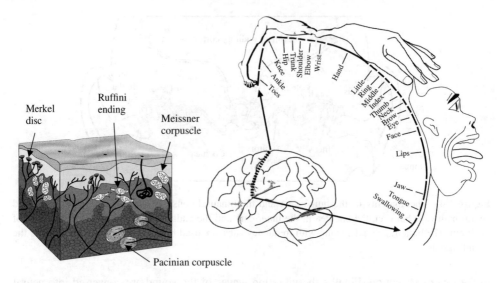

Figure 1.5 Left, anatomy of the human skin. The four principal types of mechanoreceptors that are shown convert the mechanical pressure of the tactile stimulus into neural signals. Right, the cortical homunculus, originally depicted by Wilder Penfield and refined by subsequent researchers, indicating the location and the relative portions of the somatosensory cortex dedicated to processing tactile signals from each part of the body. Source: http://www.intropsych.com/ch02_human_nervous_system /homunculus.html, reproduced with permission from R. Dewey.

assume an abundant use of the fingers for human interactions on touch-enabled interactive displays and devices!

So, as the preceding discussion reveals, there is a common theme to the neurophysiology of our perceptual processes. Our sensory systems are cleverly designed, with disproportionately large parts of the cerebral cortex assigned to the most important parts of the perceptual sensors – for example, the fovea of the eyes for central vision, hair cells in the cochlea of the ears for hearing, and the tips of the fingers for touch interaction. While we also possess other sensation mechanisms, we predominantly use vision, audition, and touch for interacting with the physical world surrounding us. Thus it is appropriate that in this book we focus on eyes, ears, and touch as the primary modalities of natural human interfaces to the interactive displays and devices.

In contrast to the biological systems, most of the computing and entertainment devices of today have rather rudimentary perceptual sensing and processing capabilities. For example, let us consider mobile phones, tablets, and laptop computers. They are typically "one-eyed" (with a single camera), like the Cyclops of Greek mythology. In addition, most of them are monaural (with a single microphone), and many are yet to get tactile sensitivity (touch screen), especially among the laptop computers.

However, this situation is expected to improve in the near future, with rapid technical advances on many fronts. Taking a page from nature's playbook, device architects and designers are now starting to consider adding "human-like" sensing and perception capabilities to computing and communications devices, to give them the abilities to "see", "hear", and "understand" human actions and instructions, and use these new capabilities to enable natural and intuitive human interactions. These developments promise to advance human-computer interactions beyond the confines of keyboard, mouse, joysticks, and remote controls, and allow the use of natural interactions built on touch, vision, and speech sensing and recognition technologies.

It is only when we attempt to implement such perceptual capabilities in machines that we realize how complex the task of sensation and perception is, despite the fact that we sense and perceive the world around us with surprising ease and casualness, at every moment of our conscious lives! In the next section, we review the key technologies for human interfaces with electronic devices – both the legacy technologies that have been widely adopted over the past decades, and also the recently emerging modalities of interactions with the displays and systems in natural and intuitive ways.

1.3 Human Interface Technologies

1.3.1 Legacy Input Devices

Before delving into the emerging natural interface technologies and the new class of applications and user experiences enabled by them, it behooves us to take a look at the history and reflect on the most successful user input technologies that have formed the backbone of human-machine interactions as we have come to know and practice them in our daily lives over the last few decades. A complete account of all human interface technologies and associated historical developments is not our intent, as it will be impossible to undertake such an endeavor within the practical limits of this chapter. A number of comprehensive reviews have been published in the past, and we point interested readers to them [13–15].

Here, we briefly cover a few of the seminal inventions and mainstream commercial implementations that have found adoption by the masses, and have defined human interactions with displays in major ways leading up to the modern era. As we look back, we appreciate the key reasons for their successes – simplicity in the technical implementations, utilizing the best of available technology ingredients within affordable price points and, above all, fulfilling user needs that helped enrich human lives and activities at the time of the particular invention and beyond.

First, the ubiquitous remote control device which has arguably defined our relationship and interactions with the television displays and shaped our content viewing behavior on it. While the concept of a remote control was described by Nikola Tesla as far back as 1898 [16], the first television display remote control was developed and commercialized by the Zenith Radio Corporation in 1950, who aptly named the device the "Lazy Bones" [17].

Televisions had been commercially available since the 1920s but they required people to walk up to them to manipulate the control knobs, while the natural content viewing position would have been sitting on a sofa in front of it. So, in retrospect, the environment was ripe for the invention of the remote control; the need was clear and, as we will see, technology ingredients were available. Zenith's "Lazy Bones" hand-held controller connected to the television with a long cable. While this solved a legitimate human need and allowed changing of television channels without having to get off the couch, the inadvertent tripping over the cable also pointed to the need to go wireless.

In came the "Flash-matic" in 1955, also from Zenith, which used a beam of light pointed to the sensors located in the four corners of the television set to control it remotely and wirelessly. The excitement on it can be gleaned from a magazine advertisement from the day that boasted, "*You have to see it to believe it!*" Despite the enthusiasm, the light-control mechanism did not perform well in bright rooms, with the ambient light occasionally changing the settings. Zenith addressed the problem by switching to using ultrasound as the means of remote communication in their next device, named the "*Space Command*". An advertisement from 1957, shown in Figure 1.6, eloquently described, "*New miracle way to tune TV from your easy chair by silent sound.*"

Modern remote controls have come a long way since then, with fashionable and sleek form factors, utilizing infrared light to control entertainment devices. In recent years they have increasingly featured motion-sensing and voice control capabilities.

Next, we turn to the invention of the computer mouse by Douglas Engelbart, in 1963, which marked the beginning of a new era of human-computer interactions. Prior to the invention and deployment of the mouse, inputs to the early computers were limited to text-based commands typed on a keyboard. Figure 1.7 shows the first mouse prototype built by Engelbart and Bill English. It consisted of two wheels which rotated in mutually perpendicular directions as the mouse was dragged on a surface, allowing relative tracking of the location of the mouse on a 2D plane [18]. The idea of such a device construction occurred to Engelbart in 1961 as he was sitting in a computer graphics conference, pondering how to make a system that would allow easy and efficient interactions with the graphical objects on the computer screen [19].

It is noteworthy that the mouse was just of one of many mechanisms for computer input that Engelbart and his team at the Stanford Research Institute experimented with, albeit the most successful one in retrospect. Even though modern versions have replaced the wheels with a

Figure 1.6 An advertisement for the Zenith "*Space Command*" remote control device from 1957. Source: www.tvhistory.tv, reproduced with permission.

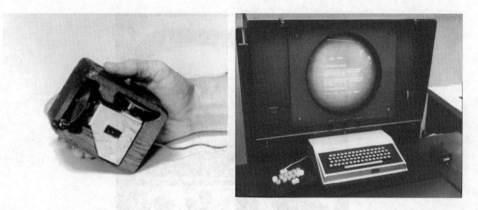

Figure 1.7 Left, the first computer mouse constructed by Douglas Engelbart and Bill English in 1963. Right, a computer workstation with an early mouse connected to it for user interactions Source: SRI International, reproduced with permission.

laser beam and replaced the cord with wireless transmission, we still refer to the device as a "mouse", which originally got its name from the resemblance to the namesake mammal due to the tail at the back to connect with the computer. The mouse, in conjunction with graphical user interfaces, made operation of computing systems easy for the user, and thus became ubiquitous along with the rapid proliferation of the personal computer over recent decades.

It is hard to overstate the scale of the impact of these "legacy" human interface devices on the proliferation of their respective host systems. As the television became the center-piece of entertainment in the homes around the world, and the personal computer became the primary vehicle for productivity and information, the remote control and the mouse became our unavoidable companions. However, despite their widespread use over decades, the human interfaces and interactions with these devices are very limited, and it is time to look forward. As we will see, recent advances spanning new sensor technologies, inference algorithms, comput-ing resources, and system integration have brought us to a point where we can start to realize natural human interfaces for interacting with electronic devices and systems. We now turn our discussion to user inputs with natural interfaces, based on touch, voice, vision, and multimodal interaction technologies.

1.3.2 Touch-based Interactions

The recent evolution of the display from merely presenting visual information to the user into an interactive interface device is largely due to the integration of touch-sensing capability into the display module, especially on the mobile devices. It took decades of development from the first report of capacitive touch screens, by Johnson in 1965 [20, 21], for the technology and usages to go mainstream with consumers worldwide.

The "touch display" constructed by Johnson, a cathode-ray-tube monitor with a capacitive touch overlay, is shown in Figure 1.8. Johnson, an engineer at the Royal Radar Establishment in England, developed the technology for use in air traffic control systems. The abstract of Johnson's paper stated, *"This device, the 'touch display', provides a very efficient coupling*

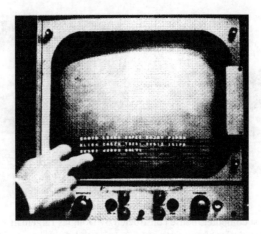

Figure 1.8 Image of the first "touch display" using the capacitive technique for sensing the contact location, reported by E. A. Johnson in 1965. Source: www.history-computer.com, reproduced with permission.

between man and machine." While Johnson's statement was specific to the device he constructed, it proved to be a profound foretelling of the devices and applications to come decades later, which would be adopted by the masses.

While the mouse and remote control devices enjoyed volume adoption for many years, interacting with the content on the display using these devices is necessarily an "indirect" manipulation experience. On the other hand, the introduction of touch-sensitive displays allowed people to "directly" interact with objects on the display by simply touching them, leading to easier and more intuitive experiences that required little or no training. In recent years, touch-screen technologies and their commercial deployment have been going through a tremendous phase of adoption and growth, thanks to the widespread proliferation of smartphones, tablets, ultrabook laptops, all-in-one desktop computers, and information kiosks of all form factors. In fact, the user experiences on these devices and systems have undergone a radical change due to the seamless integration of touch-screen technology and touch-friendly software interfaces, which have brought a new class of highly interactive applications to the users.

There are many different technical approaches to detecting touch inputs on a display device [22]. In Chapter 2, Walker provides an in-depth review of various touch technologies used in interactive displays. As described in detail in the chapter, methods for sensing contact location on the surface of a display can be predominantly categorized into capacitive, resistive, acoustic, or optical techniques.

In the capacitive approach, a layer is placed on or within the image display that stores electrical charges. In one of the implementations, called the mutual-capacitance method, when the user touches any location on the display, some part of the charge is transferred out into the user, resulting in a decrease in the electrical charge originally stored in that location. In another approach called the self-capacitance method, contact by a human body part increases the capacitance of a single electrode with respect to the ground. Electronic circuits designed to detect these changes within the touch-sensing layer are used to identify the location of the touch and provide this information to the software operating systems, applications and user interfaces.

In the resistive touch-sensing approach, when a user applies a mechanical force by touching a location on the display, two optically-transparent conductive surfaces, originally separated by a small space between them, come into contact at the touch location. The coordinates of this location are determined by voltage measurements and provided to the software for processing.

Acoustic and optical techniques involve measurements of the changes in ultrasonic waves and infrared optical waves, respectively, due to the touching of the surface of the display by the user. The specific system implementations widely vary across a wide range of products developed by a number of companies around the world.

The specific technologies covered in the chapter include projected capacitive, analog resistive, surface capacitive, surface acoustic wave, infrared, camera-based optical, in-cell integration, bending wave, force-sensing, planar scatter detection, vision-based, electromagnetic resonance, and combinations of these technologies. Walker presents the working principles of the various technologies, the associated system architectures and integration approaches, major advantages and disadvantages for each method, as well as the historical accounts, industry dynamics, and future outlook for the aforementioned touch-sensing technologies, including a discussion on various levels of integration of touch functionalities within the display. While many of the devices that are commercially available today use a touch-sensitive screen on top of the display module, recent developments and commercial products have demonstrated that touch-integrated displays without requiring separate touch-screen module to be attached on the panel provide a path to reducing the thickness, weight, integration complexity, and cost. Walker's chapter also covers such "embedded touch" technologies in detail.

The introduction of touch inputs to interactive displays and systems had a profound impact in the market. Let us take a quick look at the market size to gain appreciation for the footprint of touch-screen technology. The industry ships well over a billion units of touch-screens every year. While most of these are with mobile devices, touch screens are widely being adopted in devices of all form factors. It is just a matter of time before most displays will have touch input capabilities, especially the ones that require close-range user interactions.

1.3.3   Voice-based Interactions

Arguably, the most effective and prevalent form of communication and interaction between human beings is based on spoken language. To appreciate this, just do a "thought experiment" where you are a maverick world explorer and suddenly find yourself in a place among people who do not understand a thing you say and vice versa! Communicating with voice and speech has been a fundamental enabler of the modern human civilization and social interactions. Justifiably, there has been significant interest and efforts in the academic research community as well as the industry on developing human-computer interfaces utilizing voice input, processing, and output [23].

While we speak and understand what others are saying (mostly) without effort, making a computer understand spoken language by humans is a nontrivial task, and the effort has spanned a century. A quick look at Figure 1.9 gives us a glimpse of just a small part of the challenge, which shows a typical speech waveform recorded for the utterance of the phrase "*mining a year of speech*". When we speak sentences, we use non-uniform segmentations or temporal spacings between utterances and often generate an acoustic signal with no breaks where we perceive one, or with breaks in places that we do not perceive! We also often speak incomplete sentences in conversations, and mix meaningless utterances as "bridges" between

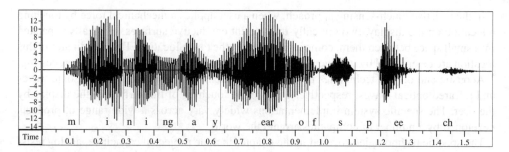

Figure 1.9 Typical speech waveform for the phrase "mining a year of speech", indicating the non-uniform temporal spacing between utterances and the nonintuitive breaks in the acoustic signal. The horizontal axis shows time in seconds and the vertical axis shows the signal strength in arbitrary units. Source: Reproduced from http://www.phon.ox.ac.uk/mining_speech/, with permission from John Coleman.

sentence segments. Broadly, the task of a speech recognition algorithm is to translate the spoken utterances into a series of text and, subsequently, to extract the meaning conveyed by them. A speech-based user interface uses these capabilities to establish a voice interaction scheme between the user and the device.

Since the early pioneering work in the 1920s and 1930s by Harvey Fletcher and Homer Dudley in Bell Laboratories on human speech modeling and synthesis [24, 25], the research on automatic speech recognition advanced steadily over decades, especially through the introduction of statistical algorithmic approaches in speech modeling in the 1980s, and the more recent advances in natural language understanding. In the 1968 epic science fiction movie *2001: A Space Odyssey*, screenwriters Stanley Kubrick and Arthur C. Clarke envisioned the HAL 9000, a computer that would be born in the 1990s and communicate with humans with fluent and natural conversational speech. While we are yet to achieve the marvelous feats that HAL was imagined to be capable of, recent developments in the domain of voice interfaces and interactions are starting to yield commercial success, with an increasing number of applications in mobile devices, consoles, and automotive usages.

In many instances, voice-based interfaces to computing devices result in simple and intuitive interactions between the humans and the devices. For example, a simple command in spoken language such as *"play the song [song title]"* would instantly pick out that specific song from among many songs stored in the device or a server and start to play it back. Similarly, a command such as *"post this picture to my Facebook page"* would instantly upload the picture that the user just captured using his or her smartphone, or one that he or she selected from the images stored in the device. *"Play the Wimbledon match saved from last night"* would find the tennis match from the media storage device and start playing it on the television screen. *"Give me the directions to SFO"* would display the map and driving directions to the San Francisco International Airport.

Accomplishing these tasks with traditional interfaces would require flipping through multiple command windows, typing in text inputs, and many clicks or taps. On the other hand, the same tasks completed via speech commands would be inherently easier and faster, if the devices are capable of understanding and interpreting voice commands and instructions with the required accuracy levels in the real-world usage conditions.

Voice recognition with natural language understanding and speech synthesis promise to expand significantly the usages for computing, communications, entertainment, and a plethora of other electronics devices and systems. It makes the use of devices possible when the hands and eyes are occupied in other tasks such as cooking, driving, shopping, gardening, jogging, etc., and potentially makes computing tasks feasible for many disabled individuals.

As we will discuss later, voice-based interactions are especially powerful when used along with other interface methods such as gesture or eye gaze tracking. In the future world of truly ubiquitous computing, when sensing and inference technologies become embedded into everything around us – from the things we wear to the things we carry to the places we live and work in – interactions based on spoken language will be crucially important. For now, from the viewpoint of interactive displays, voice interfaces seem poised to make the experiences of interacting with the content on displays of all form factors easier and more intuitive.

In Chapter 3, Breen *et al.* present an in-depth review of the fundamentals and developments in voice-based user interfaces. The authors provide a thorough discussion on the key elements of speech interfaces, including speech recognition, natural and conversational language understanding technologies, dialog management, speech synthesis, hardware and system architecture optimization for efficient speech processing, and applications to a broad array of interactive devices and systems.

1.3.4   Vision-based Interactions

As we discussed in Section 1.2, visual perception, more specifically the ability to see and understand the 3D environment, is a crucially important capability that enables us to navigate with ease in the physical world around us, and to interact with it and each other. 2D cameras and imaging applications are already ubiquitous part of computing and entertainment devices now – especially mobile phones, tablets, and laptop computers – and are increasingly appearing in all-in-one desktop computers and even in the high-end large-screen televisions.

Currently, the primary application for 2D cameras integrated within mobile devices is the capturing of digital still photographs and videos, while those in larger devices and displays are mostly used for video conferencing applications. Computer vision researchers have also developed 2D image processing algorithms that can detect, track, and recognize faces and facial expressions, understand poses and simple gestures [26–29].

A traditional 2D camera captures the projection of the 3D world onto the 2D image plane, and discards a wealth of other visual information in the 3D space in front of it. There have been significant research efforts on recovering the 3D information from single 2D images to understand human poses. Reconstructing 3D spatial information from 2D projection is an ill-posed problem with inherent ambiguities, and is a challenge even for fitting of known skeletal structures such as the human body, even though promising results have been reported for limited uses [30–32]. These approaches are in general computer intensive and often require manual user inputs, and hence are not suitable for applications that require real-time and unaided understanding of the 3D environment and general human gestural interactions in the 3D space.

In contrast, the 3D imaging pipeline in the human visual system captures and utilizes 3D visual information to enable efficient and robust cognition and interaction. Adding real-time 3D visual perception capabilities can enable truly interactive displays and systems that are capable of understanding the users and the rich natural user interactions. These include real-time 3D image sensing techniques to capture the 3D scene in front of the display; computer vision algorithms to understand the 3D images and real-time user activities in the

3D space; and appropriate user interfaces that couple human actions and instructions to the system intelligence and responses in intuitive ways.

Vision-based gesture recognition is a burgeoning field of research and development across the world, with rapidly advancing techniques reported by both academic and industrial labs, in conjunction with developments in classifying and implementing various levels of interactions based on the study of human motion behaviors [29, 33–35]. Chapter 4 provides an overview of vision-based interaction methods, including 3D sensing and gesture recognition techniques, status and outlook for implementing them in human-computer interaction applications.

Systems and applications based on 3D sensing devices have started to appear in the market, which offer richer and more robust interactive experiences than those implemented on the traditional 2D imaging methods [36, 37]. These early commercial successes promise to propel the adoption of 3D vision technologies into a wide array of devices and systems in the future, thereby making 3D user interactions pervasive. Adding real-time 3D imaging technology to electronic devices enables fine-grain user interactions and object manipulations in the 3D space in front of the display.

There are various methods for real-time 3D sensing, all of which generally output a depth-map for the scene in addition to a color image, allowing reconstruction of the 3D objects and scenes that were imaged. Three of the most prominent methods are: structured-light 3D sensing techniques, stereo-3D imaging, and time-of-flight range imaging techniques [37]. Chapters 5–7 delve into the details of each of these specific 3D imaging techniques that provide the foundation for implementing 3D interactive applications.

With real-time acquisition of 3D visual information using the techniques described above, rich human-computer interaction schemes can be implemented using 3D image recognition and inference techniques that enable interactions beyond touch screens. Figure 1.10 shows some examples that are naturally enabled by 3D gesture interactions in front of the display, rather than the use of traditional 2D inputs such as mouse or touch-screen [37]. The image on the left shows a scenario where the user is expected to reach out and "grab" a door knob, "turn" it, and "pull" it out of the plane of the display to "open" the door. The image on the right shows a "slingshot" application, where the user "pulls" a sling with the fingers, "directs" it in the 3D space, and "releases" it to hit and break the targeted elements of a 3D structure. These actions would clearly be quite difficult to implement with a mouse, keyboard, or a touch-screen, and would not be intuitive experiences for the user. However, implementations of 3D gesture interactions using real-time 3D image capture and 3D computer vision algorithms result in more natural and intuitive user experiences for this type of applications.

Figure 1.10 Examples of interactive applications and experiences enabled by real-time 3D sensing and inference technologies include manipulations of objects in the 3D space in front of the display [37].

Besides gesture interactions and object manipulations in the 3D space, real-time 3D imaging can also transform photography, video conferencing, remote collaboration, and video blogging applications. For example, the users can be segmented in the images more easily and accurately using the depth map generated by the 3D imaging devices, and subtracted from the background or placed in front of a custom background. This is illustrated in Figure 1.11.

While image processing techniques can be used to achieve this on traditional 2D images, 3D sensing devices allow cleaner segmentation and real-time applications utilizing the 3D scene information. For example, one could participate in a business meeting via a video conferencing application from the comfort of his or her home, but is shown on the screen in front of his or her office background instead!

Another category of applications that can be significantly enhanced is augmented reality, where rendered 3D graphical content is added to the captured image sequences. Beyond the traditional augmented reality applications that currently use 2D cameras, 3D imaging can augment video content with 3D models of objects and scenes for realistic visual representations, and allow the users to interact with the elements within the augmented world. Imagine applications that allow you to virtually try on clothes or jewelry in front of an interactive display equipped with a 3D imaging device, or virtually decorate your room to select appropriate furniture.

Beyond the tracking and recognition of hand and body gestures, there have also been significant developments in technologies that can detect eye gaze direction and determine where on the display the user is looking. Eye gaze plays an important role in inter-human interactions in our daily lives. Gaze is an important indicator of attention. For example, Figure 1.12 shows the regions of interest for a specific person while viewing an image.

Neurophysiological studies have shown the importance of gazes in consistent interaction with the physical world [38, 39]. While the primary function of the eye is to capture the visual information from the scene as part of the visual perception process, we also deploy eye gazes in close coordination with speech and gestures as we communicate and interact. As an example, as you say *"please give me that red ball"* and look at the red ball on the chair, it becomes

Figure 1.11 3D segmentation using a depth-sensing imaging device allows easy manipulation of the background. In this illustration, the boy is shown on the left in front of the original background, and on the right on a different background after post-processing. Note that an analysis of the inconsistent shading in the right image would reveal that the background is not original. Depth-sensing imaging devices enable real-time segmentation using the 3D scene information, for use in applications such as video conferencing or video blogging with a custom background. Source: Reproduced with permission from Sean McHugh. www.cambridgeincolour.com. (See color figure in color plate section).

Figure 1.12 An example of visual attention indicated by the directions of eye gazes. Left, an image shown to the viewer. Right, regions of interest within the image. Source: cambridgeincolour.com, reproduced with permission from Sean McHugh. (See color figure in color plate section).

obvious to the person looking at you that it is not the red ball lying on the floor that you are referring to. The person can comprehend this by simply following the direction of your eye gaze, even if you are not using your finger to point to the ball on the chair.

Researchers have long been fascinated by the prospect of incorporating this powerful interaction mechanism into user interfaces with computing systems, especially in conjunction with other interaction modalities. For example, one could simply look at an icon on the screen of the laptop computer and say "open it" or "launch it", rather than reaching out to tap on the touch screen or using the mouse to point and click, or even performing a hand gesture in free space. In Chapter 8, Drewes provides an in-depth review of eye gaze tracking technologies, systems, and applications, including the current limitations of deploying gaze tracking in human-computer interaction schemes and potential paths to mitigate these challenges.

1.3.5 Multimodal Interactions

Human perception and interactions are often multimodal – we use all our senses and combine the neural signals generated by them in order to understand the physical world and interact with it. For example, we use binaural audio signals along with frequency cues to locate the source of a sound, and then use the convergence and accommodation mechanisms in the eyes to point our binocular vision towards it, and bring the light rays emanated by the object to focus on our retina in order to see it at the same time that we hear its sound. Similarly, on other occasions, auditory perception may follow visual perception and add to it. For example, while strolling in a park, we may first see a bird, then pay attention to the sound of its chirping. In the real world, we use multiple modalities of interactions to communicate with each other. Based on intent and context, we use combinations of touch, gesture, voice, eye gaze, facial expressions and emotions to interact intuitively with fellow human beings.

In a seminal paper in 1976, aptly titled *Hearing Lips and Seeing Voices*, McGurk and MacDonald narrated their serendipitous discovery of an interaction between vision and hearing that has since come to be known as the "McGurk effect" [40]. This work demonstrated that when the auditory element of a sound uttered by a person is accompanied by the visual element of a different sound with a dubbing process, it leads to the perception of a third sound. The fusing of seeing and hearing in our perceptual processes is also evident in the ventriloquism effect, as well as in theaters where we get the illusion that the actors on the screen are speaking, even though the speaker systems are located elsewhere in the room. Neurophysiological evidence has established that the neural signals from one perceptual sensor can enhance, override, or modify those from the other as we comprehend our surroundings with multisensory perception. The interplay of different sensory areas within the human brain has been demonstrated, providing experimental evidence of connections between the receiving areas in the brain for vision, hearing, and touch [41].

Natural and intuitive human-computer interaction schemes must therefore be multimodal. The early results of combining speech recognition and position sensing, described by Bolt in his 1980 paper, demonstrated the feasibility of natural discourses between the human and the machine, such as "put that there", "make that a large blue diamond", "call that … the calendar", etc. [42]. Quek writes, *"For human-computer interaction to approach the level of transparency of interhuman discourse, we need to understand the phenomenology of conversational interaction and the kinds of extractable features that can aid its comprehension,"* and demonstrates the use of speech and gesticulation as coexpressive forms of communication [34].

In Chapter 9, LaViola *et al.* review multimodal perceptual interfaces in human-computer interaction, exploring the combination of various input modalities to constitute natural communications. The chapter examines the dominant interaction types, usability issues of various levels of multimodal integration, and methods to mitigate them towards achieving life-like natural interactions. Addressing human-factors aspects of multimodal interface schemes is crucial to the commercial success of new devices and systems incorporating multimodal interactive functionalities. Besides the input modalities covered in the preceding chapters, such as touch, gesture, speech, eye gaze, and facial expressions, this chapter also provides a discussion on integrating the emerging brain-computer interface technologies based on electroencephalography and muscle activity detection based on electromyography.

Science fiction authors have long fantasized about a future world where one controls computers, machines, and systems with brain waves, such that one simply thinks and things happen! While that future practically still remains to be realized, recent developments in brain interface technologies have demonstrated the ability to control and manipulate content on the display by signals emitted by the brain as a result of thoughts. Research and development continues in this domain, which promise to bring yet-to-be imagined interaction schemes and applications that will enrich future interactive displays and systems [43]. In Chapter 9, LaViola *et al.* consider integration of such brain-computer interfaces within the multimodal interaction schemes.

Besides multimodal interactions with the content on a display, the advances in face- and voice-based user recognition methods promise to replace the use of passwords for user identification by natural multimodal biometric authentication technologies. In our social interactions in daily lives, we use face, voice, and behavioral traits based human recognition schemes to establish the identity of people that we interact with. In contrast, a computer's ability to identify its users is still largely limited to passwords or tokens. As computing becomes pervasive and integrated in all aspects of our lives and the society, this is no longer going to be sufficient.

In Chapter 10, Poh *et al.* reviews multimodal biometrics, including technological design and usability issues, and recent developments in the field. As another example of multimodal perception, as we communicate with each other, we often use the cues present in facial expressions to understand verbal discourse. The same words, uttered with different expressions on the face, can mean very different things. Facial expressions can be voluntary to reinforce communications via specific gestures sported on the face, or involuntary indications of the inner feelings and emotions. The interpretations of one's facial expressions by other observers are dependent on context [44].

More than 150 years ago, Duchenne conducted experiments on human subjects to study how the movements of muscles produce various expressions on the face. As an example from his work, Figure 1.13 shows a number of facial expressions that were imparted on the faces via muscle contractions induced with electrical probes, and recorded using the newly available camera device [45]. The advent of digital cameras, advanced image processing techniques, and computational resources over the last decades have made more natural studies of facial expressions possible. More recently, 3D sensing and processing techniques are increasingly being used for more advanced and automated recognition of facial expressions. A brief review of the developments in vision-based facial expression recognition techniques is included in the discussion on visual sensing and gesture interactions in Chapter 4.

1.4   Towards "True" 3D Interactive Displays

Although visual displays have become ubiquitous and indispensable parts of our lives, the vast majority of them currently serve the primary role of displaying monocular (2D) visual information and are unable to reconstruct important visual cues that are salient to our 3D perception of the real world. In the recent years, stereo-3D displays have also started to gain traction in the market. The major focus for the stereo-3D displays that are commercially available currently has been on providing the stereopsis cue, where different images are presented to the left and right eyes of the viewer to create depth perception by using the binocular fusion process in our visual system. There are a number of books that cover the principles of various display technologies to reconstruct both 2D and 3D imagery [1–5].

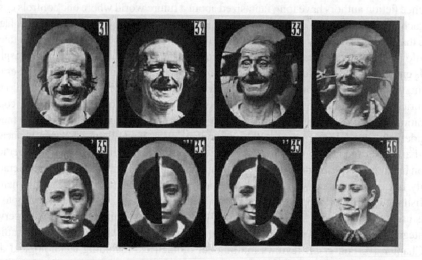

Figure 1.13   Original work of Duchenne, published in 1862, on inducing various facial expressions by muscular contractions activated by the application of electrical probes. The top figure shows the same expressions induced on both sides of the face, whereas the bottom figure shows different expressions induced on either sides of the face. Adapted from Duchenne [45]. Source: Reproduced with permission from www.zspace.com.

The ultimate goal is to construct "true" 3D interactive displays and systems that would provide life-like and immersive visual and interaction experiences to the user. The development of such displays requires careful examination of the human visual perception system and processes to reconstruct signals that are consistent with the visual cues that we utilize for sensing the 3D world in our day-to-day lives. So, how do we see and perceive the third dimension with our visual sensing and processing system?

Our 3D perception of the real world utilizes several important 3D visual cues, besides stereopsis. These include the motion parallax effect, where nearer objects appear to move faster across the view relative to objects that are further when the viewer moves; the convergence effect, where the eyes rotate inward or outward to converge on an object that is located nearer or further; the accommodation effect, where the shape of the lens in the eye is changed to focus on an object; the occlusion effect, where nearer objects partially hide objects that are further away; the linear perspective effect, where parallel lines converge at a distant point on the horizon; the texture gradient effect, where uniformly spaced objects appear to be more densely packed with distance; shadows that are consistent with 3D location of the objects and the lighting of the environment; and other cues arising from prior knowledge, such as familiar sizes, atmospheric blurs, etc. A number of these important 3D visual cues that contribute to our 3D perception are shown and explained in Figure 1.14.

It had been demonstrated that implementing motion parallax capability in a display to show unique views such that the images projected on the retina consistently varies with the head and eye movements of the viewer, in addition to stereopsis, provides a more lifelike visual experience. An example of such an implementation is a display from zSpace, illustrated

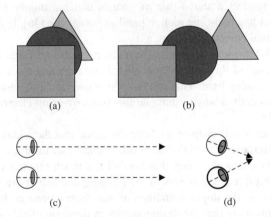

Figure 1.14 Depiction of some the salient visual cues that contribute to our 3D perception of the world, and which the future "true" 3D displays need to produce in order to provide immersive 3D visual experience. Top figures of the stacked shapes depict the binocular disparity with stereopsis cue: (a) is the image seen by the left eye of a viewer, and (b) is the image seen by the right eye. It also demonstrates the occlusion cue, as a single view is enough to indicate that the square shape is nearer to the viewer, whereas the triangle shape is further away. It also explains the motion parallax cue, since the square moves more across the visual field than the circle, which is further away, as the eye position moves from left to right. The bottom figures explain the convergence and accommodation cues: (c) depicts the case where the optical axes of the eyes are almost parallel to each other when seeing an object very far away; (d) eyes converge to see a nearer object, and the shape of the lens is accommodated to bring the image of the object to focus on the retina. Besides these, there are other 3D visual cues that are explained in the text.

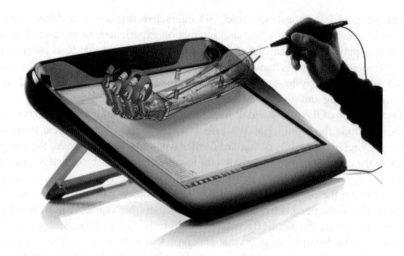

Figure 1.15   Illustration of an interactive display made by zSpace that combines motion parallax effect along with stereopsis. The system tracks the head movement of the user and displays a stereo-pair of images that are unique to the viewer's position. A stylus is used for real-time interaction with the virtual objects on the display. Source: www.zspace.com, reproduced with permission.

in Figure 1.15. This 3D display system tracks the user's head movements with infrared camera sensors and renders a stereo-pair of images that are unique to the viewer's head position, thereby providing real-time motion parallax visual cue [46]. It also includes a stylus to manipulate the virtual objects in 3D space.

Traditional stereo-3D displays also suffer from inconsistent focus cues due to the mismatch between the convergence of the eyes to focus visual attention and the accommodation of the lenses to focus the incoming light. This conflict was shown to cause visual fatigue in human observers [47], and recently a way to mitigate this issue by using electrically tunable lenses has been proposed [48].

In the earlier sections of this chapter, we have discussed that the additions of touch sensors and associated user interfaces, especially on mobile displays, are turning displays into two-way interaction devices. We have also seen that beyond the touch inputs that are limited to the 2D plane of the displays, recent progress in 3D imaging and activity recognition techniques are increasingly allowing the implementations of user interactions in the 3D space in front of the display. We anticipate that the developments in these two fields will be combined to architect systems that will have end-to-end 3D user interfaces, simultaneously showing 3D visual content and understanding 3D user inputs.

Clearly, using a 2D user input scheme such as touch or a mouse to manipulate the visual content shown on a 3D display does not provide a natural or intuitive user experience, and using a 3D interaction scheme would be more appropriate. For example, studies of subjective user experiences have demonstrated significant issues with in-plane touch-based user interaction on the 3D visual content displayed by stereo-3D displays [49], whereas users demonstrated a tendency to interact with 3D virtual objects with gestures [50]. Although there is increasing interest in the research and development of direct 3D interaction schemes for manipulating objects shown on 3D displays [51–54], further developments are still ahead towards realizing practical mainstream implementations.

The future "true" 3D interactive displays will need to present dynamic 3D visual content that provides consistent stereopsis, parallax, and focus cues to the viewers for depth perception in addition to the monocular 3D cues, and at the same time include 3D sensing and inference technologies to allow immersive interactions with the objects reconstructed in the 3D space. Chapter 11 presents an in-depth analysis of the requirements and progress towards achieving this goal. This chapter first details the fundamental requirements to reconstruct "true" 3D visual information using the principles of light field and the basics of human visual perception process. It then reviews the progress in technology developments towards realizing such "true" 3D visual displays that would provide all the important visual cues that are necessary for lifelike 3D perception. Finally, the integration of human interactions with the 3D visual content on such displays and systems is addressed, including human factors issues and potential solutions.

1.5 Summary

Visual information displays are now everywhere. They are the face of all kinds of computing, communications, entertainment, and other electronic devices and systems. As a result of the relentless drive in technology developments over the past few decades to achieve high visual quality, a wide range of sizes with thin profiles, low power consumption and affordable price points, displays have evolved to deliver stunning visual performance. Commercial shipments have skyrocketed due to the rapid consumer adoption of devices of all form factors, ranging from wearable devices to handheld smartphones to tablets and laptops to large-screen televisions and information kiosks. Now, the displays are entering a new era, morphing from a one-way visual information device to a two-way interactive device.

Human-computer interaction schemes are going through a transformation as well, with the traditional keyboard and mouse interfaces being replaced or augmented by direct and natural input methods utilizing touch, voice, and gestures. Thanks to the rapid proliferation of touch-sensing technologies, the user experiences on mobile devices have literally been transformed, with exciting new user interfaces and applications. The recent advances in 3D computer vision with real-time 3D image capture techniques and inference algorithms promise to take it one step further, by enabling rich human-computer interactions in the 3D space in front of the display. In addition, significant technology advances have been demonstrated on speech interfaces, eye gaze detection, and brain-computer interfaces. Multimodal interactions, based on multisensory perception and combination of the various input modalities, promise to make the interaction experience lifelike.

This book presents an in-depth review of the technologies, applications, and trends in the rapidly emerging field of interactive displays, focusing on natural and immersive user interfaces. In this chapter, we have provided an overview of the sensing and perception processes that are relevant to the understanding and development of interactive displays, and reviewed the natural human interface technologies. Just as the introduction of the mouse and the graphical user interface a few decades ago brought about numerous new applications on the computers, and the proliferation of the touch interface enabled another set of new applications on the smartphones and tablets over the past few years, the natural and intuitive user interfaces based on 3D and multimodal sensing and inference technologies are all set to usher in a new class of exciting and interactive applications on computing, communications, and entertainment devices. The future of the displays is interactive, and that future is already here!

References

1. Bhowmik, A.K., Bos, P.J., Li, Z. (Eds.) (2008). *Mobile Displays: Technology & Applications*. John Wiley & Sons, Ltd.
2. Lee, J.H., Liu, D.N., Wu, S.T. (2008). *Introduction to Flat Panel Displays*. John Wiley & Sons, Ltd.
3. Brennesholtz, M.S., Stupp, E.H. (2008). *Projection Displays*. John Wiley & Sons, Ltd.
4. Tsujimura, T. (2012). *OLED Displays*. John Wiley & Sons, Ltd.
5. Lueder, E. (2012). *3D Displays*. John Wiley & Sons, Ltd.
6. IHS Displays Report Portfolio, www.ihs.com 2009–2013.
7. Goldstein, EB. (2013). *Sensation and Perception*. Cengage Learning.
8. Snowden, R., Thompson P., Troscianko T. (2006). *Basic Vision: An Introduction to Visual Perception*. Oxford University Press.
9. Lee, M., Jago, J., García-Bellido, D.C., Edgecombe, G.D., Gehling, J.G., Paterson, J.R. (2011). Modern optics in exceptionally preserved eyes of Early Cambrian arthropods from Australia. *Nature* **474**, 631–634.
10. Parker, A. (2011). On the origin of optics. *Optics & Laser Technology* **43**, 323–329.
11. Penfield, W., Rasmussen, T. (1950). *The Cerebral Cortex of Man: A Clinical Study of Localization of Function*. MacMillan.
12. Dewey, R. *Psychology: An Introduction*. www.intropsych.com.
13. Jacko, J.A. (Ed) (2012). *Human-Computer Interaction Handbook: Fundamentals, Evolving Technologies, and Emerging Applications*. CRC Press.
14. Milner, N.P. (1988). A review of human performance and preferences with different input devices to computer systems. In: Jones DM, Winder R. (Eds). *Proceedings of the Fourth Conference of the British Computer Society on People and computers IV*, pp. 341–362. Cambridge University Press.
15. Buxton, W. *Human Input to Computer Systems: Theories, Techniques and Technology*. http://www.billbuxton.com/inputManuscript.html.
16. Tesla, N. (1898). *Method of and Apparatus for Controlling Mechanism of Moving Vessels or Vehicles*. US Patent 613,809.
17. A Brief History of the Remote Control. (1999). *DigiPoints: The Digital Knowledge Handbook* **3**, 4.
18. English, W.K., Engelbart, D.C., Berman, M.L. (1967). Display-Selection Techniques for Text Manipulation. *IEEE Transactions on Human Factors in Electronics* HFE-8, 5–15.
19. *Father of the Mouse*. http://dougengelbart.org/firsts/mouse.html.
20. Johnson, E.A. (1965). Touch Display – A novel input/output device for computers. *Electronics Letters* **1**, 219–220.
21. Johnson, E.A. (1967). Touch Displays: A Programmed Man-Machine Interface. *Ergonomics* **10**, 271–277.
22. Bhalla, M., Bhalla, A. (2010). Comparative study of various touchscreen technologies. *International Journal of Computer Applications* **6**(8), 975–8887.
23. Pieraccini, R., Rabiner L. (2012). *The Voice in the Machine: Building Computers That Understand Speech*. The MIT Press.
24. Fletcher, H. (1922). The Nature of Speech and its Interpretations. *Bell Systems Technology Journal* **1**, 129–144.
25. Dudley, H. (1939). The Vocoder. *Bell Labs Record* **17**, 122–126.
26. Yang, M., Kriegman, D., Ahuja, N. (2002). Detecting Faces in Images: A Survey. *IEEE Transactions on Pattern Analysis and Machine Intelligence* **24**(1), 34–58.
27. Tolba, A, El-Baz, A, El-Harby, A. (2006). Face Recognition: A Literature Review. *International Journal of Signal Processing* **2**(2), 88–103.
28. Fasel, B., Luttin, J. (2003). Automatic Facial Expression Analysis: a survey. *Pattern Recognition* **36**(1), 259–275.
29. Mitra, S., Acharya, T. (2007). Gesture Recognition: A Survey. *IEEE Transactions on Systems, Man, and Cybernetics – Part C: Applications and Reviews* **37**, 311–324.
30. Lee, H., Chen, Z. (1985). Determination of 3D human body postures from a single view. *Computer Vision, Graphics, and Image Processing* **30**, 148–168.
31. Guan, P., Weiss, A., Bălan, A., Black, M. (2009). Estimating Human Shape and Pose from a Single Image. *Int. Conf. on Computer Vision* 1381–1388.
32. Ramakrishna, V., Kanade, T., Sheikh, Y. (2012). Reconstructing 3D human pose from 2D image landmarks. *Proceedings of the European Conference on Computer Vision, Part IV* 573–586.
33. Pavlovic, V., Sharma, R., Huang, T. (1997). Visual interpretation of hand gestures for human-computer interaction: A review. *IEEE Trans. Pattern Analysis and Machine Intelligence* **19**(7), 677–695.

34. Quek, F., McNeill, D., Bryll, R., Duncan, S., Ma, X., Kirbas, C., McCullough, K.E., Ansari, R. (2002). Multimodal human discourse: gesture and speech. *ACM Transactions on Computer-Human Interaction* **9**, 171–193.
35. Wexelblat, A. (1995). An approach to natural gesture in virtual environments. *ACM Transactions on Computer-Human Interaction* **2**, 179–200.
36. Han, J., Shao, L., Xu, D., Shotton, J. (2013). Enhanced Computer Vision with Microsoft Kinect Sensor: A Review. *IEEE Transactions on Cybernetics* **43**(5), 1318–1334.
37. Bhowmik, A. (2013). Natural and Intuitive User Interfaces with Perceptual Computing Technologies. *Inf. Display* **29**, 6.
38. Pelphrey, K.A., Morris, J.P., McCarthy, G. (2005). Neural basis of eye gaze processing deficits in autism. *Brain* **128**, 1038–1048.
39. Klin, A. Jones, W., Schultz, R., Volkmar F. (2003). The enactive mind, or from actions to cognition: Lessons from autism. *Philosophical Transactions of the Royal Society of London B* **358**, 345–360.
40. McGurk, H., MacDonald, J. (1976). Hearing lips and seeing voices. *Nature* **264**, 5588.
41. Murray, M., Spierer, L. (2011). Multisensory integration: What you see is where you hear. *Current Biology* **21**, R229–R231.
42. Bolt, R. (1980). Put-That-There: Voice and Gesture at the Graphics Interface. *Proceedings of the 7th annual conference on Computer graphics and interactive techniques* 262–270.
43. Tan, D.S., Nijholt, A. (2010). *Brain-Computer Interfaces and Human-Computer Interaction.* Springer-Verlag.
44. Carroll, J., Russell, J. (1996). Do facial expressions signal specific emotions? Judging emotion from the face in context. *Journal of Personality and Social Psychology* **70**, 205–218.
45. Duchenne, G. (1862). The Mechanism of Human Physiognomy (Mecanisme de la physionomie Humaine).
46. Flynn, M., Tu, J. (2013). Stereoscopic Display System with Tracking and Integrated Motion Parallax. *International Display Workshops* **3**, D1–3.
47. Hoffman, D., Girshick, A., Akeley, K., Banks, M. (2008). Vergence-accommodation conflicts hinder visual performance and cause visual fatigue. *Journal of Vision* **8**(3), 1–30.
48. Bos, P.J., Bhowmik, A.K. (2011). Liquid-Crystal Technology Advances toward Future True 3-D Flat-Panel Displays. *Inf. Display* **27**, 6.
49. Pölönen, M., Järvenpää, T., Salmimaa, M. (2012). Interaction with an autostereoscopic touch screen: effect of occlusion on subjective experiences when pointing to targets in planes of different depths. *Journal of the Society for Information Display* **20**(8), 456–464.
50. Grossman, T., Wigdor, D., Balakrishnan, R. (2004). Multi-finger gestural interaction with 3d volumetric displays. *Proceedings of the 17th annual ACM symposium on User interface software and technology*, 61–70.
51. Alpaslan, Z., Sawchuk, E. (2004). Three-dimensional interaction with autostereoscopic displays. *Proceedings of SPIE Vol. 5291A, Stereoscopic Displays and Virtual Reality Systems XI* 227–236.
52. Blundell, B. (2011). *3D Displays and Spatial Interaction: Exploring the Science, Art, Evolution and Use of 3D Technologies.* Walker & Wood Ltd.
53. Bruder, G., Steinicke, F., Stuerzlinger, W. (2013). Effects of Visual Conflicts on 3D Selection Task Performance in Stereoscopic Display Environments. *Proceedings of IEEE Symposium on 3D User Interfaces 3DUI.* IEEE Press.
54. Zhang, J., Xu, X., Liu, J., Li, L., Wang, Q. (2013). Three-dimensional interaction and autostereoscopic display system using gesture recognition. *Journal of the Society for Information Display* **21**(5), 203–208.

2

Touch Sensing

Geoff Walker
Intel Corporation, Santa Clara, California

2.1 Introduction

This chapter is intended to provide a definitive reference on all touch technologies used in interactive displays. The objective of the chapter is to provide the reader with a substantial understanding of the operation, capabilities, advantages, disadvantages, limitations, and applications of 18 different types of touch technology. This understanding can be particularly helpful when touch is being combined with other input modalities in order to provide the user with more choices in how they can interact with a computer, as described in Chapters 1 and 9.

This chapter's scope is limited to touch technologies that operate by contact with a display screen, with the exception of stylus or finger "hover", which generally takes place within 1 cm of the touch surface. Touch on opaque (non-display) surfaces, proximity sensing, and in-air (3D) gestures are therefore excluded from this discussion. This chapter also does not include any substantial information on touchscreen manufacturing.

In covering the wide array of touch technologies and system integration details, we focus more on the breadth of information on multiple technologies rather than the depth of technical information on any one technology. In this chapter (and throughout the touch industry) the terms "touchscreen" and "touch-panel" are synonymous; the former is more commonly used in the West, while latter is more commonly used in Asia. Both terms refer identically to a touch module consisting of a touch sensor, a touch controller, and a computer interface.

This chapter categorizes all touch technologies into six basic types, each with a number of sub-types (indicated in parentheses and totaling 18), as follows: capacitive (2), resistive (3), acoustic (3), optical (5), embedded (4), and other (1). "Embedded", in this context, refers to touch capability that is fully integrated into a display by a display maker, as opposed to "discrete" touch capability which is added to a display by a touchscreen maker.

The touch industry is highly secretive; many of the 200+ companies in it are privately owned – even some very large ones. The result is that there are very few journal articles and no textbooks published by touch-technology inventors, developers or suppliers. That fact makes

Interactive Displays: Natural Human-Interface Technologies, First Edition. Edited by Achintya K. Bhowmik.
© 2015 John Wiley & Sons, Ltd. Published 2015 by John Wiley & Sons, Ltd.

this chapter somewhat different from many of the other chapters in this book. In particular, the references are much broader – they include websites, magazine and newsletter articles, white papers, patents, conference presentations, press releases, user guides, and even blogs. Also because of the lack of journal articles and textbooks on touch technologies, the historical information in this chapter tends to focus somewhat more on when a touch technology was first commercialized than when it was invented.

2.2   Introduction to Touch Technologies

Displays, from CRTs to OLEDs, have long been used as information output devices. It is only recently that displays have reached widespread use as interactive input devices, mainly due to the addition or integration of touch-sensing capability into displays. From the first report of capacitive touchscreens by Johnson in 1965 [1], it took almost three decades until touchscreens were sufficiently developed to be widely used in business-owned products (i.e., in commercial applications) such as point-of-sale (POS) terminals and airport check-in terminals [2]. The first widely visible use of touchscreens in consumer-owned products (i.e., in consumer applications) was in personal digital assistants (PDAs) in the mid-1990s, the first of which was the Apple Newton in 1993, followed by Jeff Hawkins' more-famous Palm Pilot in 1997.

The event that launched the current trend of "touch everywhere" was the introduction of the Apple iPhone in 2007 [3]. Apple's implementation of a previously obscure but exceptionally easy-to-use touch technology (projected capacitive), combined with an immersive user interface, ignited a touchscreen growth-wave that is still climbing (Figures 2.1 and 2.2) [4]. Apple's choice of touch technology also changed the dynamics of the touchscreen industry, causing the formerly dominant analog-resistive technology to be replaced rapidly by projected-capacitive (p-cap) technology (Figure 2.3) [4].

Microsoft Windows 7, launched in July 2009, marked the initial appearance of touch in consumer all-in-one (AiO) home desktop computers. The following year Apple launched the

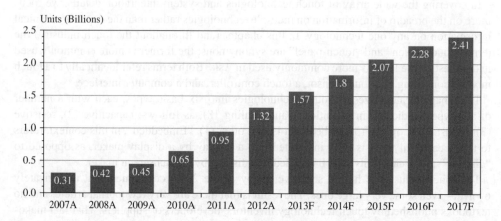

Figure 2.1 Touch module shipments in billions of units, 2007–2012 actual and 2013–2017 forecast. Source: Data from [4].

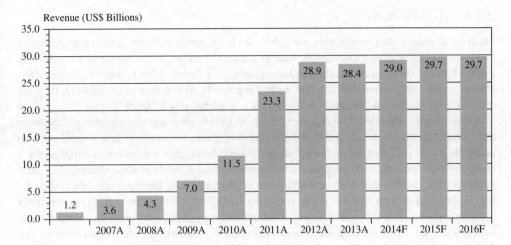

Figure 2.2 Touch module revenue in billions of US$, 2007–2013 actual and 2014–2017 forecast. Note: The author believes that the 103% increase between 2011 and 2012 is the result of a change in the market research analyst responsible for creating the report, rather than an actual change in market size of that magnitude. Source: Data from [4].

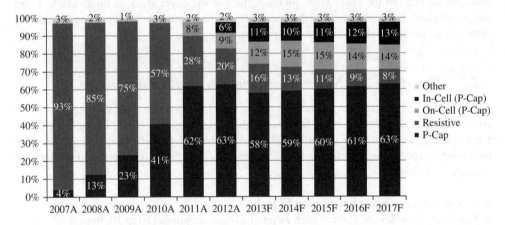

Figure 2.3 Touch technology by share of unit shipments, 2007–2012 actual and 2013–2017 forecast. Analog resistive was dominant with a 93% share in 2007; in 2012 that share had shrunk to 20%, with projected capacitive (in both discrete and embedded forms) taking 78%. DisplaySearch forecasts that in 2017 p-cap's share will increase to 90%, very close to the level of dominance that analog resistive had in 2007. Source: Data from [4]. (See color figure in color plate section).

iPad (April 2010), which was the first consumer device with 100% touch penetration (all tablets have touch capability, but not all mobile phones have touch capability). Microsoft Windows 8, launched in August 2012, marked the metamorphosis of Windows from a desktop-oriented operating system (OS) into a "touch-first" OS. As this book is being written in late 2013, the ramifications of that change are still reverberating throughout the personal computer (PC) and touchscreen industries.

2.2.1   Touchscreens

From the average user's perspective, a touchscreen is a computer display screen that can detect and respond to something touching it, such as a finger, a stylus, or the corner of a credit card. From a technology perspective, the display and whatever detects what is touching it are different electronic subsystems that must be treated separately. When they are combined, they are more properly called an "interactive display" or sometimes just a "touch display".

In this chapter, the term "touchscreen" is used to describe just the electronic subsystem that detects the user's touch and translates that touch into information that a computer can understand and use. For the majority of today's products, this subsystem is supplied by a company that specializes in touchscreens (usually called a touch-module maker). The integration of the touchscreen and the display can be performed by the touch-module maker, by the display-maker, by a systems integrator, or by an OEM/ODM device-maker (for consumer electronic devices, an Original Equipment Manufacturer (OEM) is usually the company that brands a device, while an Original Design Manufacturer (ODM) is usually the company that designs and/or manufactures a device).

Regardless of the touch technology, a touchscreen consists of three basic components: a sensor, a controller, and a computer interface. These are shown as a conceptual block-diagram in Figure 2.4. For all types of touchscreens except embedded, the sensor and the protective display cover-glass are one unit. The actual sensing elements may be underneath the cover glass, on the edge of the cover glass, on the surface of the cover glass, or immediately above the cover glass. For embedded touchscreens, the sensing elements are integrated inside the display, and the cover glass serves only a protective function.

2.2.2   Classifying Touch Technologies by Size and Application

Most touch technologies have specific applications in which they work best. Or, as Bill Buxton, one of the world's most famous touch researchers, has said, "Everything is best for something and worst for something else" [5]. Table 2.1 classifies each of the 18 touch technologies covered in this chapter in two dimensions. The first dimension is the device type and size, as follows:

- Mobile devices such as tablets (2 to 17 inches).
- Stationary commercial devices such as point-of-sale terminals (10 to 30 inches).
- Stationary consumer devices such as all-in-one home desktop computers (10 to 30 inches).
- All devices larger than 30 inches (generically called "large-format" displays).

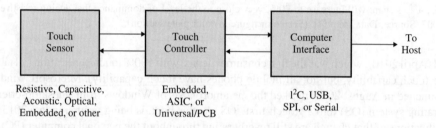

Figure 2.4   Conceptual block diagram of a touchscreen for any touch technology.

Table 2.1 In this table, the 18 touch technologies covered in this chapter are classified by device type, size, and status. Within each of four device-type and size categories, each touch technology is shown as in active current use (A), in current use but destined to disappear (L), just beginning to appear in the market (E), or not used (blank). The touch technology numbers (1 through 18) are used throughout this chapter. (See color table in color plate section)

Technology number	Touch technology	Mobile (2″–17″)	Stationary commercial (10′–30′)	Stationary consumer (10′–30′)	Large-format (>30″)
1	Projected Capacitive	A	A	A	A
2	Surface Capacitive		L		
3	Analog Resistive	A	A	L	
4	Digital Multi-touch Resistive (DMR)	E			
5	Analog Multi-touch Resistive (AMR)	E		L	
6	Surface Acoustic Wave (SAW)		A	L	A
7	Acoustic Pulse Recognition™ (Elo)		A		
8	Dispersive Signal Technology™ (3M)				L
9	Traditional Infrared		A		A
10	Multi-touch Infrared	E		E	E
11	Camera-based Optical			A	A
12	In-glass Optical Planar Scatter Detection™ (FlatFrog)			E	E
13	Vision-Based				E
14–16	Embedded (on-cell, in-cell, & hybrid)	A			
17	Embedded (light-sensing)				E
18	Force-sensing		E		

The second dimension is the status of each touch technology in each of the four device type and size categories, as follows:

- **A**: Active, meaning widely used and accepted
- **L**: Legacy, meaning in current use but destined to disappear
- **E**: Emerging, meaning just beginning to enter the market or application
- (blank): This touch technology is not used in this size/market category

Table 2.1 can be read both vertically and horizontally. For example, reading down the Mobile column, it can be seen that p-cap, analog (single-touch) resistive, and embedded touch technologies are the primary ones used in mobile devices (A); the multi-touch forms of resistive and infrared are not fully accepted yet (E); and no other touch technologies are used in mobile devices. Similarly, reading down the Stationary Commercial column, it can be seen (for example) that there are five touch technologies in active use (A) –more than any other column. This is because commercial applications have existed for more than 30 years and the associated touch technologies have evolved to become somewhat specialized for specific applications.

Reading across the Surface Capacitive row, it can be seen (for example) that the only application for the technology is in stationary commercial devices (e.g., casino gaming machines), and that it will eventually disappear (i.e., it is a legacy technology). Similarly, reading across the In-glass Optical row, it can be seen that the technology is very new (still emerging) and currently applies only in two categories – stationary consumer devices such as all-in-one home desktop computers, and larger devices such as point-of-information displays.

Note that In-glass Optical (technology number 12) is the only one whose name is currently in flux. The underlying technology was originally named Planar Scatter Detection™ by its developer (FlatFrog), but the market is beginning to use a more descriptive name.

2.2.3   Classifying Touch Technologies by Materials and Structure

Another method of classifying the 18 touch technologies described in this chapter is by the materials they include and by how they are structured. The most fundamental material in many touchscreens is a transparent conductor, typically indium tin oxide (ITO). Figure 2.5 splits the technologies into those that use ITO (eight on the left) and those that do not (ten on the right). Within the technologies that use ITO, the material is either in a continuous sheet or patterned.

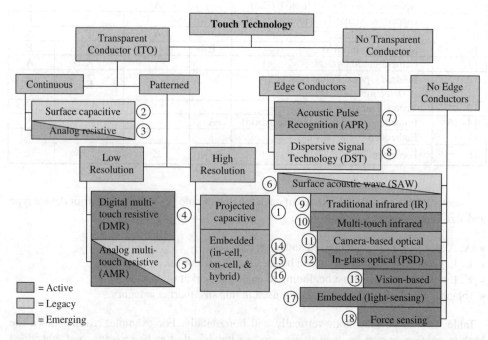

Figure 2.5   In this figure, the 18 touch technologies are classified first by their use of a transparent conductor material (typically ITO). Those technologies that use a transparent conductor are classified by whether the conductor is continuous or patterned; those that are patterned are further classified by the resolution of the patterning. Those technologies that do not use a transparent conductor are classified by their use of printed edge conductors. Note that the touch technologies are numbered to match Table 2.1. The three technologies shown in two colors (i.e., with diagonal lines) have multiple statuses in Table 2.1. (See color figure in color plate section).

If the ITO is patterned, it is either low-resolution (millimeters) or high-resolution (microns). Within the technologies that do not use ITO, the defining structure is whether they use printed edge conductors or not. Only two do so, while the remaining eight do not.

2.2.4 Classifying Touch Technologies by the Physical Quantity Being Measured

People new to the field often ask why there are so many different touch technologies. The simplest answer is that touch is an indirect measurement. If you are touching something, there is no direct method of determining exactly *where* you are touching, *how hard* you are touching, *with what* you are touching, or even the fact that it is *uniquely you* that is doing the touching. Instead, it is necessary to measure one or more of the physical quantities listed in Table 2.2 – and even then, it is still not possible to determine all four of the italicized characteristics with a single measurement. The phrase commonly used to describe this conundrum is "there is no perfect touch technology".

2.2.5 Classifying Touch Technologies by Their Sensing Capabilities

In a 2011 article on the breadth-depth dichotomy inherent in designing touch software for platforms of varying sensing capabilities [6], Daniel Wigdor at the University of Toronto proposed the taxonomy shown in Figure 2.6. In the left half of the figure, he suggested three types of sensed objects: touches (number of touches and users), stylus (level of support), and imagery (as found only in vision-based touch technology). In the right half of the figure, he suggested four types of sensed information: contacts (from different parts of the body or different users),

Table 2.2 The 18 touch technologies depend on measuring nine different physical quantities. In order to determine the combination of *where* you are touching, *how hard* you are touching, *with what* you are touching, and that it is *uniquely you* that is doing the touching, multiple measurements with multiple touch technologies are required

Technology number	Touch technology	Physical quantity being measured
1, 14–16	Projected Capacitive, Embedded (capacitive)	Capacitance
2	Surface Capacitive	Current
3–5	Resistive (all forms)	Voltage
6	Surface Acoustic Wave	Ultrasonic wave amplitude
7,8	Acoustic Pulse Recognition and Dispersive Signal Technology	Bending waves
9–12	Infrared, Camera-Based Optical, and In-Glass Optical (PSD)	Absence or reduction of light
13	Vision-based	Change in image composition
17	Embedded (light-sensing)	Presence of light
18	Force-sensing	Force

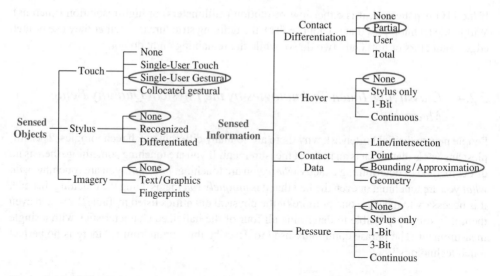

Figure 2.6   Daniel Wigdor's proposed taxonomy for classifying touch technologies based on the objects they can sense and the type of information they can deliver about those objects. The circles represent the characteristics of p-cap, as used in the iPhone, iPad, and similar mobile devices. Source: Wigdor, D., 2011. Reproduced with permission from the Society for Information Display.

hover (level of support), contact data (information about the contacting object), and pressure (level of support). Each of the 18 touch technologies described in this chapter can be characterized in terms of these seven capabilities. The figure can be used to characterize a touch technology such as p-cap, as used in the iPhone and iPad, as follows:

- Single-user only, with gesture recognition capabilities.
- Passive stylus recognized, but not differentiated from a finger.
- No imaging capability.
- No ability to differentiate contacts from different body parts or different users.
- No hover capability.
- Approximates the size of a contact with a bounding rectangle.
- No pressure-sensing capability.

2.2.6   The Future of Touch Technologies

Even though touch technologies have been in existence for half a century and in widespread use for a quarter of a century, they are not yet all mature and fully commoditized. One reason that has already been mentioned is that "there is no perfect touch technology." Another reason is that the touch industry is highly competitive and is strongly motivated by the creation of unique intellectual property. There is a continuous stream of startup companies developing improved forms of touch (e.g., new ways of measuring bending waves that better meet the needs of consumer applications, or entirely new ways of measuring touch-force), improvements in touch processing (e.g., reducing touch latency), developing new materials (e.g., new forms of printable transparent conductors that have the potential of cutting the cost of a p-cap touch module in half), and many other innovations. Some examples of where touch is going in the future include the following:

- Application in an ever-widening range of products from, one-inch wearable devices to 200-inch video walls.
- Full integration of touch and stylus.
- Embedded touch becoming lower cost and higher yield, competing even more aggressively with discrete touch.
- P-cap undergoing enhancements, such as the ability to use a #2 pencil as a touch object.
- Touch technologies that include more of the sensing capabilities listed in Figure 2.6.
- Seamless integration of on-screen (2D) touch with in-air (3D) gestures and other modes of interactivity.
- More use of touch on opaque objects, with the capability of making any object touch-aware.
- Lower cost, especially in touch for larger displays.
- Enhanced software-development environments that make it easier and faster to create a touch user-experience where "it just works" (i.e., touch that is robust, reliable, always performs in exactly the same way, and does not require the user to think about it).

2.3   History of Touch Technologies

Touch has a rich and varied history, which is not very surprising since there are six different basic touch technologies, each with several individual variations. Table 2.3 presents the history of touch from 1965 to the present (almost 50 years!). The table is divided into the six basic touch technologies. Within each technology, the companies or organizations that were most significant in the invention and/or commercialization of the technology are listed in chronological order, along with a brief explanation of their significance.

2.4   Capacitive Touch Technologies

2.4.1   Projected Capacitive (P-Cap) (#1)

The history of p-cap is less clear to a general audience than that of many other touch technologies, partially because the extreme significance of Apple's use of it in the first iPhone tends to obscure all previous uses of the technology. The basic concept of sensing touch by measuring a change in capacitance has been known since at least the 1960s. In fact, the first transparent touchscreen, invented in 1965 for use on air traffic system control terminals in the United Kingdom, used what today is known as mutual-capacitive p-cap [1]. The second published application of mutual-capacitance p-cap was in 1978; the application was in the control room of the CERN proton synchrotron [7].

Surface-capacitance touch technology (with an unpatterned touch-sensor) was commercialized by MicroTouch Systems in the mid-1980s [8]. During the mid-1990s, several US companies developed transparent capacitive touchscreens with patterned touch-sensors using indium tin oxide (ITO, the foundation of today's p-cap). Two of these were Dynapro Thin Films and MicroTouch Systems, both of which were later acquired by 3M (in 2000 and 2001 respectively)

Table 2.3 This table presents the most complete list that has ever been published of the significant companies in the history of the six basic types of touch technology. For each significant company, an explanation of the reason for their significance and the year in which it occurred are shown

Company	Significance	Year
Capacitive		
Royal Radar Establishment, UK (E.A. Johnson)	First published application of transparent touchscreen (mutual-capacitance p-cap on CRT air-traffic control terminals) [1]	1965
CERN (Bent Stumpe)	Second published application of mutual-capacitance p-cap (in the control room of the CERN proton synchrotron) [7]	1977
MicroTouch Systems (acquired by 3M Touch Systems in 2001)	First commercialization of surface-capacitive [8]	Mid-1980s
Dynapro Thin Films (acquired by 3M Touch Systems in 2000)	First commercialization of mutual-capacitive p-cap (renamed as Near-Field Imaging by 3M)	Mid-1990s
Zytronic (first license from Ronald Binstead, an inventor in the UK)	First commercialization of large-format self-capacitive p-cap [9]; first commercialization of large-format mutual-capacitive p-cap [9]	1998; 2012
Visual Planet (second license from Ronald Binstead)	Second commercialization of large-format self-capacitive p-cap [9]	2003
TouchKO (acquired by Wacom in 2007)	Invention of reversing ramped-field (surface) capacitive (RRFC™) [16]	2004
Apple	First use of mutual-capacitive p-cap in a consumer electronics product (the iPhone™) [3]	2007
Resistive		
Westinghouse Electric	Invention of first transparent analog-resistive touchscreen (3-wire); never commercialized [20]	1967
Sierracin/Intrex	First to commercialize digital (matrix) resistive; probably first to commercialize 4-wire analog resistive [21]	1973; 1979
Elographics (acquired by Raychem in 1986, who was acquired by Tyco Electronics in 1999, who spun off Elo Touch Solutions in 2012)	First to invent and commercialize 5-wire analog resistive [18] [19]	1977–1982
JazzMutant (renamed as Stantum in 2007)	First commercialization of digital multi-touch resistive; first commercial product with multi-touch interface [29]	2005
JTouch	First use of analog multi-touch resistive in a consumer-electronics product	2008

Table 2.3 (*continued*)

Company	Significance	Year
Acoustic		
Zenith (SAW IP acquired by Elographics/ Raychem in 1987, which was acquired by Tyco Electronics in 1999, which spun off Elo Touch Solutions in 2012)	Invention of surface acoustic wave (SAW, by Robert Adler, who also invented the ultrasonic TV remote-control "clicker" in 1956) [33] [34]	1985
SoundTouch Ltd. (acquired by Elo Touch Solutions in 2004)	Simultaneous invention of sampled bending-wave touch (by Tony Bick-Hardie, renamed as Acoustic Pulse Recognition™ (APR) in 2006 by Elo Touch Solutions) [40]	Early 2000s
Sensitive Object (acquired by Elo Touch Solutions in 2010)	Simultaneous invention of sampled bending-wave touch (originally named ReverSys™; incorporated into APR by Elo Touch Solutions) [41]	Early 2000s
NXT PLC (licensed to 3M Touch Systems in 2003)	First commercialization of real-time bending-wave touch (named Dispersive Signal Technology™ (DST) by 3M Touch Systems) [42]	2006
Optical		
University of Illinois	First use of infrared touch (in the PLATO IV computer-assisted instruction system) [43]	1972
Sperry Rand	Invention of first form of camera-based optical touch using CCDs	1979
Hewlett-Packard	First use of infrared touch in a commercial product (in the HP-150 microcomputer) [44]	1983
Carroll Touch (acquired by AMP in 1984, which was acquired by Tyco Electronics in 1999, which spun off Elo Touch Solutions in 2012)	Broadly commercialized use of infrared touch	1980–1999
Poa Sana	First invention of waveguide infrared touch [48]	1997–1999
SMART Technologies	Simultaneously commercialized current form of camera-based optical touch using CMOS	2003
NextWindow (acquired by SMART Technologies in 2010)	Simultaneously commercialized current form of camera-based optical touch using CMOS; supplied Hewlett-Packard, who created the first consumer computer with optical touch (the TouchSmart™ all-in-one (AiO))	2003; 2007
Perceptive Pixel (founded by Jeff Han and acquired by Microsoft in 2012)	First viral public exposure of multi-touch (in Jeff Han's TED conference videos of vision-based touch using projection)	2006

(*continued overleaf*)

Table 2.3 *(continued)*

Company	Significance	Year
Optical (continued)		
Microsoft	First commercialization of vision-based touch using projection (in Microsoft Surface v1.0)	2007
RPO (announced in 2007; assets liquidated in 2012)	Second invention of waveguide infrared touch [46] [47]	2007–2012
PQ Labs	First commercialization of multi-touch infrared [49]	2009
FlatFrog	Invention of in-glass optical touch (Planar Scatter Detection™, first shipped in 2012) [55]	2007
Baanto	First commercialization of PIN-diode optical touch [53] [54]	2011
MultiTouch	First commercialization of vision-based touch using integrated cameras (in MultiTaction®) [59]	2011
Samsung	First commercialization of vision-based touch using in-cell light-sensing (in SUR40, used in Microsoft Surface™ 2.0, which was renamed as Microsoft PixelSense™ in 2012) [60] [77]	2012
Embedded		
Planar	First to publish a technical paper on in-cell light sensing [66]	2003
Toshiba Matsushita Display	First to claim invention of in-cell light-sensing [74]	2003
Samsung	First commercial product with any form of in-cell embedded touch (in ST10 digital camera using pressed capacitive) [64] [65]; first commercial product with on-cell mutual-capacitive p-cap (on an OLED display in the S8500 Wave™ smartphone)	2009 2010
Sharp	First commercial product with in-cell light-sensing (in PC-NJ70A netbook)	2009
IDTI	Second commercial product with in-cell light-sensing (21.5-inch LCD monitor) [76]	2010
Sony (now part of Japan Display Inc.)	Invention of hybrid in-cell/on-cell mutual-capacitive p-cap (first used in Sony Xperia P™ and HTC EVO Design 4G™ smartphones) [71]	2012
Synaptics	Developed first touch controller for hybrid in-cell/on-cell capacitive (with Sony) [69] [70]	2012
Apple	First commercial product with in-cell mutual-capacitive p-cap (in the iPhone-5) [72]	2012

Table 2.3   (*continued*)

Company	Significance	Year
Other Touch Technologies		
IBM	First commercial product using force-sensing touch (In TouchSelect™ overlay)	1991
MyOrigo (sold to its management in 2004; restarted as F-Origin in Finland in 2005, which went bankrupt and was restarted in the USA in 2006, which was acquired by TPK in 2009)	Only current mature supplier of force-sensing touch in 2013 (excluding several emerging startups) [81]	2009
QSI (who spun off Vissumo in 2008, which went bankrupt in 2009); QSI was acquired by Beijer Electronics in 2010	First successful commercial product using force-sensing touch (in touch terminals used in toll booths) [79] [80]	2008

Note: References in this table are duplicates of references in the body of the chapter; there are no unique references that appear only in this table.

to form 3M Touch Systems. Dynapro Thin Films' p-cap touchscreen technology, renamed as "Near Field Imaging" (NFI), became 3M's first p-cap product in 2001. Also in 1994, an individual inventor in the UK named Ronald Peter Binstead developed a form of self-capacitance p-cap using micro-fine (25 micron) wire as the sensing electrode [9]. He licensed the technology to two UK companies: Zytronic in 1998, and Visual Planet in 2003; both are still selling it today.

P-cap remained a little-known niche technology until Apple used it in the first iPhone in 2007 [3]. Apple's engaging and immersive user-interface was an instant hit, causing most other smartphone manufacturers to begin adopting the technology. Over the next five years, p-cap set a new standard for the desirable characteristics of touch in the minds of more than one billion consumers, as follows:

- Multiple simultaneous touches ("multi-touch", initially used only for zoom).
- Extremely light touch with flick/swipe gestures (no pressure required).
- Flush touch-surface (bezel-less).
- Excellent optical performance (especially compared with analog resistive).
- Extremely smooth and fast scrolling.
- Reliable and durable touch surface.
- Fully integrated into the device user-experience so that using it is effortless and fun.

2.4.1.1   P-cap Fundamentals

There are two basic kinds of p-cap: self-capacitance and mutual capacitance. Both are illustrated in Figure 2.7. Self-capacitance (shown in Figure 2.7A) is based on measuring the capacitance of a *single* electrode with respect to ground. When a finger is near the electrode, the capacitance of the human body increases the self-capacitance of the electrode with respect to ground. In contrast, mutual capacitance (shown in Figure 2.7B) is based on measuring the capacitance between a *pair* of electrodes. When a finger is near the pair of electrodes, the capacitance of the human body to ground "steals" some of the charge between two electrodes, thus reducing the capacitance between the electrodes [10].

The key difference between self- and mutual capacitance is how the electrodes are measured, not the number or configuration of electrodes. Regardless of how they are configured, the electrodes in a self-capacitance touchscreen are measured individually, one at a time. For example, even if the electrodes are configured in a two-layer X-Y matrix, all the X electrodes are measured and then all the Y electrodes are measured, in sequence. If a single finger is touching the screen, the result is that the nearest X electrode and the nearest Y electrode will both be detected as having maximum capacitance. However, as shown in Figure 2.7C, if the screen is touched with two or more fingers that are diagonally separated, there will be multiple maximums on each axis, and "ghost" touch points will be detected as well as "real" touch points (ghost points are false touches positionally related to real touches).

Note that this disadvantage does not eliminate the possibility of using two-finger gestures on a self-capacitive touchscreen. Rather than using the ambiguous *location* of the reported points, software can use the *direction of movement* of the points. In this situation it does not matter that four points resulted from two touches; as long as pairs of points are moving toward or away from each other (for example), a zoom gesture can be recognized. For this reason, and because self-capacitance can be lower cost than mutual capacitance, the former is often used on lower-capability mobile products.

In contrast, in a mutual-capacitive touchscreen, each electrode *intersection* is measured individually. Generally, this is accomplished by driving a single X electrode, measuring the capacitance of each Y (intersecting) electrode, and then repeating the process until all the X electrodes have been driven. This measurement methodology allows the controller to unambiguously identify every touch point on the touchscreen. Because of its ability to correctly process multiple touch points (moving or not), mutual capacitance is used in preference to self-capacitance in most higher-capability mobile devices today.

2.4.1.2   P-cap Controllers

In every case, the measurement of electrode capacitance is accomplished by a touch controller. Figure 2.8 illustrates the basic structure of a controller for a mutual-capacitance touchscreen. A sensor driver excites each X electrode, one at a time. An analog front end (AFE) measures the capacitance at the intersection of each Y electrode and the excited X electrode; the analog values are converted to digital by an analog-to-digital converter (ADC). A digital signal processor (DSP) runs highly sophisticated algorithms to process the array of digital capacitance data and convert it into touch locations and areas, along with a variety of related processing such as "grip suppression" (the elimination of undesired touches near the edge of the screen resulting

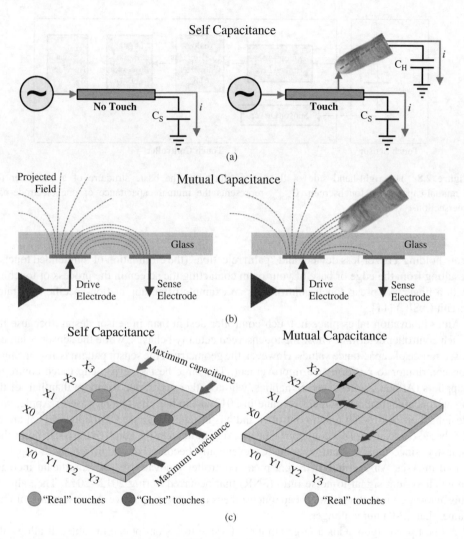

Figure 2.7   These figures illustrate the difference between self-capacitance and mutual-capacitance. **A:** Self-capacitance involves the capacitance of a single electrode to ground (C_S); a touching finger adds capacitance due to the human body capacity to ground (C_H). **B:** Mutual capacitance involves the capacitance between two electrodes; a touching finger reduces the capacitance between the electrodes by "stealing" some of the charge that is stored between the two electrodes. Adapted from 3M Touch Systems **C:** Self-capacitance measures the capacitance of each electrode on each axis, with the result that it allows "ghost points" because it cannot distinguish between multiple capacitance peaks on an axis (a total of 12 measurements on the 6×6 matrix shown). Mutual capacitance measures the capacitance of every electrode *intersection*, which allows detecting as many touches as there are intersections (36 in the 6×6 matrix shown). Source: Adapted from Atmel. (See color figure in color plate section).

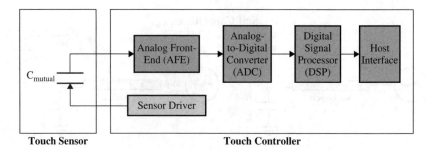

Figure 2.8 The right-hand side of this figure illustrates the basic structure of a controller for a mutual-capacitance touchscreen. C_{mutual} represents the mutual capacitance of one electrode-pair intersection.

from holding a bezel-less device) and "palm rejection" (the elimination of unintended touches resulting from the edge or base of your palm contacting the screen in the process of touching with a finger). A p-cap touch controller is an example of an application-specific integrated circuit (ASIC) [11].

More innovation takes place in touch-controller design than in sensor design, because the touch controller determines how the touchscreen actually behaves, while the sensor is largely just a source of capacitance values. However, the geometry of the sensor pattern is also an ongoing contributor to performance improvement. The three best known p-cap touch-controller suppliers (Atmel, Cypress, and Synaptics, who together accounted for more than half of the p-cap touch-controller shipment revenue in 2012, excluding the special case of Apple, who designs their own p-cap touch-controller and has it manufactured by Broadcom) [12] are all US-based companies. This could be taken as a sign of the relative youth of the p-cap controller industry, since most system-level ASICs eventually become commoditized, with suppliers based in Asia. An example of recent p-cap controller innovation is the significant increase in touch-system signal-to-noise ratio (SNR) that occurred during 2012–2013. The value of this innovation is that it allows p-cap touchscreens to support a passive stylus with a 2 mm tip, rather than just a human finger.

A fine-tipped stylus adds a large amount of value to a smartphone or tablet. It allows the user to *create* data (drawings, notes, etc.), rather than just *consume* media. In Asia, it is highly desirable to be able to write Kanji characters on a smartphone, and finger-writing is impractical because the tip of your finger obscures what you are writing. A fine-tipped stylus is also excellent as a pointing device for use with software that was not designed for touch (e.g., legacy Windows applications running on a Windows-8 tablet in desktop mode).

2.4.1.3 P-cap Sensors

A p-cap sensor is comprised of a set of transparent conductive electrodes that are used by the controller to determine touch locations. In self-capacitance touchscreens, transparent conductors are patterned into spatially separated electrodes in either a single layer or two layers. When the electrodes are in a single layer, each electrode represents a different touch coordinate pair and is connected individually to a controller. When the electrodes are in two layers, they are usually arranged in a layer of rows and a layer of columns. The intersection of each row and column represents unique touch coordinate pairs. However, as noted in the previous section, in self-capacitance, each electrode is measured individually rather than measuring each intersection with other electrodes, so the multi-touch capability of this configuration is limited.

In a mutual-capacitance touchscreen, there are almost always two sets of spatially separated electrodes. The two most common electrode arrangements are:

1. a rectilinear grid of rows and columns, spatially separated by an insulating layer or a film or glass substrate; and
2. an interlocking diamond pattern consisting of squares on a 45° angle, connected at two corners via a small bridge.

When the interlocking diamond pattern is used on two spatially separated layers, the processing of each layer is straightforward. However, this pattern is most commonly applied in a single *co-planar* layer in order to achieve the thinnest possible touchscreen. In this case, the bridges require additional processing steps in order to insulate them at crossover points.

Figure 2.9 illustrates the stack-up of a typical mutual-capacitance touchscreen. To keep this and all similar drawings in this chapter as easy to understand as possible, several simplifications have been made, as follows:

(a) The electrode pattern shown (rows 3 and 5) is a spatially separated rectilinear grid rather than the more common interlocking diamond; row 3 shows the end-views of the Y-electrodes, while row 5 shows a side view of one X-electrode.
(b) The common use of optically clear adhesive (OCA) has been omitted; for example, the space between rows 2 and 3 is typically filled with OCA.
(c) The touchscreen is shown using a glass substrate; in many mobile devices (particularly larger ones) the substrate is often two layers of polyethylene terephthalate (PET) film, one for each set of electrodes.
(d) All the layers below the thin-film transistor (TFT)-array glass in the LCD (e.g., bottom polarizer, brightness enhancement films, backlight, etc.) have been omitted.

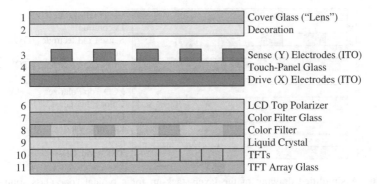

1	Cover Glass ("Lens")
2	Decoration
3	Sense (Y) Electrodes (ITO)
4	Touch-Panel Glass
5	Drive (X) Electrodes (ITO)
6	LCD Top Polarizer
7	Color Filter Glass
8	Color Filter
9	Liquid Crystal
10	TFTs
11	TFT Array Glass

Figure 2.9 A simplified drawing of the layer stack-up for a typical mutual-capacitive touchscreen is shown on top of a simplified drawing of an LCD display. The touch sensor substrate (row 4) is assumed to be a separate piece of glass with ITO on each side.

One of the key aspects of Figure 2.9 is that it shows the touchscreen adding a fourth sheet of glass to the stackup. All LCDs use two sheets of glass, and essentially every mobile device adds a third sheet of glass (or plastic) as a protective and decorative covering over the LCD. Adding a fourth sheet of glass is generally considered to be undesirable, since it adds weight, thickness and cost to the mobile device. There are two basic methods of eliminating the fourth sheet of glass:

1. The method used by the touchscreen industry, generally called "one-glass solution" (OGS), but also known by a variety of company-specific names, such as "sensor on lens".
2. The method used by the LCD industry, called "embedded touch". These methods are in direct competition.

Figure 2.10 illustrates the one-glass solution, in which the touchscreen electrodes are moved to the underside of the decorated cover-glass ("lens") [13]. In this solution, the touchscreen manufacturer either purchases the decorated cover-glass from an appropriate supplier, or vertically integrates and acquires the equipment and skills necessary to manufacture the cover-glass. The touchscreen manufacturer then builds the touch module (sensor plus controller), using the decorated cover-glass as a substrate and sells the entire assembly to a mobile device OEM/ODM (the touchscreen manufacturer may also buy the LCD specified by a device OEM/ODM and integrate the two together in order to provide more value to the OEM/ODM). The advantage of the one-glass solution to the end user is that the mobile device is lighter and thinner due to the elimination of the fourth piece of glass. The advantage of the one-glass solution to the touchscreen manufacturer is that they continue to derive revenue from the production of touch modules instead of forfeiting revenue to the LCD industry.

Figure 2.11 illustrates the simplest form of embedded touch (called "on-cell"), in which the fourth piece of glass is eliminated by moving the touchscreen electrodes to the top of the color-filter glass, underneath the LCD's top polarizer. Note that an on-cell configuration has exactly the same functionality as the p-cap configurations in Figure 2.9 and Figure 2.10; only the location of the electrodes is different. The advantage of the on-cell solution to the end user is exactly the same as the one-glass solution – the mobile device is lighter and thinner, due to

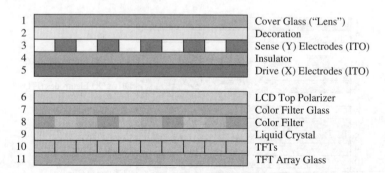

1	Cover Glass ("Lens")
2	Decoration
3	Sense (Y) Electrodes (ITO)
4	Insulator
5	Drive (X) Electrodes (ITO)
6	LCD Top Polarizer
7	Color Filter Glass
8	Color Filter
9	Liquid Crystal
10	TFTs
11	TFT Array Glass

Figure 2.10 A simplified drawing of the layer stack-up for a typical "one-glass solution" (OGS) mutual-capacitive touchscreen is shown on top of the same LCD as in Figure 2.9. The touch sensor is constructed on the underside of the display cover-glass (row 1). This configuration eliminates the separate sheet of glass used for the touch sensor in Figure 2.9.

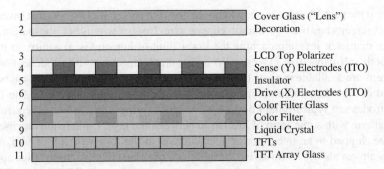

Figure 2.11 A simplified drawing of the layer stack-up for a typical on-cell embedded touchscreen is shown. In this configuration, the touch sensor is constructed on top of the color-filter glass (row 7), underneath the display's top polarizer (row 3). The functionality of this touchscreen is the same as in Figures 2.9 and 2.10; only the location of the sensor layers is different.

the elimination of the fourth piece of glass. The advantage of the on-cell solution to the LCD manufacturer is that it increases their revenue due to the added value of touch functionality (but the touchscreen manufacturer loses revenue).

One other factor in embedded touch's favor is that with the touch sensor integrated into the LCD, it makes sense to consider integrating the touch controller and the display driver together into a single ASIC, or at least establishing a direct connection between the two chips to enable cooperation. Manufacturing yield can be a significant issue with on-cell embedded touch, since depositing the electrodes on the top surface of the color-filter glass substantially increases the value of that one piece of glass; if either the color-filter deposition or the touch-electrode deposition is defective, both must be discarded. Product-line management also becomes a more complex issue for the LCD manufacturer, since instead of shipping (for example) 50 million identical LCDs to a device manufacturer, the LCD manufacturer may be required to ship 10 different models, each in quantity five million with a uniquely decorated cover glass.

It is commonly believed that integrating the touch capability into the LCD should automatically make touch better and cheaper. It should be apparent from the above that on-cell embedded touch is not necessarily an automatically better solution than the one-glass solution. There are factors to be considered on both sides, and some of those factors are more business and operational-related than technical. Competition between touch-module manufacturers and LCD manufacturers will remain a major factor in the progression of all forms of embedded touch. In their Q2-2013 touch forecast, DisplaySearch estimates that all forms of embedded touch will account for only 26% of all p-cap touchscreen unit shipments by 2017 [12].

2.4.1.4 P-cap Sensors Made with Wire Instead of ITO

In all of the above discussion about p-cap touch-sensors, the transparent conductors are assumed to be made of ITO. However, as touchscreens become larger, ITO becomes an increasingly difficult material to use due to (a) relatively high sheet resistivity 50–200 ohms per square ($\Omega/\square$), which slows down the touch-sensing process, and (b) lower manufacturing yields, which substantially increases the cost of the touchscreen. With rare exceptions, touchscreens made with ITO are never found above 32 inches.

For at least the past 10 years, the conductor material of choice for large-format (over 32-inch) p-cap touchscreens has been 10-micron copper wire. Copper wire is not transparent, but at 10 microns in diameter, it is approaching the lower limit of human visual acuity, so it is barely visible. Both self-capacitive (1–2 touch) and mutual-capacitive (10 + touches) wire-based touchscreens are available in sizes from 40 inches to over 100 inches. In most cases, the substrate used in these large-format touchscreens is a plastic film (usually PET). The 10-micron wire electrodes are typically laid down by a manufacturing robot in two layers following a zig-zag pattern, with some form of insulation between the layers. While the touchscreen sensor may be shipped to an integrator or device maker in the form of a roll of film, the film is eventually almost always laminated to the back side of a sheet of glass for the final application. One of the fundamental reasons for this is that the top surface of all LCDs is much too soft (with a pencil hardness of only 2H or 3H) to avoid accidental damage from touches.

2.4.1.5 P-cap Touch Modules

The term "touch module" applies only to discrete touchscreens, since embedded touchscreens are an integrated part of the display. The previous sections have focused on p-cap touch-controllers and sensors; these are the primary components of a p-cap touch module. The next most important component of a touch module is the flexible printed circuit (FPC) that connects the sensor to the controller. Typically, the touch controller is mounted (along with a few passive components) on the FPC, close to the sensor in order to minimize noise pick-up. The other end of the FPC is usually inserted into a connector on the device's main board.

A p-cap touch module is usually attached to the display in only one of two ways: "air bonding" or "direct bonding". In the first case, a double-sided adhesive gasket is applied around the top perimeter of the display, the touch module is aligned with the display, and the two pieces are pressed together. This leaves an air gap between the display and the touch sensor; the gap can range from 0.25 mm to more than 1 mm, depending on the size of the display. The advantage of air bonding is that the process is low cost with high yield; the disadvantages are that it results in additional reflecting surfaces that can significantly degrade image quality in high ambient light, and it makes the assembly slightly thicker.

In direct bonding, the entire top surface of the display is coated with an optically clear adhesive (dry or liquid). After alignment, the touch module is pressed against the display. There are multiple types of adhesive that are commonly used; the curing method depends on the type. The advantages of direct bonding are that the optical performance is always higher than air bonding, parallax error is usually reduced, and the durability of the top surface is increased (e.g., it can withstand a higher ball-drop specification). The disadvantage is that the process is high cost with low yield.

Most applications of p-cap touchscreens today are in consumer-owned devices. According to DisplaySearch, in 2013 more than 92% of these were smartphones and tablets [12]. The remaining consumer devices include notebook PCs, all-in-one desktop PCs, portable media players, portable game consoles, e-book readers, portable navigation devices, and cameras. Again according to DisplaySearch, in 2013 less than 1% of all p-cap touchscreens were in business-owned (commercial) devices [12]. The main reason for this disparity is that essentially the entire p-cap touch-module industry is focused on the 92% (smartphones and tablets). This means that the industry is generally uninterested in the smaller quantities and more demanding performance and environmental specifications that commercial applications always involve – even though businesses are willing to pay more per device.

In contrast, applications of wire-based large-format touchscreens (part of the 1%) usually involve interaction with the public. One of the best-known applications is "through store-window" retailing, where closed retailers engage potential customers outside of business hours by letting them interact with (for example) a product selection application through the store's windows. Other applications include in-store digital signage, public information kiosks, such as mall directories, and vending machines.

The advantages and disadvantages of p-cap touch technology are shown in Table 2.4 below.

2.4.2   Surface Capacitive (#2)

Surface capacitive was invented and commercialized by MicroTouch Systems, a company founded in 1982 and acquired by 3M in 2001 to form part of 3M Touch Systems. Because it lacked the easily damaged plastic top surface used in analog-resistive touchscreens (the dominant touch technology at that time), surface capacitive was perceived in the 1990s as the solution to more demanding touch applications.

As shown in Figure 2.12, a surface-capacitive touchscreen sensor consists of a uniform sheet of transparent conductor deposited on top of a sheet of glass. The most common transparent conductor used in surface conductive touchscreens is antimony tin oxide (ATO) deposited to produce a highly uniform sheet resistivity of $1,200-2,000 \ \Omega/\square$. Lower-cost versions of the technology sometimes use ITO or pyrolytic tin oxide (TO) instead, with lower sheet-resistivity. The conductive coating is surrounded by, and connected to, linearization-pattern electrodes made of screen-printed silver frit that are, in turn, connected to the touchscreen flex tail (the

Table 2.4   Advantages and disadvantages of p-cap touch technology

P-cap advantages	P-cap disadvantages
Unlimited, robust multi-touch (if properly implemented)	High cost (mostly in the sensor; ITO-replacement materials will help reduce the cost)
Extremely light touch (zero pressure)	Touch-object must have some capacitance to ground (or be an active stylus)
Enables flush touch-surface (bezel-less)	Challenging to integrate (requires substantial "parameter tuning" for each new implementation)
Very good optical performance (especially compared with analog resistive)	Difficult to scale above 32 inches with invisible (ITO) electrodes
Extremely smooth and fast scrolling (if properly implemented)	No absolute pressure-sensing; only relative finger-contact area
Durable touch surface unaffected by scratches and many surface contaminants (protected sensor)	
Can be made to work with running water on the surface (but rarely done in 2013 consumer products!)	
Can be made to work through extremely thick glass (≈ 20 mm)	
Can be sealed to NEMA-4 or IP65 standards	

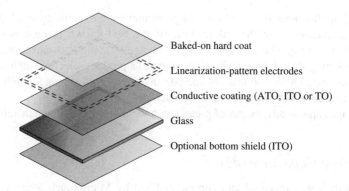

Baked-on hard coat

Linearization-pattern electrodes

Conductive coating (ATO, ITO or TO)

Glass

Optional bottom shield (ITO)

Figure 2.12 The construction of a typical surface-capacitive touch sensor is shown. The touch sensor consists of a uniform transparent-conductive coating on top of a sheet of glass. The conductive coating, surrounded by linearization-pattern electrodes, is protected by a baked-on dielectric hard coat.

purpose of the linearization electrodes is to correct the inherent non-linearity (bow) in the electric field associated with the properties of electrical currents flowing between corners of a rectangular conductive surface). The conductive coating and linearization electrodes are covered by a baked-on, transparent, dielectric hard coat typically made of silicon dioxide; this layer often also includes anti-glare (AG) functionality. The hard coat also almost always includes anti-stiction functionality which reduces the static friction between a finger and the surface; this makes dragging objects (e.g., cards in a video-poker game) much easier.

Also shown in Figure 2.12 is an optional bottom shield typically made of ITO; its purpose is to shield the conductive layer from the electromagnetic interference (EMI) emitted by the display. Since the bottom shield increases the cost of the touchscreen and reduces the transmissivity (i.e., decreases image brightness), it is often seen as undesirable. Equivalent EMI reduction is often currently accomplished through firmware in the touchscreen controller.

Surface capacitive uses a uniform electrostatic field established across the conductive coating by applying an AC signal to the four corners of the conductive coating. An AC signal (typically 1–2 volts in the range of 30–100 KHz) is required because the dielectric hard-coat prevents a DC drive signal from coupling to the user's finger. All four corners are driven with exactly the same voltage, phase, and frequency. When a user's finger contacts the top coating, a small amount of electrical energy is capacitively coupled from the conductive coating to the user, causing a small amount of current to flow through each corner connection. The controller identifies a touch by comparing a known "baseline" current in the no-touch state with the change in current when a user touches the screen. The touch location is identified by measuring the amount of current supplied to each corner, and the magnitudes of these currents are proportional to the proximity of the touch location to the corners (the equivalent circuit of a surface-capacitive touchscreen is shown in Figure 2.13). The controller electronics measure these currents, converts them to DC, filters them to remove noise, amplifies them, converts them to digital via an analog-to-digital converter (ADC), calculates the touch location, adds appropriate characterization information and outputs the location coordinates to the host computer [14].

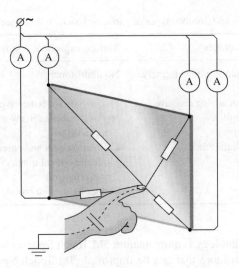

Figure 2.13 Equivalent circuit for a surface-capacitive touchscreen is shown. The circled 'A' symbols represent the measurement of current flowing through each corner connection. Source: By Mercury13 (Own work) [CC-BY-SA-3.0 (http://creativecommons.org/licenses/by-sa/3.0)], via Wikimedia Commons.

The primary applications for surface-capacitive touchscreens are in regulated gaming (casinos), point-of-sales/retail terminals, point-of-information/self-check-in kiosks, and medical equipment. Regulated gaming involves a very long approval cycle for any hardware or software changes, which will slow down any possible transition from surface capacitive to p-cap. However, surface capacitive is not as durable as p-cap, because the ITO layer is on the top surface of the glass, covered only by a protective coating. Anecdotal evidence points to scratches (presumably by diamond rings) as the number one cause of failure of surface capacitive touchscreens in casino applications, which means that the greatly increased durability of p-cap is likely to be very appealing in casino applications. Another factor that will tend to drive p-cap into casino applications is the casinos' desire to attract younger customers. According to a survey reported by Harrah's Entertainment in 2006 [15], the average age of a casino patron in the USA is 46. In order to attract younger customers, casino operators believe that games must be made more interactive and exciting; p-cap's multi-touch capability can help accomplish this.

Surface-capacitive's advantages and disadvantages are shown in Table 2.5 below.

Surface capacitive is a single-touch technology. A "simulated gesture" capability similar to that in analog resistive was developed for surface capacitive by a 3M Touch Systems competitor around 2009, but it has had little effect because surface capacitive is used almost exclusively in commercial applications, where the demand for multi-touch has so far been much less than in consumer applications. However, it is likely that the demand for multi-touch in some commercial applications will change in the near future. Many users of commercial applications (e.g., casino game-players, public-access kiosk users, and medical equipment users) are likely to have smartphones and/or tablets with p-cap touchscreens, so they have a built-in expectation of multi-touch. The providers of casino games, kiosk software, and medical equipment are likely to respond to that expectation by enhancing their products with multi-touch functionality. This, in turn, will drive surface capacitive out of those applications and replace it with p-cap.

Table 2.5 Advantages and disadvantages of surface-capacitive touch technology

Surface-capacitive advantages	Surface-capacitive disadvantages
Excellent drag performance with extremely smooth surface	No multi-touch
Much more durable than analog resistive	Finger-only (or tethered pen)
Resistant to contamination	Not as durable as many other glass-based touch technologies
Highly sensitive (very light touch)	Calibration drift and susceptibility to EMI
	Moderate optical quality (85–90% light transmission)
	Cannot be used in mobile devices

Surface capacitive technology is quite mature; 3M Touch Systems has refined it to the point where there is not much more that can be improved. 3M Touch Systems has maintained a majority share of the market ever since the acquisition of MicroTouch Systems in 2001 but, according to DisplaySearch, the total surface-capacitive market in 2013 was only about $45M, so it is not a significant factor in the overall 2013 touchscreen market of $31B [12].

Correctly sensing the future of touch, 3M Touch Systems has shifted its focus from surface capacitive to projected capacitive, as evidenced by the almost total absence of surface capacitive in 3M Touch Systems' booths at trade shows in 2013. As the market for surface capacitive shrinks, the few remaining Asian competitors are starting to exit the market, which will accelerate the rate of decline. The bottom line is that surface-capacitive touch technology is entering its end-of-life phase; within 5–7 years, the technology will be an historical curiosity.

2.4.2.1 Reversing Ramped Field Capacitive

Standard surface-capacitive technology is inappropriate for mobile use because it requires a very stable reference ground in order to establish the baseline current for the "no touch" condition. CapPLUS™, a variation of surface capacitive employing "reversing ramped field capacitive" (RRFC™) technology, very cleverly eliminates the restriction against mobile use [16]. RRFC technology was invented by Touch Konnection Oasis (TouchKO), a small company founded in 1996 in Texas; the company was acquired by Wacom in 2007.

In standard surface-capacitive, the conductive substrate is covered by a single, flat electrostatic field. In RRFC, four ramp-shaped fields are used instead, as shown in Figure 2.14. This is accomplished by applying an AC voltage on two adjacent corners of the conductive substrate and a DC voltage on the opposite two corners; this produces a voltage ramp across the sensor and a corresponding ramped electrostatic field. The touch controller repeats this for all four of the possible corner combinations in sequence, making four measurements of the change in current caused by a touching finger (two for X and two for Y). These measurements are shown conceptually as four perpendicular cylinders in Figure 2.14. The signals captured during the measurements are then subjected to additional digital signal processing that compensates for factors such as grounding changes, metal bezels, EMI, variations in skin dryness or finger size, thin gloves, etc. This allows the touch-event signal to be independent of all environmental capacitance effects except those due to the finger-touch.

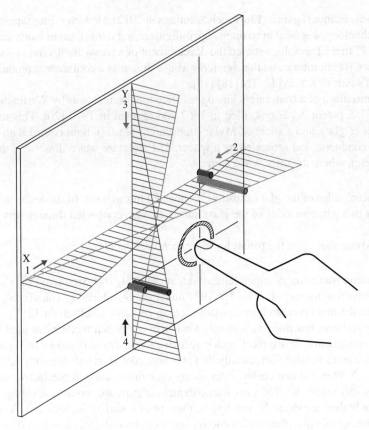

Figure 2.14 Reversing Ramped Field Capacitive (RRFC) touch technology uses four ramp-shaped fields (two voltage ramps and two electrostatic field ramps, shown as shaded triangles) instead of the single, flat electrostatic field used in legacy surface-capacitive. Source: Reproduced with permission from Wacom.

The result is a significantly improved technology that solves most, if not all, of the problems of legacy surface capacitive. Unfortunately, there are two very big disadvantages, as follows:

- RRFC is still a single-touch technology; many surface-capacitive applications other than traditional kiosks are strongly trending towards multi-touch technologies.
- Wacom is a sole-source supplier for RRFC; unless there is an overwhelming market-driver (e.g., such as there is for Wacom's pen-digitizer used in tablets such as Samsung's Galaxy Notes), OEM/ODMs tend to avoid single-source suppliers.

2.5 Resistive Touch Technologies

2.5.1 Analog Resistive (#3)

The invention of the analog-resistive touchscreen has often been attributed to Elographics in 1975 [17]. (Elographics was founded in 1971; they were renamed as Elo Touch Systems in

1986 and then renamed again as Elo Touch Solutions in 2012). However, Elographics' original resistive technology was used in an *opaque* pen-digitizer, not a *transparent* touchscreen; it was not until 1977 that Elographics started the development of a transparent version (curved to fit the face of a CRT monitor), and that version was not shown as a commercial product until the 1982 World's Fair in Knoxville, TN [18] [19].

The first invention of a transparent analog resistive touchscreen was by Westinghouse, documented in US patent 3,522,664, filed in 1967 and granted in 1970 [20]. This touchscreen used a sheet of glass and a sheet of Mylar (transparent plastic), both coated with a uniform transparent conductor and separated by a spacer. The structure was a three-wire touchscreen (now obsolete), where the three wires were:

1. two adjacent sides of the glass substrate, connected through sets of diodes;
2. the other two adjacent sides of the glass substrate, also connected through sets of diodes; and
3. the top Mylar sheet (see the patent for more technical details).

This invention was never commercialized; it is more likely that Sierracin/Intrex's four-wire analog-resistive touchscreen, first sold in 1979 under the brand name "TransTech", may have actually been the first commercially available analog-resistive touchscreen. [21]

An analog resistive touchscreen is simply a mechanical switch mechanism used to locate a touch. The construction of a typical analog-resistive touchscreen is shown in Figure 2.15. A glass substrate and a flexible film (usually PET) are both coated on one side with the transparent conductor ITO. With the two coated sides facing each other, the two conductive surfaces are separated by very small (50–250 µm), transparent, insulating spacer-dots. A voltage is applied across one or both of the sheets (depending on the type of resistive touchscreen). When a finger presses on the flexible film, the two conductive surfaces make electrical contact. The resistance

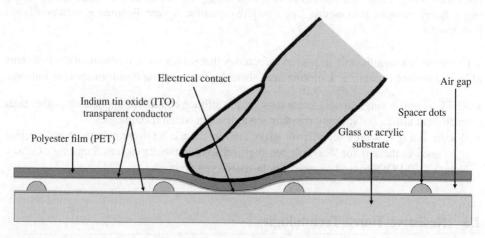

Figure 2.15 An analog-resistive touchscreen is a mechanical switch mechanism used to locate a touch. Two conductive layers are separated by tiny insulating spacer dots; when the two layers are pressed together, an electrical contact is made. The touch location is calculated from the ratio of voltages on the conductive layers. Adapted from Elo Touch Solutions.

of the ITO creates a voltage divider at the contact point, and the ratio of the voltages is used to calculate the touch position.

2.5.1.1 Analog-Resistive Variations

Resistive touch technology has three key variations:

(a) the number of "wires";
(b) the layer construction; and
(c) the options.

The number of wires refers to the number of connections to the sensor; the three common types are 4-wire, 5-wire, and 8-wire.

In a 4-wire touchscreen (Figure 2.16), connections are made to bus bars on the left and right (X) edges of one conductive sheet, and bus bars on the top and bottom (Y) edges of the other. To determine the X position of the touch, the controller applies a voltage across the X connections and measures the voltage at one of the Y connections. The controller then reverses the process, applying voltage across the Y connections and measuring the voltage at one of the X connections to determine the Y location [22].

In a 5-wire touchscreen (Figure 2.17), the X and Y voltages are applied to the four corners of the lower conductive sheet and the upper sheet is used only as a contact point (wiper). To determine the X position, the controller applies a voltage to the two right-hand X-axis corners and grounds the two left-hand X-axis corners. The coversheet (the fifth wire) is used as a voltage probe to measure the X position. The controller then reverses the process, applying a voltage to the top two Y-axis contacts and grounding the bottom two Y-axis connections. Again,

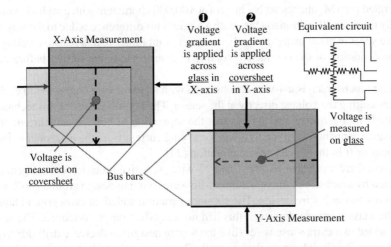

Figure 2.16 In a four-wire touchscreen, a voltage gradient is applied between the two X-axis bus bars on the glass and the resulting voltage is measured on the coversheet. The voltage gradient is then applied between the two Y-axis bus bars on the coversheet and the resulting voltage is measured on the glass.

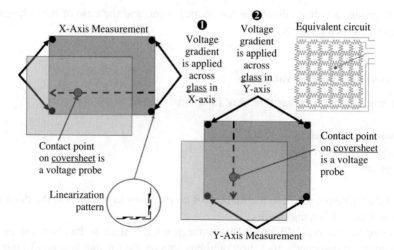

Figure 2.17 In a five-wire touchscreen, a voltage gradient is applied along the X-axis of the glass, and the coversheet (wire #5) is used as a voltage probe. Then the voltage is applied along the Y-axis of the glass, and again the coversheet is used as a voltage probe.

the coversheet is used as a voltage probe to measure the Y position. A 5-wire touchscreen is always ready for a touch; when waiting for a touch, the four corners are driven with the same voltage, and the coversheet is grounded through a high resistance. When there is no touch on the screen, the voltage on the coversheet is zero. When the screen is touched, the controller detects the increased current flow to the coversheet and starts the measurement process, as previously described [23].

The key difference between 4-wire and 5-wire touchscreens is service lifetime; 4-wire is typically rated for 1M touches with a finger (or 100,000 characters with a stylus), while 5-wire is typically rated for 30 million touches with a finger. This difference is due to the way the upper conductive sheet is used; using it as only a contact point rather than a resistive voltage divider allows the condition of the conductive coating to deteriorate much further before ceasing to function.

An 8-wire touchscreen is a 4-wire touchscreen with an extra wire connected to each bus bar to allow measuring the voltage directly at the sensor. The key advantage of this technique, generally called "four-terminal sensing", is that the separation of voltage and current eliminates the impedance contribution of the four wires carrying current from the controller to the sensor. This reduces drift in the touchscreen calibration [24].

In the past, there were also "6-wire" and "7-wire" resistive touchscreens; in general, these were created by touchscreen manufacturers who were trying to design around Elo Touch Solutions' patents on the 5-wire design. The six-wire variation added an extra ground layer on the back of the glass substrate; however, this had no real effect on performance. The seven-wire variation added two extra sense wires (like the 8-wire design) to decrease drift due to environmental changes, but it did not work very well. These unusual products performed essentially the same as 5-wire touchscreens.

Resistive touchscreens utilize seven different layer constructions, as follows:

1. Film/film
2. Film/glass
3. Film/plastic
4. Film/film/plastic
5. Film/film/glass
6. Glass/film/glass
7. Glass/glass

The first term in the construction is the top layer (i.e., five of the constructions use PET film as the top layer); the last term in the construction is the substrate. The first two of the constructions account for 80% of the market in terms of units, with the majority of the suppliers of those constructions located in Taiwan and China [12]. The first construction is used mostly in mobile devices (especially mobile phones), while the second construction is used in both mobile devices and commercial applications. The third construction is used mainly in products where glass breakage cannot be allowed (e.g., children's toys). The fourth construction is a film/film touchscreen attached to a rigid plastic substrate for improved durability; the fifth construction is the same as the fourth, except with a plain-glass substrate for rigidity (often used with digital resistive); the sixth construction is known as "armored", since it eliminates the durability issues of the PET top layer; the seventh construction is used mainly in automotive applications, due to its environmental robustness.

Resistive touchscreens are available with a large number of options, many more than for any other touch technology. Common options include the following (see Section 2.13 for more details):

- Hard coating – for durability.
- Anti-reflection coating – to reduce diffuse reflections.
- Anti-glare coating – to change specular reflections into diffuse reflections.
- Anti-fingerprint coating – to prevent fingerprint oils from adhering to the top surface.
- Anti-pollution (or "anti-corruption") coating – to prevent substances such as permanent-marker ink from adhering.
- Anti-microbial coating – to reduce bacterial adhesion on medical devices.
- Ruggedized substrate – for durability.
- Surface armoring – laminating micro-glass on top of film/glass construction for durability.
- High transmissivity/low reflectivity – to improve visibility in outdoor use.

2.5.1.2 Analog-Resistive Characteristics

Analog resistive is a single-touch technology – i.e., it does not support real multi-touch. As noted in the p-cap (#1) section of this chapter, consumer expectation set by more than one billion smartphones and tablets is that touchscreens must support multi-touch. In 2008, a resistive-controller enhancement, sometimes called "simulated gestures", was developed as a marketing workaround for the lack of multi-touch. Today, many standard resistive controllers include a simulated gesture capability [25].

There are several methods of implementing simulated gestures; one is to measure the current consumed by the sensor during operation. With a single touch, the current is normally constant and thus not monitored but, with two contact points, the two conductive sheets become resistors in parallel, which increases the current consumption. This allows analog resistive to support a few simple two-finger gestures, such as zoom and rotate, but it cannot pass standard multi-touch tests such as the Microsoft Windows Touch Logo.

Simulated gestures are important in marketing touchscreens because they allow low-end analog-resistive touchscreens to appear to be similar to p-cap touchscreens in at least one respect. In reality, the user experience with resistive simulated gestures is very different, not only because of the limited gesture capability, but also because most resistive touchscreens require significantly more touch force than p-cap, which makes it difficult to consistently press hard enough while moving two fingers at the same time.

The advantages and disadvantages of analog-resistive touch technology are shown in Table 2.6 below.

The four disadvantages shown in Table 2.6 above are all in direct conflict with the new *de facto* standard established by p-cap. These disadvantages are causing analog resistive to rapidly lose market share to p-cap in consumer-electronics applications; according to DisplaySearch, analog-resistive had only a 16% share of consumer unit-shipments in 2012, and 73% of that was in mobile phones [12].

The situation is quite different in commercial applications where, according to Display-Search, analog resistive had an 88% share of unit shipments in 2012 [12]. The major commercial applications for resistive include automotive, factory equipment, retail/point-of-sales (POS), kiosks for point-of-information (POI) and self-check-in, and office equipment such as copiers and printers. The reasons for the continuing strength of resistive technology in commercial applications are as follows:

- Resistive has been the standard for more than 30 years, so many applications have adapted to its disadvantages.
- There is much less demand for multi-touch in commercial applications, although it is starting to grow in some segments.

Table 2.6 Advantages and disadvantages of analog-resistive touch technology

Analog-resistive advantages	Analog-resistive disadvantages
Works with finger, stylus, or any non-sharp object ("touch with anything")	No multi-touch (only "simulated gestures")
Lowest-cost touch technology: $1 or less per diagonal inch of screen dimension	Poor optical quality (up to 20% of the display's emitted light can be lost to layer reflections
Widely available from around 100 suppliers (it is a commodity)	Poor durability (easily damaged PET top surface)
Can be sealed to IP65 or NEMA-4 environmental standards	Relatively high touch-force
Resistant to screen contaminants	
Low power-consumption	

- Commercial applications are mostly point-and-click, with almost no use of swipe gestures, so the touch force is less critical.
- To meet the rapidly increasing demand for flush-bezel in commercial applications, most resistive suppliers have implemented a "flush appearance" version of 5-wire [26].
- There is significant demand for stylus use, and resistive works very well with any type of passive stylus.
- There has not been anything in the commercial touch world remotely like the "iPhone Big Bang" that changed everything in the consumer touch world.

Resistive touch-technology has nowhere to go but down in both consumer and commercial applications. Its primary advantages in consumer applications are its low cost and its stylus capability. However, p-cap will incorporate both of these advantages within five years, which will drive resistive's market share of consumer applications well down into single digits.

In commercial applications, resistive will lose share mainly to p-cap. The rate at which that will happen depends on:

(a) how quickly the cost of p-cap falls;
(b) how quickly more p-cap suppliers sign up to meet the more specialized needs of commercial applications;
(c) how quickly demand increases in each application for p-cap's key capabilities (multi-touch, high optical-performance, flush bezel, and very light touch). For example, the demand for flush bezel is likely to increase much faster in customer-facing applications, such as healthcare and point-of-information, than in point-of-sale or factory-equipment applications. Similarly, the demand for multi-touch is likely to increase much faster in casual and casino gaming (as entertainment establishments try to capture younger customers) than in point-of-sale, where it is difficult to imagine any need for multi-touch (for example) on a fast-food restaurant order-terminal.

DisplaySearch expects resistive's share of unit shipments in commercial applications to drop only moderately, from 88% in 2012 to 72% in 2017 [12].

2.5.2   Digital Multi-touch Resistive (DMR) (#4)

A form of resistive touchscreen called "matrix resistive", in which the sheets of transparent conductor (ITO) are segmented into grids of rows and columns, was actually the first resistive touchscreen technology to be developed. Sierracin/Intrex first started selling ITO-coated PET film in 1973; according to an employee who worked there at the time, Sierracin/Intrex created a matrix-resistive tick-tack-toe game as a trade-show demo of their film [27]. This quickly led to customers developing products that incorporated matrix-resistive touchscreens. Of course, the row-and-column matrix technique was not unique at the time; it had been used much earlier with opaque (metal) conductors in membrane-switch panels.

In a digital-resistive touchscreen, the ITO coatings on the two layers (substrate and cover-sheet) are patterned into horizontal and vertical strips positioned at right angles to each other, as shown in Figure 2.18. When the touchscreen is pressed, one or more intersections of the ITO coatings make electrical contact, and each intersection forms an independent switch. The pitch (spacing) of the strips depends on the desired switch-matrix layout; there

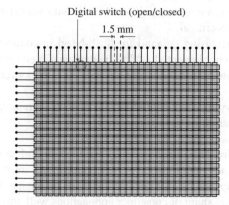

Digital switch (open/closed)

1.5 mm

Figure 2.18   A digital-resistive touchscreen is made up of two layers (substrate and cover-sheet) coated with a transparent conductor (ITO) that is patterned into horizontal and vertical strips positioned at right angles to each other. When the touchscreen is pressed, one or more intersections of the ITO coatings make electrical contact, completing a circuit. This specific example is a representation of the touch sensor in Stantum's iVSM (see the text below).

is no requirement for symmetry, so a matrix can be any number by any number (e.g., 4 rows by 12 columns). Most digital-resistive touchscreens are custom-designed and do not require controllers [28].

In the 1970s, digital resistive touchscreens were used mostly in commercial applications such as factory automation, copying machines, fax machines, calculators, and cash-dispensers. Once 4-wire and 5-wire analog-resistive touchscreens started to spread in the 1980s, digital-resistive touchscreens became less popular, due to their lower resolution and inability to handle writing or drawing.

Digital resistive was a single-touch technology until JazzMutant, a French supplier of music controllers founded in 2002, launched the Lemur™ in 2005. While multi-touch had been used in research projects since 1982 [5], the Lemur music controller was actually the first commercial product to use a multi-touch interface [29]. When JazzMutant decided to market their multi-touch technology separately in 2007 (the same year during which the first iPhone was announced), they renamed the company as Stantum.

Stantum's technology is digital multi-touch resistive (DMR), branded as "Interpolated Voltage-Sensing Matrix" (iVSM™). With the exception of the addition of a sophisticated multi-touch controller, the technology uses the same basic transparent switch-matrix concept that was introduced to the market more than 30 years earlier. The key differences are as follows:

- The pitch of the ITO vertical and horizontal strips is only 1.5 mm, much narrower than had been used previously. While this allows the controller to do interpolation between intersections to achieve much higher resolution, it also greatly increases the number of connections to the controller (e.g., 400 connections for a ten-inch screen).
- The touch-activation force is relatively light at 8–15 grams.
- The controller, which supports up to ten simultaneous touches, is optimized for both finger-touch and stylus-touch. This means that it incorporates "palm rejection" (the ability to ignore any touches other than the tip of the stylus), which is critically important for effective stylus use.

Because they were a small startup in France, Stantum decided to use a licensing business model rather than become a touchscreen hardware supplier. Stantum initially licensed their controller to two ASIC manufacturers (ST Microelectronics and Sitronix) and partnered with Gunze USA, a touchscreen manufacturer focused on commercial applications. Between 2009 and 2011, Stantum had some limited success in commercial and military applications, where the combination of finger and stylus is in more demand. In 2012, Stantum partnered with Nissha Printing in developing a version of iVSM called "Fine Touch Z", which includes a layer of Peratech's transparent pressure-sensing material between the two substrates, greatly increasing the pressure-sensing capability of the touchscreen [30]. While keeping a lower profile than in previous years, Stantum continues to work with their partners on designing products for commercial applications such as tablets for K-12 education. While Stantum is not the only supplier of digital multi-touch resistive technology, they are without doubt the most well-known.

The advantages and disadvantages of digital multi-touch resistive technology are listed in Table 2.7.

2.5.3 Analog Multi-touch Resistive (AMR) (#5)

When the iPhone in 2007 ignited the seemingly insatiable desire for multi-touch everywhere, the analog-resistive touchscreen industry created a hybrid of digital and analog resistive touch-technology called "analog multi-touch resistive" as a lower-cost alternative to p-cap. JTouch in Taiwan, in 2008, was the first touchscreen supplier to commercialize the technology. Generically, this technology is usually known by the acronym AMR; however, some touchscreen suppliers have branded their own version of the technology. For example, Touch International calls it "multi-touch analog-resistive sensor" (MARS). Some suppliers simply advertise "matrix-resistive touchscreens" which requires checking the datasheet to determine whether it is analog or digital. One method of making the determination is to look at the connections coming from the edges of the sensor. If there are many connections on all four edges, then it is analog. If there are many connections on only two edges, then it is digital.

In this technology, shown in Figure 2.19, each conductive surface is patterned into strips so that each overlapping intersection of strips forms a square; each square acts as a miniature

Table 2.7 Advantages and disadvantages of digital multi-touch resistive touch technology

Digital multi-touch resistive advantages	Digital multi-touch resistive disadvantages
True multi-touch	Poor optical quality (up to 20% of the display's emitted light can be lost to layer reflections
Works with finger, stylus, or any non-sharp object ("touch with anything")	Poor durability (easily damaged PET top surface)
Lower cost than p-cap	Touch-force low but still higher than p-cap
Simple and familiar resistive technology	High number of sensor connections
Can be sealed to IP65 or NEMA-4 environmental standards	Limited number of suppliers
Resistant to screen contaminants	Custom sensor design usually required
Low power-consumption	

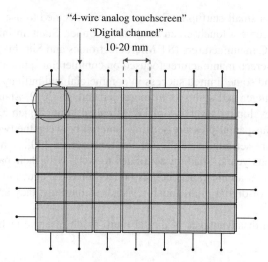

Figure 2.19   In an analog multi-touch resistive (AMR) touchscreen, the normally uniform conductive layers are patterned into strips so that each intersection of strips forms a square, typically 10–20 mm on each side. Each square acts like an independent four-wire analog touch screen.

four-wire touchscreen. That is, the location of a touch within any square is determined by the same analog voltage-divider method used in single-touch resistive. The technology can handle up to as many separate touches as there are squares. However, if two touches occur in a single square, the two touches are averaged and a single touch is reported – exactly the same as in a four-wire touchscreen.

The physical construction of an AMR touchscreen is very similar to that a four-wire analog resistive touchscreen, with the exception of the patterning of the ITO layers. AMR touchscreens are often available only in film-glass construction, although sometimes glass-film-glass is offered for higher durability. AMR touchscreen controllers are available from standard sources such as Texas Instruments [31]. Some touchscreen manufacturers, such as AMT (Apex Material Technology), make their own AMR controllers [32].

AMR was initially targeted as a low-cost multi-touch solution for all-in-one (AiO) consumer desktop computers running Windows 7. In order to minimize the number of sensor connections in the typical 22-inch AiO touchscreen (i.e., to reduce cost), the squares were typically sized at 10–20 mm wide. Unfortunately, this meant that if the user held two fingers closely together and dragged them across the screen, the touchscreen output randomly switched between one or two touches, depending on the location of the touches. In many cases, the consumer user interpreted this behavior as an indication that the touchscreen was defective. In addition to this fundamentally flawed user-experience, AMR in the consumer electronics market also had the following problems:

- It was not significantly lower cost than p-cap.
- It was quite difficult to make properly, especially in larger sizes.
- It had all the same fundamental limitations of resistive (relatively high touch force, low optical performance, and low durability).

After only a few months on the market, one major OEM withdrew their AMR-based AiO products; essentially all other AMR-based AiO products failed to get any meaningful level of consumer acceptance within a year or two. The resistive touchscreen industry learned from this experience; in 2013 the only AMR touchscreens on the market were:

(a) smaller;
(b) made with squares small enough so that it was impossible to get two fingers on one square; and
(c) targeted at specialized commercial and military applications.

In summary, although the technology was developed to compete with the multi-touch capability of p-cap at a lower cost in consumer products, it failed badly and has now become an insignificant niche technology.

The advantages and disadvantages of analog multi-touch resistive technology are listed in Table 2.8.

2.6 Acoustic Touch Technologies

2.6.1 Surface Acoustic Wave (SAW) (#6)

Surface acoustic wave (SAW) in its current form was invented in 1985 by Dr. Robert Adler, a well-known inventor at Zenith [33]. (Dr. Adler is best known for being the co-inventor of the ultrasonic TV remote control, first sold in 1956 [34].) Zenith sold SAW touchscreen technology in 1987 to Elo Touch Solutions which, at the time, was owned by Raychem and known as Elographics. After the sale, Robert Adler continued consulting for Elo, actively contributing to the commercialization of SAW technology into the 1990s.

As shown in Figure 2.20, a SAW sensor is relatively simple, consisting of a piece of ordinary soda-lime glass, four piezo transducers, and four wave-directing partial-reflectors made of low-temperature glass frit that is screen-printed on the surface and then fired. The piezo

Table 2.8 Advantages and disadvantages of analog multi-touch resistive touch technology

Analog multi-touch resistive advantages	Analog multi-touch resistive disadvantages
Multi-touch, but without two touches on the same square	Poor optical quality (up to 20% of the display's emitted light can be lost to layer reflections
Works with finger, stylus, or any non-sharp object ("touch with anything")	Poor durability (easily damaged PET top surface)
Simple and familiar resistive technology	Touch-force low but still higher than p-cap
Can be sealed to IP65 or NEMA-4 environmental standards	High number of sensor connections (with a pitch tight enough to handle two fingers held together)
Resistant to screen contaminants	Limited number of suppliers
Low power consumption	

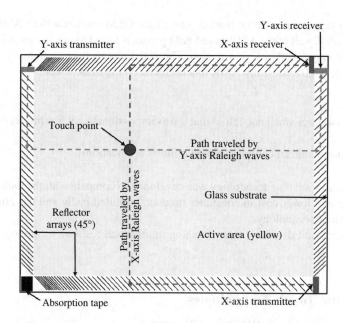

Figure 2.20   A surface acoustic wave (SAW) touch sensor consists of a glass substrate, two transmitting transducers (one per axis), two receiving transducers, and four 45-degree reflector patterns. Rayleigh waves travel from a transmitting transducer down a reflector, across the screen, and up the opposite reflector to a receiving transducer. Adapted from Elo Touch Solutions.

transducers are configured in pairs, one for the X-axis and one for the Y-axis. The X and Y transmitting transducers send bursts of ultrasonic Raleigh waves across the surface of the glass, aimed down the X and Y transmitting reflectors. Frequencies in the range of 4–10 MHz are known to work, but for historical reasons, most SAW touchscreens today operate at 5.53 MHz. The transmitting reflectors consist of an array of ridges at a 45-degree angle; as the Rayleigh waves hit the ridges, some of them are reflected across the screen. The spacing between adjacent ridges is an integral multiple of one wavelength of the propagating wave on the glass. This prevents destructive interference of the wave train as it travels down the reflector array and is partially reflected at each ridge. A matching set of receiving reflectors on the opposite edges of the screen direct the waves towards the X and Y receiving transducers.

The transit time of any given Rayleigh wave from transmitting transducer to receiving transducer depends on the length of the path; waves reflected by the beginning of the reflector take less time than waves reflected by the end of the reflector. The use of a "time of flight" method such as this is possible when the medium behaves in a non-dispersive manner, i.e., when the velocity of the waves does not vary significantly over the frequency range of interest. In this way, the physical location of the touch on the screen is mapped into the time domain. When a human finger or other soft (sound-absorbing) object touches the glass, it absorbs a portion of specific X and Y Rayleigh waves.

As shown in Figure 2.21, the touch location is determined by measuring where the reduction in wave amplitude occurs in the time domain for the X and Y waves. By measuring the amount of amplitude reduction, an indication of the touch pressure in the Z-axis can be obtained – although this is rarely done in practice.

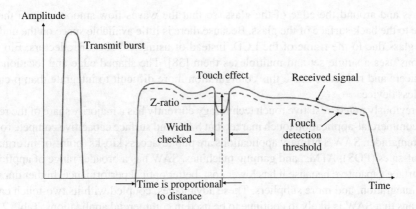

Figure 2.21 In a SAW touchscreen, a transmit burst of Rayleigh waves in the X or Y axis produces an amplitude response curve in the time domain; the location of a touch on an axis is determined by the time-domain location of an amplitude dip. Source: Adapted from Elo Touch Solutions.

SAW controllers are adaptive. They continually monitor the received waveform in the "no-touch" condition in order to ignore most forms of contamination on the touch surface, adjust for changes in environmental conditions, and change the noise threshold when necessary.

SAW touchscreens are available in sizes from 6 to 52 inches. However, because of the relatively high attenuation (absorption) of Raleigh waves in soda-lime glass, screens larger than about 24 inches typically use borosilicate or barium glass, due to its lower attenuation (this type of glass adds up to 30dB to the signal-to-noise ratio of a 42-inch SAW touchscreen system [35]). However, as the touchscreen size approaches 42 inches, optical touch-technologies become more cost-effective so, even though it is technically possible to build a 52-inch SAW touchscreen, it is becoming increasingly rare in actual applications.

SAW was originally a single-touch technology, but the two largest suppliers, Elo Touch Solutions and General Touch (which together account for the majority of the market) both developed two-touch versions in 2009 and 2010 respectively. Elo's approach is to add a second set of reflectors at angles other than 45° (e.g., 15° or 75°) in order to provide an additional source of touch location data [36]. The primary drawback of two-touch SAW is that SAW requires a significant amount of force to register a touch – between 20 and 80 grams, depending on the implementation. Even for a single-touch application, this is much more force than p-cap requires (which is essentially zero). Maintaining sufficient pressure with two fingers while performing a gesture such as zoom or rotate, or having to press hard to perform a swipe gesture, are not very good user experiences. Some models of all-in-one Windows 7 consumer desktop computers from Lenovo and Samsung were marketed with two-touch SAW, but there has not been any significant consumer-market penetration beyond that. Windows 8 has ended any consideration of SAW in the consumer market, since the Windows 8 touch specifications require a minimum of five touches [37].

Another problem with standard SAW is that it requires a bezel to cover the reflectors around the border of the glass. Both Elo Touch Solutions and General Touch have developed a bezel-less (alternatively called "zero-bezel", "flush", or "edge-to-edge") two-touch version. Elo Touch Solutions' approach is to move the transducers and reflectors to the underside of the glass and around the edge of the glass, so that the waves flow smoothly from the front surface to the back surface of the glass. Because there is little available space on the underside of the glass due to the frame of the LCD, instead of using two sets of reflectors, Elo Touch Solutions uses a single set and multiplexes them [38]. The shaped edge and location of the transducers and reflectors make this configuration more difficult to integrate than p-cap in a bezel-less device.

As previously noted, resistive touch technology currently has a majority share of the revenue in the commercial-application touch market, but SAW and surface capacitive compete for most of the remainder. SAW's primary applications are public-access kiosks (point-of-information), point-of-sales (POS), ATMs, and gaming machines. SAW has a broader range of applications than surface capacitive because it has lower cost, better optical performance, higher durability, easier integration, and more suppliers. These advantages, coupled with its two-touch capability, means that SAW is likely to continue to be used in commercial applications. Table 2.9 lists the advantages and disadvantages of SAW touchscreens.

2.6.2   Acoustic Pulse Recognition (APR) (#7)

Both acoustic pulse recognition (APR) and dispersive signal technology (DST, covered in the next section) make use of bending waves. Bending waves are a form of mechanical energy created when an object impacts the surface of a rigid substrate. Bending waves differ from surface waves in that they travel through the full thickness of the substrate, rather than just on the surface of the material; one advantage that results from this difference is superior scratch resistance.

When an object such as a finger or stylus touches the substrate, bending waves are induced that radiate away from the touch location. As the waves travel outward, they spread out over time due to the phenomena of dispersion, where the velocity of a bending wave propagating through solid material is dependent upon the wave's frequency. An impulse caused by a touch contact generates a number of bending waves within the substrate, all at different frequencies. Because of dispersion, these bending waves propagate out to the edges of the glass at different speeds, rather than in a unified wave front. The result is that sensors at the edges or corners of the substrate receive a wave formation that does not resemble the original impulse at all; the

Table 2.9   Advantages and disadvantages of surface acoustic wave (SAW) touch technology

SAW advantages	SAW disadvantages
High optical performance due to plain-glass substrate	No multi-touch (> 2 touches)
Finger, gloved hand, and soft-stylus activation	Very sensitive to surface contamination, especially by water
Very durable; can be vandal-proofed with tempered or chemically strengthened (CS) glass	Relatively high activation force (20–80 g typical)
Relatively easy to integrate; available in waterproof and/or dustproof versions	Requires a soft (sound-absorbing) touch object

wave formation is further modified by reflections from the internal surfaces of the substrate. The net result is a chaotic mass of waves, all interfering with one another throughout the substrate. The key difference between APR and DST touch technology is how that chaotic mass of waves is processed.

In acoustic pulse recognition (APR) touchscreens, the glass substrate is "characterized" in advance by tapping the substrate in thousands of locations, using a robot. The bending-wave "signature" of each location is sampled and stored in a lookup table in non-volatile RAM that is associated with the particular substrate. In operation, bending waves produced by a touch are sensed by four piezoelectric transducers located asymmetrically on the perimeter of the substrate (see Figure 2.22). The asymmetry helps ensure that the signatures are as complex as possible; a high level of complexity helps differentiate the signatures. A controller processes the output of the four transducers to produce the signature of the current touch and then compares it with the stored samples in the lookup table; interpolation between samples is used to calculate the correct touch location [39].

The concept of APR was developed in the early 2000s by Tony Hardie-Bick, an individual inventor in the UK with his own company, SoundTouch Ltd. Elo Touch Solutions acquired the assets of SoundTouch around 2004. After some development for commercialization, the technology was announced in 2006. It was designed to be a more-durable, low-cost replacement for analog resistive, before Apple made multi-touch capability a market requirement in 2007.

A touch technology based on the same fundamental idea of sensing bending waves was developed simultaneously and independently by Sensitive Object in France (it was branded "ReverSys") [40]. Sensitive Object's intellectual property (IP) and Elo Touch Solutions' IP did not infringe each other, but they were closely interleaved, so the two companies executed a cross-license agreement in 2007, shortly after the launch of both products. The two companies continued to develop separately after the agreement was signed, because the purpose of the

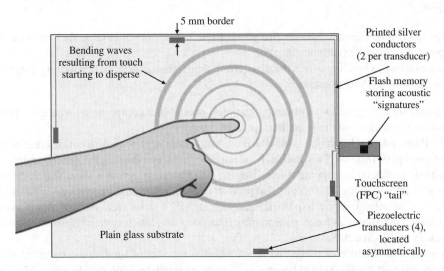

Figure 2.22   An Acoustic Pulse Recognition (APR) touch sensor consists of a piece of glass with four piezoelectric transducers on the back surface of the glass. When a finger or any object contacts the glass, bending waves are produced in the glass substrate and sampled by the transducers; a controller determines the touch location. Adapted from Elo Touch Solutions.

agreement was to avoid a lawsuit over the existing IP, not to share IP going forward. One of Sensitive Object's key innovations was a method of quickly characterizing a substrate in just a few steps, versus Elo Touch Solutions' method of tapping the substrate thousands of times with a robot arm. Elo Touch Solutions purchased Sensitive Object in January of 2010 for $62M [41]. The combination of the two companies' IP produced a very strong portfolio.

However, even with its strong portfolio, APR has a number of fundamental limitations due to its reliance on bending waves. The most important of these limitations is that APR does not have a "hold" function (equivalent to keeping a mouse button depressed). When the touching object stops moving, bending waves are no longer produced. This means that the commonly used sequence of drag-pause-drag on the Windows desktop does not work, since the APR driver must issue an automatic mouse-up at the beginning of the pause. This essentially limits the technology to commercial use (i.e., non-Windows user interfaces) rather than consumer use.

Another key limitation of APR is that it requires a "tap" to produce enough bending waves to be detected. If a shy or uncertain user "sneaks up" on an APR touchscreen and presses it without a distinct tap (even if the user presses very hard), the touch will not register. After the lack of "touch-and-hold" and the need for a distinct tap, the next most important limitation is that APR is fundamentally a single-touch technology. In a world increasingly dominated by multi-touch technologies, single-touch is becoming less and less relevant.

There are three additional limitations due to APR's reliance on bending waves. The first limitation is that APR touch technology is not deterministic. Touching the same exact location many times produces a "cloud" of points surrounding the target coordinates. This means that drawing with a stylus will not produce exactly the same result every time. This is quite different from analog resistive, where touching the same exact location always produces the same coordinates.

The second limitation is that APR's bending-wave detection algorithms cannot be optimized for both:

1. the intermittent bending waves that result from a series of rapid taps, such as occur in a point-of-sales (POS) application; and
2. the continuous bending waves that result from drags, such as occur when writing a signature in the same POS application.

Optimizing for "general-purpose applications" results in a compromise, where the performance of rapid taps and drags are both sub-optimum.

The third and final limitation is that the mounting (clamping) of an APR touchscreen is critical to good performance. This can easily be understood by considering the difference in sound vibrations produced by tapping a free-hanging sheet of glass versus a sheet that is clamped tightly on all four sides. This means that product manufacturers and system integrators worldwide must be trained on how to properly integrate an APR touchscreen. For this reason, APR is not currently available as a component, only integrated into touch systems (finished products) produced by Elo Touch Solutions.

Given the dominance of p-cap and the substantial number of limitations described above, APR as it is today is unlikely to become a mainstream touch technology. However, Elo Touch Solutions is gaining some limited market share with APR in POS applications where none of the above limitations matter. Beyond touchscreens on displays, Elo Touch Solutions may be able to exploit various other non-traditional application niches through the incorporation of

ReverSys' ability to make any rigid object into a touch-sensing surface. For example, the back case of a smartphone could be made touch-sensitive in this way.

Since APR and DST are so similar, the advantages and disadvantages of both touch technologies are combined in Table 2.10.

2.6.3 Dispersive Signal Technology (DST) (#8)

Dispersive Signal Technology™ is 3M Touch Systems' trade-marked name for their version of a touch technology based on sensing bending waves.

The key difference between Elo Touch Solutions' APR (described in the previous section) and 3M Touch Systems' dispersive signal technology (DST) is that, instead of comparing the bending waves produced by a touch with stored characterization samples, DST analyzes the bending waves in real time to calculate the touch location. Figure 2.23 presents a graphic representation of the effects of bending waves on a glass substrate. The third graphic is representative of a wave pattern that APR would sample and compare; the fourth graphic represents the result of processing the pattern through DST's real-time algorithms.

As noted in the description of APR, the transmission velocity of bending waves through the substrate changes with frequency, causing the signal to be dispersed or spread. Upon receiving the signal, DST's approach is to re-integrate the spread signal by applying processing that allows for the differences in delay vs. frequency, then applying correlation processing between the four sensors before ultimately triangulating the original touch co-ordinates. In effect, it is a form of spread-spectrum technique, which is inherently tolerant of signal reflections and interference [42].

NXT PLC (New Transducers Ltd.) in the UK licensed their DST core technology exclusively to 3M Touch Systems in 2003. NXT (which renamed itself as HiWave Technologies PLC in 2010, and then again as Redux Labs in 2013), is known as the creator of the first flat-panel loudspeaker. In this device, piezoelectric transducers mounted on the periphery of a rigid substrate are driven with an audio signal, causing the substrate to function as a speaker

Table 2.10 Advantages and disadvantages of Elo Touch Solutions' Acoustic Pulse Recognition (APR) and 3M Touch Systems' Dispersive Signal Technology (DST), both of which are based on sensing bending waves

APR & DST advantages	APR & DST disadvantages
Works with finger, stylus, or any touch object	No "touch-and-hold"
High optical quality due to plain-glass substrate	No multi-touch
Very simple sensor (glass + four piezoelectric sensors)	Requires enough touch-object velocity (a tap) to generate bending waves
Resistant to surface contamination; works with scratches and extraneous objects on the surface	Mounting method in bezel is critical
Can easily be made bezel-less (flush surface)	Non-deterministic operation ("cloud of points")
	Difficult to optimize for both rapid taps (intermittent bending waves) and drags (continuous bending waves)

  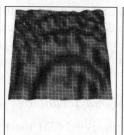 

| 1: Initial touch contact | 2: Dispersion progressing with reflections beginning | 3: Highly complex dispersion pattern with many reflections | 4: Pattern after DST algorithm processing |

Figure 2.23   This figure is a graphic representation of the effect of bending waves on a glass substrate. The third graphic is typical of a pattern that APR would sample and compare offline; the fourth graphic is the result of processing the pattern through DST's real-time algorithms. Source: Adapted from 3M Touch Systems.

diaphragm. NXT had realized (and patented) the reverse idea, that vibrations (bending waves) in the substrate could be sensed by the transducers and used to locate the source of the waves (the touch location).

3M Touch Systems and NXT did a substantial amount of joint development to commercialize DST technology, with 3M pre-announcing their first DST product in 2004 and actually launching the product in 2006. The initial launch was not successful; 3M Touch Systems withdrew the first product from the market for more than a year, finally re-launching it in 2007. Since 3M Touch Systems' bread-and-butter product at the time was surface-capacitive touchscreens in sizes from 5.7 to 32 inches, 3M Touch Systems targeted DST at large-format displays between 32 and 55 inches in order to avoid cannibalizing any of their surface-capacitive business. In contrast, Elo Touch Solutions focused on sizes under 32 inches – not for competitive reasons, but because they were unable to make APR perform well enough in digital signage applications (Elo's APR-based digital-signage products were withdrawn from the market in 2012).

Applications of DST are similar to those for camera-based optical and traditional infrared; interactive information and digital signage are the primary focus. DST has most of the same basic limitations as APR, summarized in Table 2.10.

Around the end of 2011, 3M Touch Systems stopped all further development on DST. Without further development, the technology will eventually become uncompetitive. Although 3M Touch Systems is currently continuing to sell DST into existing interactive information and digital-signage applications, it is likely that the technology will disappear from the market within five years.

2.7   Optical Touch Technologies

2.7.1   Traditional Infrared (#9)

The first widely recognized example of an infrared touchscreen appeared in 1972 in the PLATO IV educational system at the University of Illinois [43]. In this system, a 16-by-16 grid infrared touchscreen was overlaid on an orange plasma bit-mapped display in order to provide finger-selectable functions. One of the first commercial implementations of an

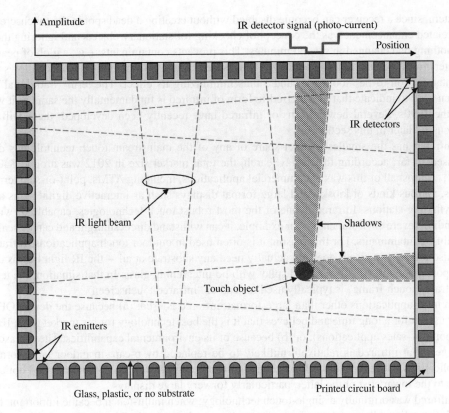

Figure 2.24 A traditional infrared touch sensor consists of a frame with infrared LEDs on two adjacent sides and infrared photo-detectors on the two opposite sides. The LEDs create a grid of infrared light; a touch is recognized when an infrared-opaque object blocks the light beams. Adapted from Elo Touch Solutions.

infrared touchscreen was in 1983 in the HP-150, HP's first touch microcomputer (it had a nine-inch CRT and ran the CP/M operating system) [44]. During the 1980s and 1990s, Carroll Touch was considered to be the leading supplier of infrared touchscreens. AMP acquired Carroll Touch in 1984. In 1999, Tyco International acquired AMP and then later the same year acquired Raychem, who had acquired Elo Touch Solutions (Elographics) in 1986. In this way, Carroll Touch became part of Elo Touch Solutions in 1999.

As illustrated in Figure 2.24, a traditional infrared touchscreen has IR LEDs along two adjacent sides of the screen and IR photo-detectors along the other two sides. Each IR LED is pulsed in sequence, emitting light that is received by the opposing photo-detectors (the sequenced pulsing is why this technology is sometimes called "scanning infrared"). A grid of IR light beams in the X and Y directions is therefore formed just above the surface of the screen. When a finger or any IR-opaque object breaks the beams, a controller calculates the location of the touch.

A minor but significant enhancement to infrared that was invented in the early 1990s by Elo Touch Solutions was the concept of "one transmitter, multiple receivers". Instead of establishing a one-to-one correspondence between transmitting and receiving elements, each transmitting LED is seen by up to five receivers. This increases the robustness of the touch system, since a receiver can become disabled without creating a dead spot in the touchscreen. A related enhancement was the concept of checking for stationary objects and ignoring them if nothing had changed in several minutes. This prevents contaminants (e.g., a glob of peanut butter) from creating a dead spot on the touchscreen; the use of multiple receivers allows the contamination to be "looked around", thus minimizing its effect. The term "traditional" is often used to indicate that the referred-to form of infrared is fundamentally the same as it was in the 1990s. Several newer forms of infrared have recently been developed; these will be discussed later in this section.

Infrared has the smallest market share of any of the mainstream touch technologies discussed so far; according to DisplaySearch, the total market size in 2012 was around $40M [12]. Almost all of this was in commercial applications, including ATMs, point-of-sale terminals, various kinds of kiosks, and large-format displays such as interactive digital signs and wayfinding stations. Infrared is one of the most robust touch technologies, capable of withstanding severe environments – for example, it can withstand direct sunlight and can be sealed against contaminants. For this reason, it is often used in outdoor touch applications. Infrared is also unique because it does not actually need any substrate at all – the IR light beams can be positioned directly above the display with no intervening glass. In that situation, the term "infrared touch-frame" is typically used instead of "infrared touchscreen".

In most applications other than large-format, infrared is used: (a) because the device OEM has used it for a long time and believes that it is the best technology for his market (e.g., IBM in point-of-sales applications); or (b) because of its environmental capabilities. These reasons suggest that infrared is relatively unlikely to be replaced by p-cap. In indoor large-format applications, however, infrared is facing significant competition from camera-based optical, due to the lower cost of the latter, particularly for very large displays.

Infrared was originally a single-touch technology; when multi-touch became important, the major suppliers all started supporting some degree of two-touch. Because there are only two available axes of information (X and Y), two touches can not be uniquely resolved without additional information (this is the same problem as ghost touches in self-capacitive). This limited degree of multi-touch support is sometimes called "1.5 touches". In the late 2000s, Elo Touch Solutions developed a clever method of adding a third axis of information using diagonal light beams (they called the additional dimension "U") [45]. This allowed the resolution of two touches most of the time, except for the special case when the two touches were exactly in line with the diagonal beams (occluding each other). Unfortunately, Elo Touch Solutions never put the technology into mass production due to its high cost.

Traditional infrared is likely to continue to exist as a unique technology, particularly in applications where environmental resistance is critical. Market share in small-to-medium sizes is likely to remain relatively constant during the next five years, while market share in large format is likely to decline due to the growth of camera-based optical.

The advantages and disadvantages of traditional infrared touch technology are listed in Table 2.11.

2.7.1.1 Waveguide Infrared

Beginning around 2000, an Australian startup named RPO began developing polymer optical waveguides targeted at the "last mile" telecommunications market. When the "last mile" market collapsed in 2002 because of extreme overexpansion of fiber, RPO regrouped and

Table 2.11 Advantages and disadvantages of traditional infrared touch technology

Traditional infrared advantages	Traditional infrared disadvantages
Scalable to very large sizes (over 100 inches)	Mostly single-touch; limited support for two touches ("1.5 touches")
Activation by any IR-opaque object with zero touch force	Pre-touch (touch activation before actually contacting the surface of the screen)
High optical performance due to plain glass substrate	Profile height (IR emitters and receivers project above the touch surface); printed circuit board (PCB) must totally surround screen
Easy to integrate; available with or without a substrate; even available as build-it-yourself frame segments	Bezel must be designed to include an IR-transparent window
Very durable; can be vandal-proofed with tempered or chemically strengthened (CS) glass	Extremely difficult to prevent false touches (from shirtsleeves, insects, etc.)
Can easily be environmentally sealed for use outdoors	Relatively low resolution and accuracy
Can be made immune to interference from high levels of ambient IR (e.g., 75K lux)	Minimum touch-object size typically > 5 mm (no fine-tip stylus)
	Relatively high cost compared with camera-based optical (scales as a function of perimeter)

looked for a new application. In 2004, they decided to use the optical waveguides they had developed in a variation of traditional IR touchscreens [46]. They named their technology Digital Waveguide Touch (DWT™). As illustrated in Figure 2.25, their concept was to use a single light source with two sets of optical waveguides transmitting the light in the X and Y directions, with a second pair of waveguides collecting the light and guiding it into a single multi-pixel photo-detector. The waveguides were produced on fabrication ("fab") equipment similar to that for LCDs using photolithography; this allows very high resolution, so the waveguide channels could be as small as 10 μm each.

As often happens, the difficulty and restrictions of implementing the concept in the real world resulted in a somewhat more complex design, shown in Figure 2.26.

Constrained by limited border space, RPO used only a single pair of waveguides to collect the light and guide it into the receiving photo-detector (a line-scan CMOS sensor containing 100+ pixels each for X and Y). In order to distribute the light across the substrate, RPO used two lensed IR-LEDs, one for X and one for Y. The substrate itself was used as an optical waveguide (making use of total internal reflection (TIR)), and a parabolic reflector in the opposite two sides made a 180° change in the direction of the light and helped spread the light out across the substrate. The white lines in Figure 2.26 show the path of one light ray from the LED on the left side of the touchscreen into the optical sensor, in both a top view and an edge view. The result was a low-cost, high-performing IR touchscreen optimized for the 3-inch to 15-inch (mobile) size range [47].

Where the technology fits best is on devices with reflective screens (e.g., e-readers that use E-ink's electrophoretic display). IR's ability to operate without any additional layers on top of the screen is an excellent match with a reflective screen's need to effectively use every photon of light that is available (RPO's glass substrate, a form of waveguide, can be placed

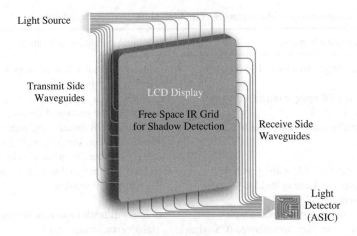

Figure 2.25 Conceptual diagram of RPO's waveguide-infrared touch technology. Source: RPO.

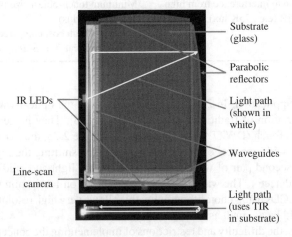

Figure 2.26 RPO's actual construction of a 3.5-inch waveguide-infrared touchscreen. Source: Photo by RPO; annotation and arrows by the author.

underneath the e-reader display). However, like all touch technologies, waveguide infrared has some fundamental limitations in this application, as follows:

- Multi-touch is limited to two touches because there are only two sources of positional information; ghost touches are minimized in firmware, but not completely eliminated.
- A bezel is required to protect the waveguides and reflectors. Total height is only about 1.5mm, but it is still not zero, like the bezel on today's smartphones and tablets.
- The technology is relatively sensitive to debris on the screen, because the waveguide channels are only 200 μm above the surface.

RPO announced the technology in 2007, showed improved performance in 2008, showed larger sizes in 2009, and showed the technology working in a 13.3 inch notebook in 2010 – all at the Society for Information Display's (SID's) Display Week conferences. RPO was counting on a partnership with a very large LCD/consumer–electronics manufacturer (one big customer). When the partnership suddenly derailed at the end of 2010, RPO was not sufficiently prepared with an alternative source of funding to support the manufacturing ramp required by the consumer-electronics market. After a total of $55M investment over more than ten years, RPO went into liquidation in 2011. The sale of the assets (patents) took place in 2012. It is unclear if this technology will ever be put to productive use again.

There is one further historical aspect to this technology that is worth mentioning. A very similar waveguide-infrared touch technology was invented and patented, well before RPO even started thinking about theirs, by Poa Sana (Swahili for "very cool"), a startup located in Silicon Valley, California. Poa Sana's first patent was filed in 1997 and issued in 1999 [48]. Between 1997 and 2002, Poa Sana made little progress towards commercialization of the technology, spending most of the $3.5M they raised on R&D. In 2003, National Semiconductor, which was at the time exploring entering the touchscreen market, acquired the rights to Poa Sana's IP. After spending several years working with the technology and assessing market opportunities for it, National Semiconductor concluded that the future of the technology was not bright enough, so they gave Poa Sana's IP back to its founders. There were never any lawsuits filed between RPO and Poa Sana, because there was never enough revenue produced by either company to justify a lawsuit.

There are probably three basic reasons why waveguide infrared technology failed at the time that it did, as follows:

- The application for which the technology was best (e-readers and other products with reflective displays) was still a fairly small niche market in 2010 (Remember, "everything is best for something and worst for something else" [5]. Any technology must excel in at least one application in order to be successful).
- The fundamental nature of the technology could not support true multi-touch and bezel-less designs, both of which became essential for consumer electronics products after the introduction of the iPhone in 2007.
- The waveguide technology limited the touchscreen size to less than about 14 inches, preventing its application in many other potential markets.

Any technology has an optimum time and place, so it may be that waveguide infrared's time just has not come yet. At least one company (Nitto Denko) has filed and/or received recent patents in this technology area, so it may not be totally dead.

2.7.2 Multi-touch Infrared (#10)

Multi-touch infrared touch-technology is based on a new "imaging" method of using essentially the same physical IR LED emitters and photo-detector receivers that are present in traditional infrared touchscreens. The new method supports up to 32+ simultaneous finger-touches; the main difference is in how the controller manages the emitters and receivers. Instead of simply looking for pairs of interrupted light beams, the controller in most implementations of this new technology uses as many of the available receivers as possible to capture the shadows of all objects touching the screen created by each individual IR emitter.

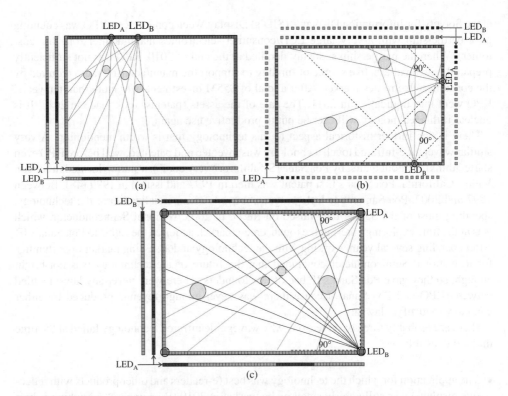

Figure 2.27 Multi-Touch Infrared touch-screens typically use an "imaging" approach. There are currently three different architectures: 28A by PQ Labs, 28B by Image Display Systems (PulseIR™), and 28C by unknown. The architectures are similar, in that each IR LED emitter flashes, and many or all of the IR photo-detectors on two or three opposite sides record their level of light intensity, producing a one-pixel-high "image" showing the shadows of all the objects between the LED and the photo-detectors.

There are three different arrangements of IR emitters and detectors that are used in this imaging method. The first two are identified by their creator: (A) PQ Labs, who invented the category and launched their first product in January 2009, and (B) Image Display Systems; the third arrangement (C) cannot be attributed to a specific creator at this time.

As illustrated in Figure 2.27A, B and C, one IR LED emitter flashes, and many or all of the IR photo-detectors on the two opposite sides record their level of light intensity, producing a one-pixel-high "image" showing the shadows of all the objects between the emitting LED and the photo-detectors. Each pixel in the image (i.e., data from a specific photo-detector) typically is rendered in a gray-scale, which is useful in delineating the edges of moving objects, since real-world shadows do not have sharp edges when the object is moving. Each time an IR LED flashes, the "image-capture" process is repeated. Doing this extremely rapidly and then combining the image sequences in a mathematical array [49] allows a relatively large number of shadow-creating objects to be tracked simultaneously.

The hardware used in all three architectures is relatively similar. The difference in the user experience produced by different multi-touch infrared products is determined mostly by the quality of the algorithms used to interpret the "shadow-image" data, reject ghost-points, handle occlusion, and track moving and non-moving objects.

There are only a few suppliers of this technology so far; the best-known ones are PQ Labs (the creator of the category), Citron (DreamTouch™ brand), Image Display System (PulseIR™ brand), TimeLink, and ZaagTech.

Two key characteristics of this technology's use of resources are the need for very high-speed sequencing and a large amount of continuous image processing. Because the maximum resolution of which this technology is capable is related to the spacing between the IR photo-detectors (generally spaced on 5 mm centers), line-drawing with this type of touchscreen can show evidence of "stair-stepping" or "jaggies". Also, as the touch objects get larger or closer to the IR LEDs, their shadows become larger which, in effect, reduces the amount of data that each image is capable of holding.

Oddly, the primary problem with this technology is the lack of a clear application. This technology seems to be driven more by mainstream enthusiasm about multi-touch than by any actual application need. Current commercial infrared touchscreen applications rarely need more than two touches, and nobody has defined *any* real applications for 20–40 touches. Another impediment is that the problem of identifying which touch belongs to which user does not yet have an elegant (practical) solution. Multi-player games on large horizontal displays (gaming tables) are probably the best opportunity for multi-touch infrared, but it is not clear that this technology is fast enough or has enough resolution for that application. Large displays used in multi-person gaming in casinos today (e.g., roulette) typically use wire-based p-cap because it is fast, the touchscreen surface can be made flush, it is unaffected by light-blocking objects on the surface, and it is readily available from suppliers such as Zytronic [50].

Multi-touch infrared technology is not very suitable for interactive whiteboard applications because of inadequate resolution, slow speed and large minimum touch-object size. Interactive whiteboard applications generally require a stylus and very rapid recognition of the stylus being lifted ("pen-up") by less than one millimeter. It is very difficult to meet the pen-up requirement with any form of infrared in which the IR emitters are positioned above the surface of the screen. While there are implementations of multi-touch infrared technology in interactive whiteboard products (particularly from Asian competitors of PQ Labs), the resulting user experience is generally unsatisfactory.

The advantages and disadvantages of multi-touch infrared touch technology are listed in Table 2.12.

The current status of multi-touch infrared touch technology can be best assessed by visiting the website of PQ Labs (the clear leader of the category). At the time this book is being written in 2013, the website's Demo page [51] includes seven videos averaging 2.5 minutes each; in all of those 18 minutes of video, there are almost no applications shown that use more than two touches. The most compelling application shown is an air-hockey game with two players, each using two paddles.

On the other hand, further study of the PQ Labs website reveals marketing claims regarding the elimination of many of the disadvantages of the technology category. For example, they claim that they distribute the controller processing load by adding up to 10 "Light Processors" (some undefined form of CPU) that execute most of the multi-touch processing in the touchscreen frame, thus optimizing touch speed and accuracy. PQ Labs claims a minimum touch-object size of only 1.5 mm (compared with the typical > 5 mm specification) – except that the claim does not appear in the product specifications, only in the marketing description of

Table 2.12   Advantages and disadvantages of multi-touch infrared touch technology

Multi-touch infrared advantages	Multi-touch infrared disadvantages
Scalable multi-touch from 2 to 32+ (only the controller changes)	No clear application for a high number of touches in the large-format market
Most of the same advantages as traditional infrared (scalable size, broad activation, zero touch force, high optical performance, durable, sealable)	Most of the same disadvantages as traditional infrared (pre-touch, profile height, PCB surround, bezel design-complexity, false touches, low resolution and accuracy; minimum touch-object size)
Object-size recognition (a by-product of the imaging method that captures a view of each object from multiple vantage points)	Performance is often not as good as traditional infrared (slower response, stair-stepped lines, more jitter, etc.)
	More difficult to make immune from high levels of ambient IR, which makes it less useful outdoors
	Disproportionately higher cost than traditional infrared (may be due to temporary "market pricing")

the product. PQ Labs claims to have "optimized handwriting algorithms and a unique White-board Mode" which allows their products to support clear writing of complex graphics such as mathematical formulas. They claim immunity to "bow and twist" and immunity to "adverse lighting conditions" – without providing either a "minimum planarity requirement" specification or a "maximum ambient IR lux" specification.

The above paragraph provides an excellent illustration and example of one of the fundamental problems that occurs throughout the touch industry – insufficient specifications with an excess of marketing claims.

2.7.3   Camera-based Optical (#11)

Although camera-based optical touch only came to prominence in 2009 with the launch of Windows 7, the technology has existed for more than 30 years. In 1979, Sperry Rand Corp. was the first to patent the concept of using two infrared linear image sensors (they were charge-coupled devices (CCDs) at the time) to locate the position of a touch on the top surface of a display. SMART Technologies in Canada and NextWindow in New Zealand both developed the first commercial complementary metal-oxide-semiconductor (CMOS) based optical touch systems independently in the early 2000s. SMART used the technology in a few of their products during the 2000s, but did not start making significant use of it until 2010.

Hewlett-Packard was the first to use optical touch in a desktop product, launching the TouchSmart™ consumer all-in-one computer in 2007 with NextWindow's touch technology. SMART filed a lawsuit against NextWindow for patent infringement in April of 2009 and then licensed their camera-based optical technology to Pixart in June of 2009. Pixart immediately began supplying optical sensors to Quanta for the launch of Windows 7 in October 2009; Quanta became NextWindow's primary competitor. SMART acquired NextWindow in April of 2010, thus ending the lawsuit and lessening the financial impact of Quanta as a competitor. Combining the optical-touch IP of both companies made more sense than the possibility of one company invalidating one or more of the other's patents as a probable outcome of the lawsuit.

Camera-based optical is a form of light-blocking IR touch ("camera" here refers to an assembly that typically includes an image sensor, a lens, an IR filter, a housing and a cable). In the most common form of camera-based optical touchscreen (illustrated in Figure 2.28A), a peripheral backlight is provided via IR LEDs in the corners of the screen, with a retro-reflector around the periphery of the screen (a retro-reflector is a material that returns light in the direction from which it came, regardless of the angle of incidence). As a result of the retro-reflectors, light is radiated from the edges of the screen across the surface of the screen. CMOS line-scan or area imagers (cameras) are placed in two or more corners of the screen; when a finger touches the screen, the peripheral light is blocked and a shadow is seen by the cameras.

Note that even if the camera uses an area imager rather than a one-pixel-high line-scan imager, it is not seeing a grey-scale image of the touching finger; it is simply seeing the presence or absence of light. A controller processes the data from the cameras and uses triangulation to determine the location of the touching finger [52].

Figure 2.29B shows a graph of light intensity seen by one 512-pixel optical sensor. The sharp dip in the graph at pixel 358 is the result of a touch on the screen (i.e., a point where all of the peripheral backlight is blocked). The moderate dip around pixel 250 is the junction of the two edges of the screen (i.e., the bottom and right-hand edges, as seen by a camera in the upper left corner); this is the most distant point from the camera. The sharp peak around pixel 270 is the one point where the retro-reflector is sending light directly back into the camera.

The majority of camera-based optical touchscreens used in desktop products in 2009–2012 had only two CMOS sensors, mainly for cost reasons. Using triangulation, two cameras are required in order to calculate the X and Y locations of a single touch point. If two simultaneous touch points can be seen by both cameras (i.e., each camera sees two distinct shadows), then there are four potential touch points – two real touch points and two "ghost" points (false touches positionally related to real touches). This is the same problem that exists with self-capacitive p-cap, traditional IR and single-touch SAW – all touch systems where information can be obtained only from two axes. Distinguishing real points from ghost points

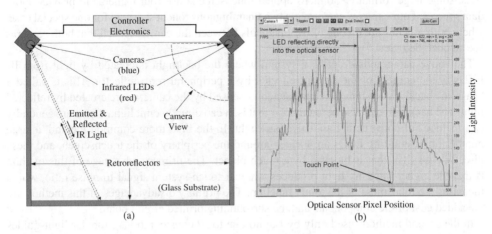

Figure 2.28 Camera-based optical touch uses a backlight created by IR LEDs in the corners of the screen and a peripheral retro-reflector. CMOS line-scan sensors (cameras) are placed in two or more corners of the screen; when an IR-opaque object touches the screen, the peripheral light is blocked and a shadow is seen by the cameras. Source: Adapted from NextWindow.

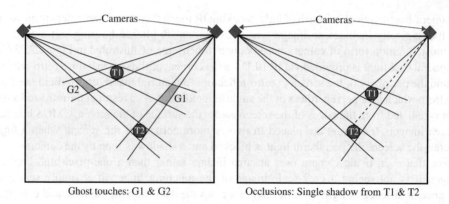

Ghost touches: G1 & G2   Occlusions: Single shadow from T1 & T2

Figure 2.29 Ghost touches (G1 and G2, shown in green on the left) and occlusions (a single shadow from touch-objects T1 and T2 on the right) become a problem when trying to uniquely identify two touches in a two-camera optical touchscreen because there are not enough sources of positional data.

in optical touch requires the application of sophisticated algorithms operating on multiple sets of points over time.

Another situation in which advanced algorithms is important is when the position of the two simultaneous points is such that one of the cameras cannot distinguish between them (i.e., one touch-point occludes the other). Figure 2.29 A & B illustrates both ghost points and occlusion.

Much of an optical touchscreen's controller processing time in a two-camera optical touch system is used running algorithms to eliminate ghost points and compensate for occlusion. In fact, the quality of the multi-touch experience in a two-camera optical touch system depends largely on the sophistication of the algorithms, not the quality of the hardware. For these reasons, some large format (> 30 inch) optical touchscreens use four cameras to provide more data sources. Four cameras can provide two unambiguous touches, except for one special case, where both touches are on one of the diagonals between the cameras, such that both cameras see an occluded view.

The above paragraphs describe a system where the backlight is created by light from IR LEDs in the corners of the screen reflected by a peripheral retro-reflector. This is called a "passive" backlight system, since the shadows sensed by the cameras are created by reflected light. The backlight can also be "active" – that is, it can directly emit light. There are generally two methods of constructing an active backlight. In the first, more common method, a large number of IR-emitting LEDs are located around the periphery of the touchscreen, and these directly emit light that is blocked by a touch-object. The primary advantage of this method is that the higher-intensity light produces a higher touch-system signal-to-noise ratio, which increases the robustness of the touch function. The primary disadvantages of this method are the added cost of the components and the surrounding printed circuit board.

In the second method (used only by Lumio due to IP considerations), tubular light-guides around the periphery of the screen redirect IR light produced by an LED at each end of a waveguide segment, so that it spreads out across the surface of the screen. The primary advantages of this method are lower cost and much lower profile height (3–4 mm versus 6–10 mm); the primary disadvantages are lower light intensity and its single-source nature.

Applications for camera-based optical touch in 2013 were in two main areas:

1. desktop all-in-one touch computers and touch monitors;
2. large format interactive information, digital signage, conference and training rooms, and large interactive LCDs replacing whiteboards in some educational applications.

The desktop application area developed mainly because Microsoft's Windows 7 Touch Logo specification was written around the capabilities of camera-based optical, which at the time was the lowest-cost technology capable of supporting two touches. The Windows 8 Touch Logo specification is written around p-cap, with a minimum requirement of five simultaneous touches. NextWindow has been able to meet the Windows 8 Touch Logo specification by using six cameras – one in each corner, with two additional cameras trisecting the top edge of the screen. At the present time, no other camera-based optical suppliers (other than SMART Technologies, NextWindow's parent, which only sells complete systems and not touchscreens) have announced products that meet the Windows 8 Touch Logo specification.

Because of the high cost of p-cap in desktop (15 to 30 inch) sizes, camera-based optical is seen as a desirable alternative for Windows 8 consumer products. However, PC OEM/ODMs generally prefer to have multiple sources so, until suppliers other than NextWindow emerge, penetration of camera-based optical in Windows 8 desktop products will be limited.

The primary competition for camera-based optical in large-format applications is traditional IR; secondary competition comes from wire-based p-cap, SAW, and Dispersive Signal Technology (DST) from 3M (the latter two technologies are limited to a maximum size of about 52 inches). Camera-based optical's primary advantage over traditional IR is its scalability, which translates into lower cost for larger touchscreens. A traditional IR touchscreen of any size must have a printed circuit board completely surrounding the screen, while a camera-based optical touchscreen can use printed retroreflectors which can be attached to a plastic or metal bracket. The latter is much lower in cost. A secondary advantage of camera-based optical over traditional IR is higher resolution and speed.

In large format applications, both optical touch and traditional IR have individual strengths, so it is likely that both will continue to exist in the large format market for a number of years. Over time, camera-based optical will overtake traditional IR because the hardware is simpler and more capability can be added through software. The advantages and disadvantages of camera-based optical touch technology are shown in Table 2.13 below.

2.7.3.1 PIN-Diode Optical

PIN-diode optical is a variation of camera-based optical that is currently available only from Baanto due to IP considerations. This optical touch-technology uses PIN diodes (diodes with a p-i-n semiconductor structure) as light sensors, instead of CMOS cameras, as described above. PIN diodes have the following characteristics which simplify the structure of an optical touch-screen [53]:

- PIN diodes directly read light intensity and operate at up to 10,000 frames per second, which supports very high-performance touch systems.
- PIN diodes do not require a lens and can be configured with a field-of-view (FOV) approaching 180 degrees (see Figure 2.30), which eliminates reliance on sensors placed in the corners of a touchscreen and eliminates optical aberrations that require computational correction.

Table 2.13 Advantages and disadvantages of camera-based optical touch technology

Camera-based optical advantages	Camera-based optical disadvantages
Scalable to very large sizes (up to 120 inches)	Profile height (camera modules project above the touch surface); increases with larger screens
Activation by any IR-opaque object with zero touch force (stylus independence)	Pre-touch (touch activation before actually contacting the surface of the screen), but less severe than traditional IR
Multi-touch (2–5 touches is typical, but 40 touches has been achieved using 20 cameras)	Relatively high degree of substrate planarity (e.g., ±2 mm) is required, especially with retro-reflectors as the backlight source
Relatively high resolution and accuracy, sufficient to meet the Win8 requirements	Extremely difficult to prevent false touches (from shirtsleeves, insects, debris on the screen, etc.)
Object-size recognition (a byproduct of capturing a view of each touch-object from multiple vantage points)	Multi-touch capability is inherently lower than p-cap due to many fewer data sources; two-touch performance with two cameras is very weak
High optical performance due to plain-glass substrate	Accuracy is non-uniform across screen, especially with only two cameras
Lower cost than traditional or multi-touch IR	Minimum touch-object size depends on screen size; can be as large as 7 mm for very large screens
Available with or without glass substrate	More sensitive to ambient IR than traditional IR (less-developed solutions)
	Side-by-side screens can interfere with each other due to IR light from one being detected by the other

- PIN diodes do not require exposure control, since sensor performance is not affected by illumination changes due to the distance or velocity of a touch-object.
- PIN diodes have an infinite depth of field, which means that position-detection algorithms do not change with the position of a touch-object.

PIN diodes operate entirely in the analog domain, which allows the touchscreen controller to use the information about the shadows created by touch-objects in a different manner than more digitally-oriented image processing technology (Baanto's technology is branded ShadowSense™). Some of the capabilities that this enables, and which are not generally found in other types of camera-based optical touchscreens, include the following [54]:

- Selectable touch-area (for providing palm-rejection, ignoring rain drops on the screen, or setting the minimum required finger-pressure).
- Selectable "dwell time" (minimum number of frames a touch-object must be present before reporting a valid touch event, which allows rejecting brief accidental touches).
- Selectable shadow density (allows the infrared opacity of the touch-object to become a criterion in declaring a valid touch).
- Greater ability to reject high levels of ambient infrared (up to 100K lux).
- Easier scaling to very large sizes (Baanto's largest touchscreen to date is on a 266-inch video wall).

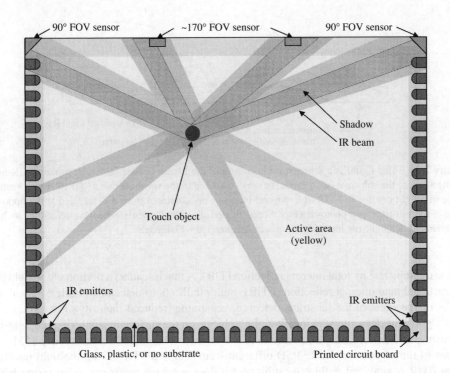

Figure 2.30 An example of a PIN-diode optical touchscreen, which uses two 90° field-of-view (FOV) sensors in the corners, and two 170° FOV sensors along the top edge. The other three edges of the touchscreen contain IR LEDs. The light rays are drawn to illustrate that a sensor may see both the shadow of a touch-object and the peripheral light, depending on the size of the touch-object. Source: Adapted from Baanto. (See color figure in color plate section).

The backlight for Baanto's PIN-diode optical touchscreens is active, utilizing 940 nm IR LEDs spaced on 5 mm centers around the periphery of the touchscreen. The touchscreen's use of the ratio between fully illuminated, partially shadowed, and fully shadowed touch-events makes the touchscreen tolerant of different LED power levels around the periphery, eliminating the need to use "binned" (matched) LEDs. Also, since the controller algorithms use the ratio between readings, changes in the total power received by the sensors do not impact the position calculation.

2.7.4 In-glass Optical (Planar Scatter Detection – PSD) (#12)

Planar Scatter Detection (PSD) is a unique form of in-glass optical touch invented at FlatFrog, a startup founded in early 2007 in Sweden that shipped its first product in May 2012. The core of FlatFrog's touch technology could be termed "optical waveguide analysis". The "waveguide" is the touch substrate, which can be any dimensionally stable transparent material. It does not have to be rigid or flat, which is a very unusual characteristic for an optical touch system. The basic functionality of FlatFrog's system is shown in Figure 2.31. In a PSD touch-sensor, light is injected into the edges of an optical substrate by multiple IR LEDs and remains confined

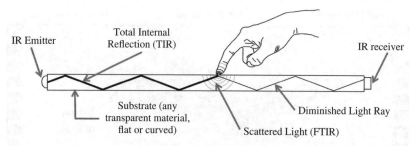

Figure 2.31 This figure is a schematic illustration of the Planar Scatter Detection principle. IR light is injected into the substrate (waveguide); it stays confined in the substrate due to TIR. A touch scatters some of a light ray due to FTIR; the reduced-intensity ray continues until it is detected at the opposite edge of the substrate by a photo-detector. Sophisticated algorithms calculate the location of the touch by analyzing all the light-ray intensities. Source: Adapted from FlatFrog.

inside the substrate by total internal reflection (TIR). A touch scatters a portion of the light due to frustrated total internal reflection (FTIR); multiple IR photo-detectors, interleaved with the LEDs on the edges of the substrate, detect the remaining (reduced-intensity) TIR light. Complex algorithms determine the location of all objects on the surface by analyzing the light-ray intensities and doing a one-dimension to two-dimension (1D-to-2D) reconstruction [55].

One of the aspects that make PSD different from standard FTIR is that the light resulting from FTIR is analyzed within the substrate; it does not have to escape, as in vision-based touch. Another is that a touch only consumes a small amount of a given light ray, so that multiple touches can be located in a straight line with enough light left over to be still sensed at the edge. Like traditional IR, PSD requires a circuit board on all four edges of the substrate, but the number of IR emitters and receivers is slightly less than required for a traditional IR touchscreen (7–8 mm component spacing versus 5 mm spacing). Unlike traditional IR, a PSD touchscreen has a totally flush bezel, since nothing projects above the surface of the display.

The only product incorporating PSD that is currently shipping in 2013 is a 32-inch LCD touch-display assembly from FlatFrog. In other words, at the present time, this touch technology is not available as a component. However, Intel Capital has invested in FlatFrog, and Intel is working directly with FlatFrog on the commercialization of the technology as a component. It is likely that by 2014, PSD touch-technology will be available as a component for all-in-one computer (e.g., 23-inch) displays. FlatFrog intends to use a licensing business model in high-volume consumer electronics applications, and a product-sales business model in low-volume commercial large-format applications. PSD may be able to offer some serious competition for traditional IR and camera-based optical in large-format, and potentially even for p-cap starting from laptop-size and up. The advantages and disadvantages of PSD touch-technology are listed in Table 2.14.

2.7.5   Vision-based Optical (#13)

Vision-based touch here refers to the use of "computer vision" to detect and process touch *in contact with a surface*. While the same term is also used (more commonly, perhaps) to describe the detection and processing of in-air gestures seen by 2D and/or 3D cameras, the

Table 2.14   Advantages and disadvantages of in-glass optical (PSD) touch technology

In-glass optical (PSD) advantages	In-glass optical (PSD) disadvantages
Very robust multi-touch (40+ touches on 32-inch screen, closest in user experience of any touch technology to p-cap)	IR emitters and receivers require printed circuit board around perimeter (9 mm wide), plus one driver ASIC per 12 pairs of components
Edge-to-edge (bezel-less) or with bezel (like p-cap)	Cannot meet Windows pen-digitizer interface specifications due to lack of hover, so not optimum for stylus applications
Practical size range is 14–84 inches (better than p-cap)	False touches are possible from soft objects that cause FTIR
High resolution (400 dpi) and accuracy, sufficient to meet the Win8 requirements (like p-cap)	Sensitive to ambient IR (may be improved in future); dust or smoke on touch surface can affect performance due to change in FTIR
High optical performance due to plain glass or plastic substrate (better than p-cap)	Emerging touch-technology; cost-competitiveness and volume product capability are still to be proven
Very light touch (similar to p-cap); 10 bits of pressure-sensitivity due to changes in optical properties of finger as pressure increases (better than p-cap)	As of 2013, directly replacing p-cap requires more space (3 mm) between cover-glass and LCD and cannot be direct-bonded; direct bonding requires component PCB to be outside LCD frame, which increases display border-width
Works with finger, glove, passive soft-tip stylus (any soft object capable of causing FTIR)	
Insensitive to EMI/RFI (better than p-cap)	
Capable of operating at up to 1,000 Hz data rate (good for large screens, better than p-cap)	
Lower cost than p-cap in desktop sizes; driver ASICs and firmware/software are only unique components	

latter technology is excluded from discussion in this chapter since it does not involve contact with a display. Computer vision also implies heavy use of image-analysis software to determine touch locations and other information about what is contacting the touch surface.

2.7.5.1   Projection

There are currently three methods of producing vision-based touch:

1. projection;
2. multiple wide-angle cameras behind an LCD; and
3. light-sensing in-cell.

The projection method used in vision-based touch is usually rear projection, with a camera located next to the projector (as shown in Figure 2.32). Frustrated total internal reflection

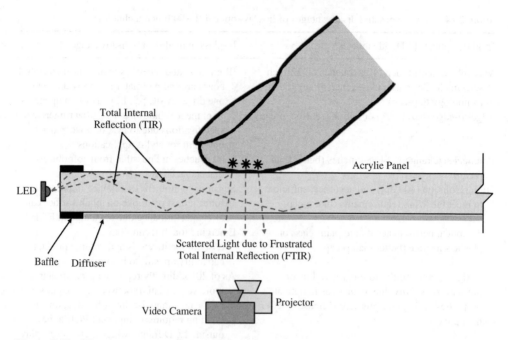

Figure 2.32 The most common method of producing vision-based touch utilizes rear projection with a camera located next to the projector. Frustrated total internal reflection is most commonly used to generate light resulting from touches on the projection surface. Source: Adapted from Perceptive Pixel.

(FTIR, also illustrated in Figure 2.32) is the most commonly used method of producing "blobs" of IR light (bright luminescent objects) resulting from touches on the projection surface [56]. Microsoft Surface 1.0, launched in 2007 (and many of the similar "touch tables" that followed in the next four years) are the best examples of rear-projection vision-based touch. The primary advantage of this method is that a system can be assembled at very low cost [57]; the primary disadvantage is the physical size of a rear-projection system, plus the relatively low image quality that results from rear projection.

There are three other methods of supplying the light that produces the IR blobs in the projection method of vision-based touch:

1. Diffused illumination (DI);
2. laser light-plane (LLP); and
3. diffused surface illumination (DSI) [58].

Diffused illumination refers to providing a uniform distribution of IR light across the underneath of the touch surface. This is usually achieved via one or more IR emitters located at some distance from the screen, and it is the method that was used in Microsoft Surface 1.0. Laser light-plane refers to using lasers to create a thin (1 mm) plane of IR light just above the touch surface; when a finger breaks the plane, an IR blob is produced. Typically, two or four lasers are used in the corners of the screen; a 120-degree line-filter is placed on each laser to spread out the laser beam. Diffused surface illumination uses a special acrylic to distribute the IR light evenly across the surface. The acrylic contains small reflective particles; when an IR LED shines its light into the edge of the acrylic, the particles redirect the light and spread it to the surface of the acrylic. The effect is similar to diffused illumination, but with more uniformity.

2.7.5.2 Integrated Cameras

The only current product that uses multiple wide-angle cameras integrated into an LCD to produce vision-based touch is MultiTaction™ from MultiTouch in Finland (see Figure 2.33). In this touch-display product, the cameras are integrated into the LCD's backlight. The primary advantages of this method are the relative thinness of the display (8 inches) compared with projection, and the high performance that can be achieved. The primary disadvantages are the cost, complexity, and thickness of the solution. Some of the more valuable characteristics of the MultiTaction product are as follows [59]:

- Immunity to external lighting conditions (achieved by reading both ambient light and the reflected light from the IR emitters embedded in the backlight).
- Unlimited number of touch points and users (the touchscreen software also identifies hands, not just touch points).
- Object recognition using 2D markers and/or generic shape-recognition.
- Operation with an IR-emitting stylus (creates clear differentiation between finger and stylus).
- Modular touch-displays can be formed into multi-user interactive walls.

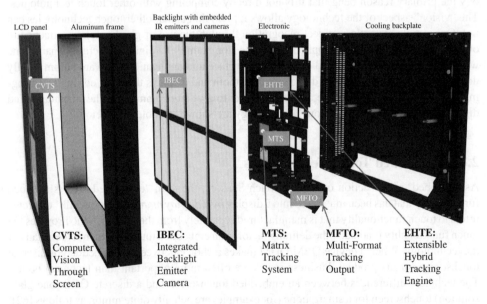

Figure 2.33 The MultiTaction vision-based touch system from MultiTouch in Finland uses IR emitters and IR cameras integrated into the backlight of a standard LCD, supported by sophisticated object-tracking software running on an embedded processor. Source: Adapted from MultiTouch.

2.7.5.3   Embedded In-cell Light-sensing

The technology of embedded in-cell light-sensing is discussed in more detail in the Embedded Touch Technologies section that follows. The Samsung SUR40 touch-display that is used in Microsoft Surface 2.0 is the only current example of embedded in-cell light-sensing technology used to produce vision-based touch in a commercially available product (Surface 2.0 was launched in 2011 and renamed in 2012 as Microsoft PixelSense in order to make the "Surface" name available for use in consumer tablets). The primary advantage of embedded in-cell light-sensing is that touch is fully integrated into the display without adding any thickness. The key disadvantages of in-cell light-sensing (as implemented in the SUR40) are extreme sensitivity to ambient IR (such that the SUR40 cannot be used in most brightly lit environments [60]) and the increase in touch latency apparently caused by the time required to process the data from the embedded light sensors.

2.7.5.4   Vision-Based Summary

Applications for vision-based touch currently fall into two categories:

1. A common platform for touch research in many universities due to the high multi-touch capability and low cost of self-fabrication [57]; and
2. A new platform for traditional commercial applications, such as product-focused touch-tables in retail stores, and interactive video walls in public venues.

Although vision-based touch is still in its infancy, it is here to stay as at least a niche technology the primary reason beng that it is not directly competing with other touch technologies. The "vision" aspect of the technology allows it to do things that other touch technologies can not. One simple example is object recognition through the use of graphic tokens attached to objects [61]. This enables an application where (for example) a digital camera or smartphone with an attached token is placed on the touch surface and application software automatically downloads the photos from the device via Bluetooth and displays them for editing or arranging on the display – without the need for the user to give any commands. The advantages and disadvantages of vision-based touch-technology are shown in Table 2.15 below.

2.8   Embedded Touch Technologies

As described in the section on P-Cap Touch Sensors, the term "embedded" refers to touch functionality that has been integrated into a display *by a display manufacturer*, while "discrete" refers to touch functionality that is manufactured separately from the display. Who provides the touch functionality is actually the defining factor in an embedded touchscreen, not the details of the technology. When a device OEM/ODM makes a choice between an embedded or discrete touchscreen for his product, business issues are often more important than technical issues. The technical differences between an embedded touchscreen and a discrete OGS (one-glass solution) touchscreen for a smartphone (for example) are actually quite minor, as follows [62]:

Table 2.15 Advantages and disadvantages of vision-based touch technology

Vision-based advantages	Vision-based disadvantages
Ideal data source for analysis by image-processing software	Not available as a component, only as a system, and only with large-format displays
Object-recognition (not just size- or shape-recognition) through the use of graphic tokens	Cannot meet Windows pen-digitizer interface specifications due to lack of hover, so not optimum for stylus applications
Very robust multi-touch	Space required by and low optical quality of projection method
Very light touch	Cost, complexity and relative thickness of integrated camera method
Works with finger, glove, or passive soft-tip stylus (projection method); works with light-emitting stylus (integrated camera and embedded methods)	Sensitivity to ambient IR light in embedded method
Low-cost self-fabrication available (projection method)	Emerging touch-technology with limited applications in view
Sophisticated functionality enabled by multiple integrated cameras and powerful embedded image-processing capability (MultiTaction)	
Standard advantages of embedded touch (SUR40)	

- A smartphone display with embedded touch is typically about 100 to 150 μm thinner than the same display with discrete touch. Given that the thickness variation between smartphone models with embedded touch is about 1.0 mm due to other features, the 100 to 150 μm difference is insignificant to most users.
- The performance of embedded and discrete touch is roughly the same. Some display-makers are still climbing the learning curve but, in the long term, the performance is expected to be equal.
- The weight of embedded and discrete touch is the same, since they both use three pieces of glass (the two sheets of glass that make up the display, plus the required cover-glass).
- The power consumption of embedded and discrete touch is roughly the same. Over time, the higher level of integration that is possible with embedded touch should allow it to consume less power.
- The cost of embedded touch and discrete touch are surprisingly close. It is currently possible to save $2 to $4 in controller and flex-cable cost in a smartphone due to the higher level of integration possible with embedded touch. As embedded touch expands into tablet-size displays, discrete touch could actually become lower in cost due to the replacement of ITO with other materials such as metal mesh.
- Off-screen touch icons (such as the Menu icon in an Android smartphone) can easily be created with discrete touch, since the cover-glass is always larger than the active area of the display. Embedded touch, on the other hand, must use additional components (e.g., capacitive buttons) to provide off-screen touch-icons.

Embedded touch has been under development for at least 10 years. All of the original development was focused on inventing new methods of detecting touch that were an integral part of, and leveraged the design of, the LCD [63]. The main methods included the following:

- "Pressed capacitive" (also called "charge-sensing", first mass-produced by Samsung), where pressure on the display caused a change in the dielectric constant of the liquid crystal. The change in dielectric constant changed the capacitance between a pair of electrodes added into some or all pixels [64] [65].
- "Light-sensing", first mass-produced by Sharp, where IR photo-detectors were added into some or all pixels. The photo-detectors read either the shadow of the touch-object in bright ambient light or the reflected backlight from the touch-object in dim ambient light or darkness [66].
- "Voltage sensing" (also called "digital switching", first produced by Samsung), where micro-switches for X and Y location were added into some or all pixels. Pressure on the display closed the micro-switches, identifying the location of the press [67].

None of these methods were fully successful, although Samsung did ship several million units in a series of point-and-shoot digital cameras using pressed capacitive. The main reasons these methods were not successful were:

(a) insufficient signal-to-noise ratio for robust operation;
(b) the requirement to actually deflect the display surface (which eliminated the possibility of using a cover glass, thus increasing the probability of display damage); and
(c) the unreliability of pressing the display very close to the frame, where the color-filter glass has little ability to move.

Once the mutual-capacitive form of p-cap became widely adopted as the technology of choice for discrete touch in smartphones, the display industry realized that figuring out how to integrate p-cap into a display was the right way to go rather than trying to invent something entirely new. Table 2.16 summarizes the three methods that have been developed for integrating p-cap into displays as embedded touch.

From a shipment perspective, in February 2010 Samsung was the first to ship on-cell embedded touch in a high-volume consumer product, the S8500 Super AMOLED™ smartphone ("Super AMOLED" is actually Samsung's brand name for their active-matrix OLED display enhanced with on-cell embedded touch). In May 2012, Sony was the first to ship hybrid in-cell/on-cell embedded touch in the Sony Xperia P™ and the HTC EVO Design 4G™ smartphones; Synaptics developed the touch controller for these products [68]. In September 2012, Apple was the first to ship in-cell embedded touch in the iPhone-5. From an invention (patent) perspective, it remains to be seen if one company (e.g., Apple) will be credited with inventing all forms of capacitive embedded touch, or if each type of capacitive embedded touch will be credited separately.

In 2013, embedded touch is found mainly in smartphone (small) displays, the reason being that scaling embedded touch to larger displays is still under development. It is expected that, by 2015, the technology will have scaled to notebook-sized (15-inch) displays. The primary problems in scaling up embedded touch are as follows:

Table 2.16 Definitions of on-cell, in-cell, and hybrid embedded touch-technology along with the first company to ship a high-volume consumer product in each category

Method	Definition	First Shipment
On-cell	Touch sensor is an array of ITO electrodes on the top surface of an LCD's color filter glass or an OLED's encapsulation glass; functionality is the same as standard p-cap	Samsung, in 2010 OLED smartphone
Hybrid in-cell/on-cell	Touch sensor is an array of ITO electrodes with the sense electrodes on top of the color filter (outside the cell) and the drive electrodes inside the LCD cell. The drive electrodes can be on the TFT glass (in IPS (in-plane switching) LCDs) or on the underside of the color-filter glass (in non-IPS LCDs).	Sony, in 2012 Sony and HTC smartphones
In-cell	Touch sensor is physically inside the LCD cell (between the TFT glass and the color-filter glass). Touch sensor can be either an array of ITO electrodes (mutual capacitive) or light-sensing elements	Apple, in 2012 iPhone-5 (capacitive); Sharp, in 2009 netbook (light-sensing)

- Larger screens have more electrodes, since the number of sense and drive electrodes is a function of the physical dimensions of the screen, not the pixel-resolution of the screen. The electrodes are also longer. Both of these factors increase the amount of time required to complete a scan of the complete touchscreen.
- Larger screens often have higher pixel-resolution, which shortens the amount of time during which the display is electrically quiet. Embedded touch-sensing is typically done during these quiet periods (in fact, the touch controller and the display controller are usually synchronized in order to optimize this timing).

The combination of these two problems – more time required per scan, and less time in which to do it – has so far prevented scaling embedded touch beyond about seven inches in mass production, and 12 inches in the lab.

2.8.1 On-cell Mutual-capacitive (#14)

On-cell mutual capacitive is conceptually the simplest form of embedded touch. Instead of being constructed on a separate sheet of glass or on the underside of the cover-glass, a p-cap touchscreen is constructed on top of an LCD's color filter glass or an OLED's encapsulation glass (see Figure 2.11 in the P-Cap Touch Sensors section). The layer stack-ups and the resulting functionality are essentially identical to those of discrete p-cap. The most common electrode pattern used in on-cell is the interlocking diamond, because it can be manufactured in one layer with bridges.

As noted above, the first mass-production of on-cell embedded touch was in an OLED smartphone. Producing on-cell touch is actually easier (and higher yield) on an OLED display than on an LCD because the OLED encapsulation glass does not have anything on the underside. The underside of the color-filter glass in an LCD has at least the color filter material on it; if it is a non-IPS LCD, it also has the common voltage (Vcom) electrodes (made of ITO) on it. For a display manufacturer, the question then becomes which side to manufacture first. If the touchscreen side is made first, it can be annealed at a high temperature, which improves the quality of the ITO and, thus, the performance of the touchscreen, but the glass cannot be thinned as usual after sealing the LCD cell, which makes the display about 0.3 mm thicker. If the color-filter side is made first, then the glass can be thinned, but the touchscreen cannot be annealed at high temperature (doing so would damage the color-filter material), and this results in a lower-performing touchscreen [69]. Typically display-makers choose the latter because: (a) it does not interfere with the yield of the LCD production process; and (b) thinness is always seen as extremely important.

2.8.2 Hybrid In-cell/On-cell Mutual-capacitive (#15)

As the name implies, hybrid in-cell/on-cell embedded touch is half inside and half outside the LCD. The sense electrodes are deposited on top of the LCD's color-filter glass (outside the cell). Some IPS displays have a grounded anti-static shield layer made of a uniform layer of ITO on top of the color filter; if this is present, that layer is patterned into strips. It still works as a shield because all of the strips except the one being sensed at any given moment are grounded. This method of anti-static shielding is in the process of being replaced by the presence of a conductive layer in the display's top polarizer. In this situation, the sense electrodes are added as an entirely new layer on top of the color-filter glass.

The drive electrodes are created by grouping and repurposing the LCD's Vcom electrodes to act as part of a touch-sensing system in addition to their normal display-updating function [70]. In an IPS display, these electrodes are located on the TFT glass. In a non-IPS display, these electrodes are located on the underside of the cover-filter glass. Since these two surfaces are only a few microns apart, there is no different in performance between the two arrangements. The number of Vcom electrodes that are grouped together to form a single embedded p-cap drive electrode depends on the pixel resolution, the physical size of the screen, and the desired electrode spacing.

For example, a 7-inch, 1280×720 LCD has an active area of 155 mm $\times$ 87 mm. If embedded p-cap drive electrodes are to be formed horizontally along the long side of the display, and an electrode pitch of around 4.8 mm is desired, then dividing the 1280 pixels into 32 groups produces p-cap drive electrodes made up of 40 grouped Vcom electrodes, at a pitch of slightly more than 4.8 mm. The sense electrodes (on top of the color-filter glass, as explained above) would run vertically along the short side of the display, yielding 18 sense electrodes at the same pitch. Figure 2.34 illustrates a hybrid in-cell/on-cell structure, using Japan Display's (formerly Sony) "Pixel Eyes" [71] as an example. The figure shows both the physical arrangement of the electrodes, as well as the stack-up (not to scale).

The method described above is actually only one of several different ways that hybrid in-cell/on-cell can be implemented. Another method, described in patents by Apple and Samsung, is to use metal overlaid on the black matrix (on the underside of the color-filter glass) as drive electrodes, instead of repurposing anything on the TFT glass.

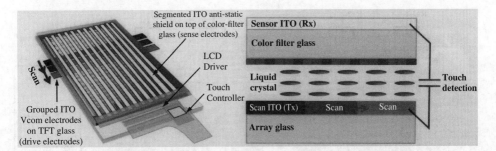

Figure 2.34 Japan Display's "Pixel Eyes" hybrid in-cell/on-cell touchscreen structure. The right side of the drawing shows that the drive electrodes are integrated into the TFT layer, while the sense electrodes are placed on top of the color-filter glass. The left side of the drawing provides more physical detail, showing that the sense electrodes are formed by segmenting the ITO anti-static shield, while the drive electrodes are formed by grouping Vcom electrodes. Also shown is an FPC containing the display and touch controllers, and connecting to both sets of electrodes. Source: Japan Display, with annotation by the author.

With on-cell embedded touch, it is quite possible to use a standard touch controller ASIC that has no connection to the LCD – after all, the touch electrodes in on-cell are only slightly closer to the LCD than they are in direct-bonded OGS, so increased LCD noise is not a big problem. But once part of the touch system moves inside the LCD, synchronization between the touch system and the LCD becomes necessary in order to deal with LCD noise.

2.8.3   In-cell Mutual-capacitive (#16)

As the name implies, in-cell embedded touch is located entirely inside the LCD cell. As noted previously, Apple was the first to ship high-volume in-cell embedded touch in the iPhone-5 smartphone in September 2012. The configuration used by Apple in the iPhone-5 is to put both the drive and sense electrodes on the TFT glass of an IPS LCD. This is accomplished using the same basic technique of grouping and repurposing the Vcom electrodes as described in the Hybrid In-Cell/On-Cell section above, except that the Vcom electrodes are grouped into two separate sets – one for drive and one for sense. In reality, this is much more difficult than it sounds.

Figure 2.35 illustrates how the grouping is done. This figure, created by BOE Technology Group's Central Research Institute, attempts to translate the drawings in Apple's patents [72] into a more easily-understandable form [73] (note, however, that the perspective is not quite right; each square is actually 54 pixels high by 126 pixels wide, so they should be drawn as rectangles). As shown in the key to the figure, the rows labeled "TX" are groups of horizontal (X) segmented Vcom electrodes, made of ITO and connected together by gate metal (in black). Each row is connected to the touch controller by touch-panel metal (labeled "TP" in the key). Since the display in the figure is 1,080 pixels high and there are 20 drive electrodes (rows), each group contains 54 Vcom electrodes. The wider columns are vertical (Y) Vcom electrodes connected via touch sense-detection metal (labeled "S/D" in the key); each of the ten columns also contains 54 Vcom electrodes. From a touchscreen perspective, the drive and sense electrodes are symmetrical. The narrow columns on each side of the wider ones are electrically

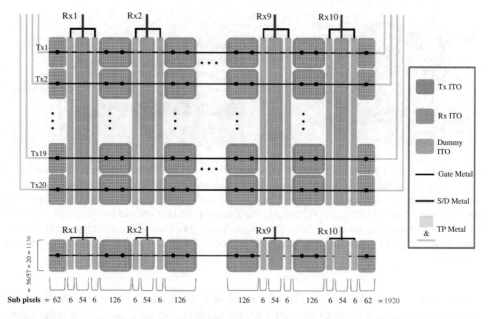

Figure 2.35 A schematic representation of how the Vcom electrodes in the TFT array of the iPhone-5 are grouped to produce both drive and sense electrodes. Source: Adapted from BOE Technology Group's Central Research Institute. (See color figure in color plate section).

connected but isolated dummy ITO (dummy ITO is often used in touch electrodes to achieve a more uniform visual appearance).

It is important to understand that the particular technique used by Apple in the iPhone-5 is only one of several different ways that in-cell embedded touch can be implemented. For example, patents by Apple and Samsung describe a method where the drive electrodes are made by segmenting Vcom, but the sense electrodes are made of metal overlaid on the black matrix (on the underside of the color filter glass). The same two companies describe in their patents another method where the sense electrodes are metal on the black matrix (as just described) and the drive electrodes are ITO stripes deposited on top of a dielectric layer over the color filter material. Sharp described, at a display conference in 2012, a method where both touch electrodes are deposited on the bare underside of the color-filter glass, before the black matrix and color-filter material are applied. LG Displays described in one of their patents an unusual self-capacitive method using just the segmented Vcom electrode.

There are two methods of synchronizing the touch controller with the display controller (TCON): (1) modifying both controllers slightly and connecting them via several lines; or (2) combining the two controllers into a single chip. Synaptics was the first touch-controller company to do both of these. The primary advantage of the second method is that it reduces the touch-system bill of materials (BOM) cost by a dollar or two, while adding some NRE (non-recurring engineering) cost for the chip development; the primary disadvantage is that the combined controller becomes unique to a particular display resolution and pixel configuration.

Clearly, the second method is only practical for very high-volume products (at least several million units).

The advantages and disadvantages of capacitive embedded touch-technology are shown in Table 2.17 below.

2.8.4 In-cell Light Sensing (#17)

In-cell light-sensing embedded touch is accomplished by adding a photo-detector into some or all pixels of an LCD (see Figure 2.36). As previously noted in Table 2.16, Sharp was the first to mass-produce a consumer product incorporating in-cell light-sensing in May 2009; the product was a netbook with a display-based touchpad. The display was a 4-inch LCD that used Sharp's CG (continuous grain) silicon technology with 854×480 pixels (245 ppi). Originally Sharp wanted to use one sensor per pixel, in order to enable the display to be also used as a scanner, but they found that doing so reduced the display aperture-ratio too much. Instead, they

Table 2.17 Advantages and disadvantages of capacitive embedded touch-technology

Embedded capacitive advantages	Embedded capacitive disadvantages
Most of the advantages of p-cap (robust multi-touch, if properly implemented in the controller algorithms; extremely light touch; allows bezel-less cover-glass; excellent optical performance; sealable; etc.)	In-cell and hybrid are practical only for very high-volume displays (millions of units); on-cell can be produced in lower quantities, but it may reduce yield of LCD manufacturing
Requires less "parameter tuning" than discrete capacitive for each new implementation in a product	Currently cannot be scaled above 12 inches; may never be as scalable as discrete
Lower cost than discrete capacitive (OGS), but not radically lower and definitely not "free"; difference may decrease when OGS starts to use ITO-replacement materials	More difficult to achieve same touch performance as discrete; display-maker may prioritize display yield over touch performance
Slightly thinner than discrete (typically $100-150\,\mu m$)	Controller manufacturers may be slower to migrate p-cap enhancements (e.g., water resistance, active stylus, etc.) to embedded versions
Power consumption may be slightly lower than discrete, especially with integrated touch and display controller	Display-maker may not be willing to produce same cover-glass variety and features as discrete touch-panel maker, or be as willing to do direct-bonding
Opportunity for touch-latency reduction, especially with integrated touch and display controller	Cannot support off-screen icons without additional components (unlike discrete)
	May not be able to work through extremely thick glass
	No absolute pressure-sensing; only relative finger-contact area (same as p-cap)

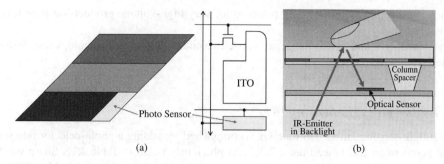

Figure 2.36 Conceptual diagram of in-cell light-sensing embedded touch. The light-sensor is an aSi photo-transistor; it is placed in the blue (lowest) sub-pixel because the sensor's peak response is in the blue-green spectral range. The drawing in the right side of the figure shows IR light emitted from an IR-LED in the backlight being reflected from a touching finger and seen by the light-sensor. Source: Adapted from [4] (left) and Samsung (right).

used one sensor per nine pixels, which produced a scanning resolution of only 27 ppi – not high enough for practical use. Even at one sensor per nine pixels, Sharp found that processing the output of the light-sensors required more CPU bandwidth than their touch controller could provide, so the performance of the display-based touchpad turned out to be very poor (only 25% of the speed of a typical capacitive touchpad).

Research in this area in the first half of the 2000s used visible-light sensors, with the expectation that the sensor would see the shadow of the touch-object in bright ambient light or the reflection of the backlight from the touch-object in dim ambient light [74] [75]. Once researchers realized, around 2006, that visible light was too restrictive (among other things, it cannot get through a black image on the LCD), the researchers switched to IR light sensors. This change also meant that IR emitters needed to be added to the backlight. Because LCDs are only moderately transparent to IR light, and the light is attenuated on both trips through the LCD, the IR emitters are required to have relative high-intensity emission. The additional power consumed by the IR LEDs in Sharp's display significantly reduced the netbook's battery life.

Between Sharp's netbook in 2009 and today (2013), there have been only two commercial products that use light-sensing embedded touch. The first is a 21-inch consumer monitor developed by IDTI (Integrated Digital Technologies, Inc.) in Taiwan [76]. Initially designed for use only with a light-pen, later versions were enhanced to support finger touch using the visible light shadow/reflection method mentioned in the previous paragraph.

The second product is the Samsung SUR40 40-inch all-in-one table computer created jointly with Microsoft for use in Microsoft Surface 2.0 [77]. This LCD includes IR emitters in the backlight, and uses one IR photo-sensor per eight pixels. In an attempt to improve the sensitivity of the touch system, Samsung used aSiGe (amorphous silicon-germanium) light sensors, which are about 15 times more sensitive than ordinary silicon light-sensors. While this definitely improved the touch sensitivity, it introduced a new problem: extreme sensitivity to ambient IR. The problem is so extreme that Samsung published a manual documenting the maximum lux that the touch system can tolerate in each type of room lighting; the number for incandescent light is only 50 lux [60]. The product includes a utility program that measures the ambient IR illumination and displays red-yellow-green areas on the screen to indicate whether the illumination level is low enough.

Samsung is currently researching the use of in-stack light-sensing embedded touch for OLED displays [78]. The concept is parallel to that described for LCDs: IR-emitting pixels in the OLED stack, with IR-detecting sensors in the active-matrix backplane.

Although research and development on in-cell light-sensing embedded touch has been going on for more than 10 years, it is still an emerging technology with unsolved challenges and no successful high-volume consumer products. Table 2.18 provides a more detailed look at the advantages and disadvantages of in-cell light-sensing embedded touch.

Table 2.18 Advantages and disadvantages of in-cell light-sensing embedded touch technology

In-cell light-sensing advantages	In-cell light-sensing disadvantages
Some of the advantages of embedded capacitive (extremely light touch; allows bezel-less cover-glass; sealable; etc.)	Practical only for high-volume displays because fundamentally it's a unique display design
May be lowest-cost form of embedded touch (only one set of sensors; easier to integrate into LCD structure with less material), although image-processing requirements may offset some of that	More difficult to achieve same touch performance as discrete or embedded capacitive; processing sensor output data requires using CPU/GPU-intensive image-processing software (same as vision-based touch)
No formal size limit (40 inches is the largest as of 2013)	Reduced optical performance (lower LCD aperture ratio due to light sensors; lower OLED light output due to IR emitters)
Slightly thinner than discrete capacitive (typically 100 to 150 μm), but same thickness as embedded capacitive	Sensitive to ambient IR; worse if more sensitive photo-detectors are used; difficult to avoid saturating photo-sensors in very bright ambient IR
Requires much less "parameter tuning" than either discrete or embedded capacitive touch-sensing	Low signal-to-noise ratio (poor touch sensitivity) at the crossover point where the IR reflected from the touch-object is equal to the ambient IR
Less sensitive to external RFI/EMI	Low signal levels from photo-sensors make touch-sensing more sensitive to internal interference (e.g., stray current from adjacent photo-sensor)
	Reduced touch sensitivity as the touch surface moves further away from the LCD (i.e., air-gap, thicker cover-glass required for larger displays, etc.)
	Optical-sensor density (ppi) is not high enough to enable using the display as a scanner
	Increased power consumption due to IR emitters
	Cannot support off-screen icons without additional components (same as embedded capacitive)

2.9   Other Touch Technologies

2.9.1   Force-sensing (#18)

Force-sensing has always been seen as the "holy grail" of touch-sensing, because the simplest possible method of detecting a touch should be to just measure the pressure (force) of the touch in multiple locations on the substrate and then triangulate to find the origin. If only it were that simple!

The earliest known commercial product based on force-sensing was IBM's "TouchSelect" touch overlays for 12″ to 19″ CRT monitors in 1991. This technology used strain gauges to mount the touchscreen. It was unsuccessful, lasting no more than about three years on the market. The next commercial incarnation of force-sensing touch in the USA was launched in 2007 by QSI, a Utah-based manufacturer of human-machine interface products and mobile data terminals. The technology, branded as InfiniTouch™, employed a clever beam-mounting method for the strain gauges that eliminated any horizontal component of the touching force [79]. In order to avoid impacting their existing business, QSI spun off the force-sensing technology in 2008 into a subsidiary named Vissumo [80]. The subsidiary was insufficiently capitalized to undertake the non-trivial task of selling a new touch-technology into a crowded market, so they ran out of money and shut down in 2009 (QSI was acquired by Beijer Electronics in 2010 and is now known under that name).

In a separate attempt at commercializing force-sensing touch, MyOrigo in Finland in 2000 developed a force-sensing touchscreen for an advanced user-interface on a proposed smartphone. MyOrigo was sold to its management in 2004 and restarted as F-Origin. F-Origin went bankrupt in Finland in 2005 and the assets were purchased by a US investor who re-formed F-Origin in the US in 2006. F-Origin further developed its force-sensing technology (branded as zTouch™) during 2007–2008, but could not get any traction in the consumer electronics market, due to the rapidly growing prevalence of p-cap. They restructured in 2009 with an ownership investment from TPK (the world's largest supplier of p-cap), and began marketing zTouch in 2010. F-Origin is currently focused on commercial applications, where the durability and environmental resistance of force-sensing touch technology are particularly valuable [81].

Force-sensing touch works by supporting the display (or cover-glass) on force sensors, typically either strain gauges or piezoelectric transducers. In order to obtain accurate measurements of the force applied to the touch surface, the display and/or cover-glass's movement must be constrained so that it only moves in the Z-direction. There are several ways of accomplishing this. Figure 2.37 illustrates the suspension spring-arm method currently employed by F-Origin.

The advantages and disadvantages of force-sensing touch-technology are shown in Table 2.19 below. Note that this analysis does not include new force-sensing methods that have not emerged yet, such as that from NextInput, which uses an array of force-sensing organic transistors under the display [82].

Earlier in this chapter, a prediction was made that multi-touch will eventually become significant in commercial applications. Given that prediction, force-sensing technology seems likely to disappear in the next five years, or at best be relegated to a very small niche.

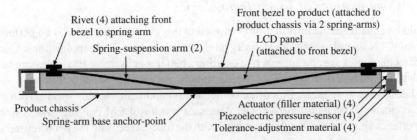

Rivet (4) attaching front bezel to spring arm
Spring-suspension arm (2)
Front bezel to product (attached to product chassis via 2 spring-arms)
LCD panel (attached to front bezel)
Product chassis
Spring-arm base anchor-point
Actuator (filler material) (4)
Piezoelectric pressure-sensor (4)
Tolerance-adjustment material (4)

Figure 2.37 In F-Origin's force-sensing touch technology, the LCD/cover-glass/front-bezel assembly is supported by two suspension spring-arms whose middle is anchored to the device housing (product chassis) and whose ends are attached to the assembly. Four piezo-electric pressure sensors are located at the corners of the assembly, placed between the assembly and the device housing. Since the assembly can only move in the z-direction due to the constraint provided by the suspension spring-arms, any pressure on the assembly (e.g., on the display cover-glass) can be detected and localized by the four piezo-electric sensors. Source: Adapted from F-Origin, with additional annotation by the author.

Table 2.19 Advantages and disadvantages of force-sensing touch technology

Force-sensing advantages	Force-sensing disadvantages
High optical performance due to plain-glass substrate	Difficult to achieve multi-touch (two touches requires eight sensors; the number of sensors increases exponentially after that)
Touch-surface independence (substrate can be any rigid material); can use 3D substrates with embedded moving objects (this is unique)	Minimum touch-force required is close to zero but not actually zero, like p-cap
Activation by any object with close to zero touch-force (stylus independence); even better than resistive	Difficult but not impossible to achieve flush (bezel-less) screen design
Pressure-sensitive; pressure can replace hover (press lightly to display choices; press harder to select); pressure can also can be used to minimize false touches	Cannot meet Win8 touch specifications due to lack of five-touch multi-touch; this limits it to commercial or non-Windows consumer applications
Inherently durable; can easily be designed to handle extended environmental conditions	Mechanical nature of touch-sensing mechanism reduces reliability
Relatively low cost due to sensor simplicity (substrate plus four piezoelectric sensors)	Most sensors add volume to the system (thickness or footprint)
No pre-touch (user must actually touch the substrate to register a touch)	
Insensitive to EMI/RFI and ambient light	
Continuous calibration filters out environmental conditions such as vibration	
Has been scaled up to 42 inches; theoretically can go higher	

2.9.2   Combinations of Touch Technologies

It should be clear from the information presented in this chapter that "there is no perfect touch technology". No single technology meets all the requirements of all applications. Combinations of technologies are one approach to creating a better touchscreen. Examples can be found in Tablet PCs, e-readers, and point-of-sale terminals. An example of each follows.

The latest Microsoft Tablet PCs often combine a p-cap finger-capable touchscreen with an electromagnetic (EM) pen-digitizer. The dominant vendor of EM pen-digitizers (Wacom in Japan) offers a single controller that can drive both the touchscreen and the pen-digitizer, which enables automatic mode-switching between finger and pen [83].

In May 2011, Hanvon announced a new method of combining technologies to accomplish the same goal of pen-and-finger operation. Hanvon combines their EM pen-digitizer with an array of pressure-sensing piezo-capacitors in the same plane as the EM sensor under the display. The piezo-capacitors, which are the same component that is used in the tip of the pen for pressure-sensing, enable sensing finger pressure through (rather than on top of) an e-reader's electronic paper display (EPD).

One of the major providers of POS terminals prefers to use traditional IR in their products. However, to minimize the problem of "pre-touch" (where the finger breaks the IR light beams and triggers a touch without actually touching the surface of the display), the provider includes a pressure-sensing piezoelectric transducer in the touchscreen mounting scheme so that touch coordinates are produced only when the user is guaranteed to be touching the screen. In this application, the transducer detects the *presence* of a touch, while the traditional IR touchscreen detects the *location* of the touch.

Combinations of touch technologies are very likely to continue to exist during the next five years, although combinations of major technologies are usually limited by the cost of the combination. Combinations of a major and minor technology are more likely to occur, as in the example above of combining traditional IR with a pressure sensor. Combinations of current touch technologies with emerging human-machine interface (HMI) technologies are also likely to occur; for example, p-cap combined with a low-cost, miniature 3D camera to enable the detection of near-field gestures in the space above the touchscreen, beyond the range of hover detection. Ultimately, it should be possible to manipulate an object on the p-cap touchscreen and then "pull" the object out of the 2D screen into the 3D space between the screen and the user, seamlessly transitioning the object manipulation from the p-cap touchscreen to the 3D camera.

2.10   Summary

This chapter has described 18 current touch technologies (plus several obsolete technologies) in a fair amount of detail. The best way of summarizing such an ungainly body of information is by predicting how the 18 touch technologies are likely to evolve in the next five to ten years, as shown in Table 2.20 below:

Table 2.20 This table presents predictions of how the 18 touch technologies described in this chapter are likely to evolve in the next five to ten years

#	Touch technology	Predicted outcome
1	Projected Capacitive	Remains #1 in consumer devices; significant growth in commercial applications
2	Surface Capacitive	Disappears from the market in 5–7 years
3	Analog Resistive	Major reduction, but never completely disappears from the market
4	Digital Multi-Touch Resistive	Niche commercial and military applications
5	Analog Multi-Touch Resistive	Niche commercial and military applications
6	Surface Acoustic Wave (SAW)	Moderate continued growth in commercial applications
7	Acoustic Pulse Recognition (APR)	Niche non-display applications (touch-sensitive surfaces and devices)
8	Dispersive Signal Technology (DST)	Disappears from the market in less than five years
9	Traditional Infrared	Moderate reduction in large format; significant reduction in commercial applications; potential growth in mobile devices with reflective displays
10	Multi-Touch Infrared	Limited growth until multi-user gaming and/or collaboration applications become common and the cost of the technology drops to consumer levels
11	Camera-Based Optical	Significant growth only over 40″
12	In-Glass Optical (PSD)	Niche applications in large format, with potential applications in consumer all-in-one desktop computers
13	Vision-Based	Niche applications in large format
14	Embedded On-Cell Capacitive	Significant growth only in high-volume consumer devices; will become the most popular form of embedded touch due to minimum interference with the LCD cell process
15	Embedded Hybrid Capacitive	Significant growth only in high-volume consumer devices; #2 in popularity
16	Embedded In-Cell Capacitive	Significant growth only in high-volume consumer devices; #3 in popularity due to the high level of modification required in the LCD cell process
17	Embedded In-cell Light-Sensing	Disappears from the market in less than five years unless there's a breakthrough in solving the current problems
18	Force Sensing	Disappears from the market in less than five years or becomes a very small commercial niche

2.11 Appendix

All touch-technology suppliers mentioned in this chapter (excluding ones no longer in business) are listed in alphabetical order in Table 2.21, along with the technologies they supply and their website URL.

Table 2.21 This table lists all the touch technology suppliers mentioned in this chapter (excluding ones no longer in business), along with the technologies they supply, and their website URL

Company	Technologies	URL
3M Touch Systems	1,2,8	www.3mtouch.com
Apex Material Technology (AMT)	1,3,5	www.amtouch.com.tw
Apple	1,15	www.apple.com
Atmel	1	www.atmel.com
Baanto	11	www.baanto.com
Citron	10	www.citron.de
Cypress Semiconductor	1	www.cypress.com
Elo Touch Solutions	1,2,3,6,7,9	www.elotouch.com
FlatFrog	12	www.flatfrog.com
F-Origin	18	www.f-origin.com
General Touch	1,3,6,9,10,11	www.generaltouch.com
Gunze USA	1,3,4	www.gunzeusa.com
IDS Pulse	10	www.idspulse.com
Integrated Digital Technologies (IDTI)	17	www.idti.com.tw
Japan Display (JDI)	16	www.j-display.com
JTouch	1,3,5	www.jtouch.com.tw
LG Displays	1,15	www.lgdisplay.com
Lumio	1,6,10,11	www.lumio.com
Microsoft	13	www.microsoft.com
MultiTouch	13	www.multitaction.com
Nissha	1,3	www.nissha.com
Peratech	4	www.peratech.com
Planar	1,3,6,10,11	www.planar.com
PQ Labs	10	www.pqlabs.com
Quanta	11	www.quantatw.com
Samsung	14,15,17	www.samsung.com
Sharp	15,17	www.sharp-world.com
Shenzhen TimeLink Technology	10	www.timelink.cn
SMART Technologies	11	www.smarttech.com
Stantum	4	www.stantum.com
Synaptics	1,16	www.synaptics.com
Texas Instruments	1,3,5	www.ti.com
TPK	1	www.tpk.com
Visual Planet	1	www.visualplanet.biz
Wacom	1,2	www.wacom-components.com
ZaagTech	10	www.zaagtech.com
Zytronic	1	www.zytronic.co.uk

References

1. Johnson, E.A. (1965). Touch Display – A Novel Input/Output Device for Computers. *Electronics Letters* **1**(8), 219–220.
   Further Reading:
   Johnson, E.A. (1967). Touch Displays: A Programmed Man-Machine Interface. *Ergonomics* **10**(2), 271–277.
   Orr, N.W., Hopkins, V.D. (1968). The Role of Touch Display in Air Traffic Control. *The Controller* **7**, 7–9.
2. Shneiderman, B. (1991). Touch screens now offer compelling uses. *IEEE Software* **2**,93–94, 107.
3. Walker, G. (2007). Touch and the Apple iPhone, *Veritas et Visus Touch Panel* **12**, 50–54, (http://www.walkermobile.com/Touch_And_The_Apple_iPhone.pdf, retrieved 10/15/13).
4. DisplaySearch, 2008–2013. *Touch-Panel Market-Analysis Annual Reports.*
5. Buxton, B. (2007–2013). *Multi-Touch Systems that I Have Known and Loved.* Microsoft (www.billbuxton.com/multitouchOverview.html, retrieved 9/25/13).
6. Wigdor, D. (2011). The Breadth-Depth Dichotomy: Opportunities and Crises in Expanding Sensing Capabilities. *Information Display* **3**, 18–23, (http://informationdisplay.org/IDArchive/2011/March/EnablingTechnologyTheBreadthDepthDichotomy.aspx, retrieved 10/15/13).
   Further Reading:
   Wigdor, D., Wixon, D. (2011). *Brave NUI World: Designing Natural User Interfaces for Touch and Gesture.* Morgan Kaufmann (Elsevier), Burlington, MA.
7. Stumpe, B. (1977). A New Principle for X-Y Touch System, CERN (http://cds.cern.ch/record/1266588/files/StumpeMar77.pdf, retrieved 10/15/13).
   Further Reading:
   Stumpe, B., Sutton, C. (2010). The First Capacitive Touch Screens at CERN. *CERN Courier* (http://cerncourier.com/cws/article/cern/42092, retrieved 10/15/13).
   CERN Bulletin (2010). *Another of CERN's Many Inventions.* BUL-NA-2010-063, (http://cds.cern.ch/record/1248908, retrieved 10/15/13).
   Beck, F., Stumpe, B. (1973). *Two Devices for Operator Interaction in the Central Control of the New CERN.* CERN (http://cds.cern.ch/record/186242/files/CERN-73-06.pdf?version=1, retrieved 10/15/13).
   Stumpe, B. (1978). *Experiments to Find a Manufacturing Process for an X-Y Touch Screen.* CERN (http://cds.cern.ch/record/1266589/files/StumpeFeb78.pdf, retrieved 10/15/13).
8. Logan, J. (1991). *The History of MicroTouch: 1982–1992, a Decade of Touch Input.* Pamphlet published by the founder of MicroTouch Systems.
9. Binstead, R. (2009). *A Brief History of Projected Capacitance Development by Binstead Designs* (http://binsteaddesigns.com/history1.html, retrieved 10/15/13).
10. Barrett, G., Omote, R. (2010). Projected-Capacitive Touch Technology. *Information Display* **3**, 16–21, (http://informationdisplay.org/IDArchive/2010/March/FrontlineTechnologyProjectedCapacitiveTouchT.aspx, retrieved 10/15/13).
    Further Reading:
    3M Touch Systems (2011). *Touch Technology Brief: Projected Capacitive Technology* (http://solutions.3m.com/3MContentRetrievalAPI/BlobServlet?lmd=1332776667000&locale=en_US&assetType=MMM_Image&assetId=1319224170371&blobAttribute=ImageFile, retrieved 10/15/13).
11. Wang, T., Blankenship, T. (2011). Projected-Capacitive Touch Systems from the Controller Point of View. *Information Display* **3**, 8–12, (http://informationdisplay.org/IDArchive/2011/March/FrontlineTechnologyProjectedCapacitiveTouchS.aspx, retrieved 10/15/13).
    Further Reading:
    Lawson, R. (2012). *Challenges and Opportunities in Touch-Controller Semiconductors, HIS.* SID-IHS Future of Touch and Interactivity Conference, Boston, MA.
12. DisplaySearch (2013). *Touch-Panel Market-Analysis Annual Report.*
13. Poor, A. (2012). *How It Works: The Technology of Touch Screens.* Computerworld (http://www.computerworld.com/s/article/9231961/How_it_works_The_technology_of_touch_screens?taxonomyId=12&pageNumber=1, retrieved 10/15/13).
14. Bauman, C. (2007). How to Select a Surface-Capacitive Touch-Screen Controller. *Information Display* **12**:, 32–36 (http://informationdisplay.org/IDArchive/2007/December/HowtoSelectaSurfaceCapacitiveTouchScreenCo.aspx, retrieved 10/15/13).

15. Harrah's Entertainment (2006). *Profile of the American Casino Gambler* (http://www.org.id.tue.nl/ifip-tc14 /documents/HARRAH'S-SurveyCasinoGambler-2006.pdf, retrieved 10/15/13).

16. Wacom (2013). *RRFC™ Reversing Ramped-Field Capacitive Touch-Technology* (http://www.wacom-components.com/english/technology/touch.html, retrieved 10/15/13).

17. Wikipedia (2013). *Touchscreen* (http://en.wikipedia.org/wiki/Touchscreen, retrieved 10/15/13).

18. Emerson, L.G. (2010). *Samuel Hurst – the 'Tom Edison' of ORNL*. OakRidger (http://www.oakridger.com /article/20101214/NEWS/312149981?tag=1, retrieved 10/15/13).

19. Elo Touch Solutions (2013). *History of Elo* (http://www.elotouch.com/AboutElo/History/default.asp, retrieved 10/15/13).

20. Westinghouse Electric (1970). *Interface Device and Display System*. US Patent 3,522,664 (http://www .freepatentsonline.com/3522664.html, retrieved 10/15/13).

21. Sierracin/Intrex (1979). TransTech product brochure.

22. Downs, R. (2005). Using Resistive Touch Screens for Human/Machine Interface. *Texas Instruments Analog Applications Journal* (SLYT209A).

23. Barrett, G. (2012). *Decoding Touch Technology: An Insider's Guide to Choosing the Right Touch for your Display*. Touch International, white paper (http://touchinternational.com/literature/choosing-touch-technology-whitepaper.html, retrieved 10/15/13).
Further Reading:
Barrett, G. (2006). *Frit and the Better Touch Screen*. Touch International, white paper (http://touchinternational .com/literature/whitepapers/FritandtheBetterTouchScreen.pdf, retrieved 10/15/13).

24. DMC Co (2011). *Technologies of Touch Screens – Analog 8-Wire Resistive* (http://www.dmccoltd.com/english /museum/touchscreens/technologies/8-wire.asp, retrieved 10/15/13).

25. Semtech (2011). *SX8677/SX8678 Haptics-Enabled Multitouch 4/5-Wire Resistive Touchscreen Controller with Proximity Sensing*. Datasheet (http://www.semtech.com/images/datasheet/sx8677_8.pdf, retrieved 10/15/13).

26. Elo Touch Solutions (2013). *AccuTouch ZeroBezel Technology Specifications* (http://www.elotouch.com /Technologies/AccuTouch/accutouch_zero-bezel_specifications.pdf, retrieved 10/15/13).

27. Fenn, J. (2013). *Turned Around a Potential Loss to Revolutionize a Business*. Professional Resume (http://www.corporatewarriors.com/john7820/fenn.doc, retrieved 10/15/13).

28. DMC Co (2011). *Technologies of Touch Screens – Digital Matrix Resistive* (http://www.dmccoltd.com/english /museum/touchscreens/technologies/Matrix.asp, retrieved 10/15/13).

29. Largillier, G. (2007). Developing the First Commercial Product that Uses Multi-Touch Technology. *Information Display* **12**, 14–18 (http://informationdisplay.org/IDArchive/2007/December/DevelopingtheFirstCommercial ProductthatUses.aspx, retrieved 10/15/13).

30. Stantum (2012). *Stantum's Newest Digital Resistive Touch-Panel* (http://www.leavcom.com/stantum_061012 .php, retrieved 10/15/13).

31. Texas Instruments (2011). *Analog Matrix Touchscreen Controller*. Datasheet (http://www.mouser.com/ds/2/405 /sbas536b-94322.pdf, retrieved 10/15/13).

32. Apex Material Technology (AMT) (2013). *Multi-Finger (MF) Touch* (http://www.amtouch.com.tw/products /advanced-resistive-touch-screen/multi-finger-mf-touch/, retrieved 10/15/13).

33. Adler, R., Desmares, P.J. (1985). *An Economical Touch Panel Using SAW Absorption*. IEEE Ultrasonics Symposium 1985, 499–502. (10.1109/ULTSYM.1985.198560).

34. Kent, J. *et al.* (2007). *Robert Adler's Touchscreen Inventions*. IEEE 2007 Ultrasonics Symposium, 9–20 (10.1109/ULTSYM.2007.18).

35. Elo Touch Solutions (1997–2001). *Acoustic Touch Position Sensor Using a Low Acoustic Loss Transparent Substrate*. US Patent 6,236,391 (http://www.freepatentsonline.com/6236391.html, retrieved 10/15/13).

36. Elo Touch Solutions (2010–2013). *Acoustic Condition Sensor Employing a Plurality of Mutually Non-Orthogonal Waves*. US Patent 8,421,776 (http://www.freepatentsonline.com/8421776.html, retrieved 10/15/13).

37. Microsoft (2013). *Windows Certification Program: Hardware Certification Taxonomy & Requirements for Windows 8.1, Device Digitizer Requirements* (http://msdn.microsoft.com/en-us/library/windows/hardware/jj134351 .aspx, retrieved 10/15/13).

38. Elo Touch Solutions (2011–2013). *Bezel-less Acoustic Touch Apparatus*. US Patent 8,576,202 (http://www.freepatentsonline.com/8576202.html, retrieved 10/15/13).
39. North, K., D'Souza, H. (2006). Acoustic Pulse Recognition Enter Touch-Screen Market. *Information Display* **12**, 22–25 (http://informationdisplay.org/IDArchive/2006/December/AcousticPulseRecognitionEntersTouchScreenMar.aspx).
40. Kent, J. (2010). New Touch Technology from Time Reversal Acoustics: A History. *IEEE International Ultrasonics Symposium Proceedings* 1173–1178.
41. Butcher, M. (2010). The $62 Million Sale of a Touch Tech Startup Adds to the Tablet Revolution. *TechCrunch* (http://techcrunch.com/2010/01/27/the-62-million-sale-of-a-touch-tech-startup-adds-to-the-tablet-revolution/, retrieved 10/15/13).
42. 3M Touch Systems (2008). *Dispersive Signal Touch Technology: Technology Profile*. White paper (http://multimedia.3m.com/mws/mediawebserver?mwsId=66666UF6EVsSyXTtmxTXoxfaEVtQEVs6EVs6EVs6E666666--&fn=DST%20Tech%20Profile.pdf, retrieved 10/15/13).
43. Wikipedia (2013). *PLATO Computer System* (http://en.wikipedia.org/wiki/PLATO_IV, retrieved 10/15/13).
44. Wikipedia (2013). *HP-150* (http://en.wikipedia.org/wiki/HP_150, retrieved 10/15/13).
45. Elo Touch Solutions (2009). *Elo TouchSystems IntelliTouch Plus Multi-Touch & Windows-7*. Communication Brief (http://www.elotouch.com/pdfs/faq_ip.pdf, retrieved 10/15/13).
46. Charters, R. (2009). High-Volume Manufacturing of Photonic Components on Flexible Substrates. *Information Display* **12**, 12–16 (http://informationdisplay.org/IDArchive/2009/December/FrontlineTechnologyHighVolumeManufacturingof.aspx, retrieved 10/15/13).
47. Thompson, M. (2009). *RPO Digital Waveguide Touch*. DisplaySearch 2009 Emerging Display Technologies Conference, San Jose, CA.
Further Reading:
Maxwell, I. (2007). An Overview of Optical-Touch Technologies. *Information Display* **10**, 26–30 (http://informationdisplay.org/IDArchive/2007/December/AnOverviewofOpticalTouchTechnologies.aspx, retrieved 10/15/13).
48. Poa Sana (1997–1999). *User Input Device for a Computer System*. US Patent 5,914,709 (http://www.freepatentsonline.com/5914709.html, retrieved 10/15/13).
49. PQ Labs (2010–2012). *System and Method for Providing Multi-Dimensional Touch Input Vector*. US Patent Application 2012-0098753 (http://www.freepatentsonline.com/y2012/0098753.html, retrieved 10/15/13).
50. Zytronic (2013). *Zytronic Reveals a New Dimension in Touchscreens for Gaming Machines*. Press release (http://www.zytronic.co.uk/assets/Uploads/ZY370-G2E-2013-Zytronic-reveals-a-new-dimension-in-touchscreens-for-gaming-machines.pdf, retrieved 10/15/13).
51. PQ Labs website, (http://www.pqlabs.com, retrieved 10/15/13).
52. Walker, G. (2011). Camera-Based Optical Touch Technology. *Information Display* **3**, 30–34 (http://informationdisplay.org/IDArchive/2011/March/FrontlineTechnologyCameraBasedOpticalTouchT.aspx), retrieved 10/15/13.
53. Baanto (2011). *ShadowSense™ Touch Detection*. White paper (http://baanto.com/uploads/Image/pdfs/whitepapers/shadowsense_touch_detection.pdf, retrieved 10/15/13).
54. Baanto (2011). *Rain and Fluid Discrimination for Touchscreens*. White paper (http://baanto.com/uploads/Image/pdfs/whitepapers/shadowsense_rain_rejection.pdf, retrieved 10/15/13).
55. Wassvik, O. (2013). *PSD: In-Glass Optical Touch for Larger Form-Factors*. DisplaySearch Emerging Display Technologies Conference, San Jose, CA.
56. Walker, G., Finn, M. (2010). Beneath the Surface. *Information Display* **3**, 31–34, (http://informationdisplay.org/IDArchive/2010/March/EnablingTechnologyBeneaththeSurface.aspx).
57. Castle, A. (2009). Build Your Own Multitouch Surface Computer. *Maximum PC* (http://www.maximumpc.com/article/features/maximum_pc_builds_a_multitouch_surface_computer?page=0,0, retrieved 10/15/13).
58. NUI Group Authors (2009). *Multitouch Technologies*. e-book, NUI Group (http://www.google.com/url?sa=t&rct=j&q=&esrc=s&frm=1&source=web&cd=1&ved=0CCwQFjAA&url=http%3A%2F%2Fnuicode.com%2Fattachments%2Fdownload%2F115%2FMulti-Touch_Technologies_v1.01.pdf&ei=C7yBUqn6N4G3igKz9YDIDQ&usg=AFQjCNGWgAwHLy64d0YfObchamgKkeby8g&bvm=bv.56343320,d.cGE, retrieved 10/15/13).

59. Anttila, H. (2012). *Multi-User Interactive Technology Advancements for Any Size Display*. IHS Touch-Gesture-Motion Conference, Austin, TX.
    Further Reading:
    MultiTouch (2011). *MultiTaction: Technology Platform for MultiTouch LCDs of Any Size*. White paper http://multitouch.s3.amazonaws.com/resources/brochures_for_print/whitepaper_MultiTaction_v1-1_USletter _print.pdf, retrieved 10/15/13).

60. Samsung Electronics (2011). *Samsung SUR40 for Microsoft Surface Venue Readiness Guide* (http://www.samsung.com/us/pdf/sur40/SUR40_Venue_Readiness_Guide.pdf, retrieved 10/15/13).

61. Microsoft (2012). *Tagged Object Integration for Surface 2.0*. White paper. (http://download.microsoft.com /download/D/7/B/D7BE282A-FCB2-4A2C-AC48-6BC8441AB281/Tagged%20Objects%20for%20Surface% 202.0%20Whitepaper.docx, retrieved 10/15/13).

62. Walker, G. (2013). Embedded Touch: The Touch-Panel Makers vs. The Display-Makers, FPD International Conference, Yokohama, Japan, (http://www.walkermobile.com/FPD_International_2013_Touch_Futures.pdf, retrieved 10/15/13).

63. Walker, G., Finn, M. (2010). LCD In-Cell Touch. *Information Display* 3, 8–14. (http://informationdisplay .org/IDArchive/2010/March/FrontlineTechnologyLCDInCellTouch.aspx, retrieved 10/15/13).

64. Lee, J. *et al.* (2007). Hybrid Touch Screen Panel Integrated in TFT-LCD. *SID 2007 Symposium Digest* 24.3.

65. Samsung Electronics (2004–2007). *Liquid Crystal Display Device Having Touch Screen Function and Method of Fabricating the Same*. US Patent 7,280,167 (http://www.freepatentsonline.com/7280167.html, retrieved 10/15/13).

66. Den Boer, W. *et al.* (2003). Active Matrix LCD with Integrated Optical Touch Screen. *SID 2003 Symposium Digest* 56.3.

67. Samsung Electronics (2007–2011). *Touch Screen Display Apparatus and Method of Driving the Same*. US Patent 8,072,430 (http://www.freepatentsonline.com/8072430.html, retrieved 10-15-13).

68. Ozbas, M. *et al.* (2012). An In-Cell Capable Capacitive Touchscreen Controller with High SNR and Integrated Display Driver IC for WVGA LTPS Displays. *SID Symposium 2012 Digest* 485–488.

69. Mackey, B. (2013). *Touch + Display, Any Way You Want It, Synaptics*. SID Display Week Conference (Session M8), Vancouver, Canada.

70. Synaptics. (2010). *Capacitive Sensing Using a Segmented Common Voltage Electrode of a Display*. US Patent Application 2010-0238134 (http://www.freepatentsonline.com/y2010/0238134.html, retrieved 10/15/13).

71. Noguchi, K. (2012). *Trend of In-Cell Touch Panel Technologies*. Sony, FPD International Conference, Yokohama, Japan.

72. Apple (2010). *Integrated Touch Screen*. US Patent 7,859,521 (http://www.freepatentsonline.com/7859521.html, retrieved 11/15/13).
    Further Reading:
    Apple (2011–2013). *Segmented Vcom*. US Patent 8,451,244 (http://www.freepatentsonline.com/8451244.html, retrieved 10/15/13).
    Apple (2009–2011). *Integrated Touch Screen*. US Patent 7,995,041 (http://www.freepatentsonline.com /7995041.html, retrieved 10/15/13).
    Apple (2010–2010). *Integrated Touch Screen*. US Patent 7,859,521 (http://www.freepatentsonline.com /7859521.html, retrieved 10-15-13).

73. Wu, C.W. (2013). *On/In Cell Touch Sensor Embedded in TFT-LCD*. BOE Technology Group, FPD China Conference, Shanghai, China.

74. Toshiba Matsushita Display (TMD) (2003). *Toshiba America Electronic Components Demonstrates First System on Glass (SOG) Input Display with Built-In Image Capture*. Press release (http://www.toshiba.com/taec/news /press_releases/2003/to-314.jsp, retrieved 10/15/13).

75. Toshiba Matsushita Display (TMD) (2005). *Toshiba Matsushita Display Announces World's First LTPS TFT LCD Prototype with Finger Shadow Sensing Input Capability*. Press release (http://www.toshiba-components.com /prpdf/5615e.pdf, retrieved 10/15/13).

76. Chen, Z.H. (2011). IDTI: In-Cell Optical Touch Panel. *Optoelectronic Notes Blog* (http://ntuzhchen.blogspot .com/2011/03/idti-in-cell-touch-panel.html, retrieved 10/15/13).

77. Samsung Electronics, 2014, Product Webpage for Samsung SUR40 with Microsoft PixelSense (formerly Microsoft Surface 2.0), (http://www.samsung.com/ae/business/business-products/large-format-display /specialized-display/LH40SFWTGC/XY, retrieved 03/18/14).

78. Samsung Mobile Display (2011–2012). *Organic Light Emitting Display Having Touch Screen Function*. US Patent Application 2012-0105341 (http://www.freepatentsonline.com/y2012/0105341.html, retrieved 10/15/13).

79. Soss, D. (2007). Advances in Force-Based Touch Panels. *Information Display* **12**, 20–24 (http://information display.org/IDArchive/2007/December/AdvancesinForceBasedTouchPanels.aspx, retrieved 10/15/13).

80. Fihn, M. (2009). Interview with Garrick Infanger from Vissumo. *Veritas et Visus Touch Panel* **3**(7/8), 80–83.

81. F-Origin (2013). *Force-Based Touch-Screen Technology* (http://www.f-origin.com/zTouch0153Technology.aspx, retrieved 10/15/13).

82. NextInput (2013). *Force-Sensitive Touch Technology* (http://nextinput.com/t/ForceTouch, retrieved 10/15/13).

83. Wacom (2013). *Touch Panels Product* (http://wacom.jp/en/products/components/displays/touch/index.html, retrieved 10/15/13).

3

Voice in the User Interface

Andrew Breen, Hung H. Bui, Richard Crouch, Kevin Farrell,
Friedrich Faubel, Roberto Gemello, William F. Ganong III, Tim Haulick,
Ronald M. Kaplan, Charles L. Ortiz, Peter F. Patel-Schneider, Holger Quast,
Adwait Ratnaparkhi, Vlad Sejnoha, Jiaying Shen, Peter Stubley and
Paul van Mulbregt
Nuance Communications, Inc.

3.1 Introduction

Voice recognition and synthesis, in conjunction with natural language understanding, are now widely viewed as essential aspects of modern mobile user interfaces (UIs). In recent years, these technologies have evolved from optional 'add-ons', which facilitated text entry and supported limited command and control, to the defining aspects of a wide range of mainstream mobile consumer devices, for example in the form of voice-driven smartphone virtual assistants. Some commentators have even likened the recent proliferation of voice recognition and natural language understanding in the UI as the "third revolution" in user interfaces, following the introduction of the graphical UI controlled by a mouse, and the touch screen, as the first and second respectively.

The newfound prominence of these technologies is attributable to two primary factors: their rapidly improving performance, and their ability to overcome the inherent structural limitations of the prevalent 'shrunken desktop' mobile UI by accurately deducing user intent from spoken input.

The explosive growth in the use of mobile devices of every sort has been accompanied by an equally precipitous increase in 'content', functionality, services, and applications available to the mobile user. This wealth of information is becoming increasingly difficult to organize, find, and manipulate within the visual mobile desktop, with its hierarchies of folders, dozens if not hundreds of application icons, application screens, and menus.

Often, performing a single action with a touchscreen device requires multiple steps. For example, the simple act of transferring funds between a savings and checking accounts using a typical mobile banking application can require the user to traverse a dozen application screens.

Interactive Displays: Natural Human-Interface Technologies, First Edition. Edited by Achintya K. Bhowmik.
© 2015 John Wiley & Sons, Ltd. Published 2015 by John Wiley & Sons, Ltd.

The usability problem is further exacerbated by the fact that there exists great variability in the particular UIs of different devices. There are now many mobile device "form factors", from tablets with large screens and virtual keyboards to in-car interfaces intended for eyes and hands-busy operation, television sets which may have neither a keyboard nor a convenient pointing device, as well as "wearable" devices (e.g., smart eyeglasses and watches). Users are increasingly attempting to interact with similar services – making use of search, email, social media, maps and navigation, playing music and video – through these very dissimilar interfaces.

In this context, voice recognition (VR) and natural language understanding (NLU) represent a powerful and natural control mechanism which is able to cut through the layers of the visual hierarchies, intermediate application screens, or Web pages. Natural language utterances encode a lot of information compactly. Saying, "Send a text to Ron, I'm running ten minutes late" implicitly specifies which application should be launched, and who the message is to be sent to, as well as the message to send, obviating the need to provide all the information explicitly and in separate steps. Similarly, instructing a TV to "Play the *Sopranos* episode saved from last night" is preferable to traversing the deep menu structure prevalent in conventional interfaces. These capabilities allow the creation of a new UI: a virtual assistant (VA), which interacts with a user in a conversational manner and enables a wide range of functionalities.

In the example above, the user does not need first to locate the email application icon in order to begin the interaction. Using voice and natural language thus makes it possible to find and manipulate resources – whether or not they are visible on the device screen, and whether they are resident on the device or in the Cloud – effectively expanding the boundary of the traditional interface by transparently incorporating other services.

By understanding the user's intent, preferences, and interaction history, an interface incorporating voice and natural language can resolve search queries by directly navigating to Web destinations deemed useful by the user, bypassing intermediate search engine results pages. For example, a particular user's product query might result in a direct display of the relevant page of her or his preferred online shopping site.

Alternately, such a system may be able to answer some queries directly by extracting the desired information from either structured or unstructured data sources, constructing an answer by applying natural language generation (NLG), and responding via speech synthesis.

Finally, actions which are difficult to specify in point-and-click interfaces become readily expressible in a voice interface, such as, for example, setting up a notification which is conditioned on some other event: "Tell me when I'm near a coffee shop."

There are also other means of reducing the number of steps required to fulfill a user's intent. It is possible for users to speak their request naturally to a device without even picking it up or turning it on. In the so-called "seamless wake-up" mode, a device listens continuously for significant events, using energy-efficient algorithms residing on digital signal processors (DSP). When an interesting input is detected, the device activates additional processing modules in order to confirm that the event was a valid command spoken by the device's owner (using biometrics to confirm his or her identity), and it then takes the desired action.

Using natural language in this manner presupposes voice recognition which is accurate for a wide population of users and robust in noisy environments. Voice recognition performance has improved to a remarkable degree over the past several years, thanks to: an ever more powerful computational foundation (including chip architectures specialized for voice recognition); fast and increasingly ubiquitous connectivity, which brings access to Cloud-based computing to even the smallest mobile platforms; the development of novel algorithms and modeling techniques (including a recent resurgence of neural-network-based models); and the utilization of massive data sets to train powerful statistical models.

Voice recognition also benefits from increasingly sophisticated signal acquisition techniques, such as the use of steerable multi-microphone beamforming and noise cancellation algorithms to achieve high accuracy in noisy environments. Such processing is especially valuable in car interiors and living rooms, which present the special challenges of high ambient noise, multiple talkers and, frequently, background entertainment soundtracks.

The recent rate of progress on extracting meaning from natural utterances has been equally impressive. The most successful approaches blend three complementary approaches:

- machine learning, which discovers patterns from data;
- explicit linguistic "structural" models; and
- explicit forms of knowledge representation ("ontologies") which encode known relationships and entities *a priori*.

As is the case for voice recognition, these algorithms are adaptive, and they are able to learn from each interaction.

Terse utterances can on their own be highly ambiguous, but can nonetheless convey a lot of information because human listeners apply context in resolving such inputs. Similarly, extracting the correct meaning algorithmically requires the application of a world model and a representation of the interaction context and history, as well as potentially other forms of information provided by other sensors and metadata. In cases where such information is insufficient to disambiguate the input, voice and natural language interfaces may engage in a dialog with the user, eliciting clarifying information.

Dialog, or conversation management, has evolved from early forms of "system initiative" which restricted users to only answering questions posed by an application (either visually, or via synthetic speech), to more flexible "mixed initiative" variants which allow users to provide relevant information proactively. The most advanced approaches apply formal reasoning – the traditional province of Artificial Intelligence (AI) – to eliminate the need to pre-define every possible interaction, and to infer goals and plans dynamically.

Whereas early AI proved to be brittle, today's state of the art systems rely on more flexible and robust approaches that do well in the face of ambiguity and produce approximate solutions where an exact answer might not be possible. The goal of such advanced systems is to successfully handle so-called "meta-tasks" – for example, requests such as "Book a table at Zingari's after my last meeting and let Tom and Brian know to meet me there", rather than requiring users to perform a sequence of the underlying "atomic" tasks, such as checking calendars and making a reservation.

Thus, our broad view of the 'voice interface' is that it is, in fact, an integral part of an intelligent system which:

- interacts with users via multiple modalities;
- understands language;
- can converse and perform reasoning;
- uses context and user preferences;
- possess specialized knowledge;
- solves high-value tasks;
- is robust in realistic environments.

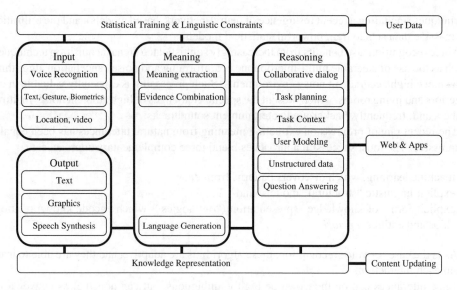

Figure 3.1 Intelligent voice interface architecture.

The elements of such a system, shown in Figure 3.1, are typically distributed across client devices and the cloud.

The reason for doing so includes optimizing computation, service availability, and latency, as well as providing users with a consistent experience across multiple clients with dissimilar characteristics and capabilities.

A distributed architecture further enables the aggregation of user data from multiple devices, which can be used to continually improve server- as well as device-specific recognition and NLU models. Furthermore, saving the interaction histories in a central repository means that users can seamlessly start an interaction on one device and complete it on another.

The following sections describe these concepts and the underlying technologies in detail.

3.2 Voice Recognition

3.2.1 Nature of Speech

Speech is uniquely human, optimized to allow people to communicate complex thoughts and feelings (apparently) effortlessly. Thus the 'speech channel' is heavily optimized for the task of human-to-human communication. The atomic linguistic building blocks of spoken utterances are called phonemes. These are the smallest units that, if changed, can alter the meaning of a word or an utterance. The physical manifestations of phonemes are "phones"; but a speech signal is not simply a sequence of concatenated sounds, like Morse code. We produce speech by moving our articulators (tongue, jaw, lips) in an incredibly fast and well-choreographed dance that creates a rapidly shifting resonance structure. Our vocal cords open and close between 100–300 times a second, producing a signal called the fundamental frequency or F0, which excites vocal tract resonances, resulting in a high bandwidth sound (e.g. 0–10 kHz).

At other times, the resonances are excited by turbulent noise created at constrictions in the vocal tract, such as the sound of [s]. The acoustic expression of a phoneme is not fixed but, rather, the realization is influenced by both the preceding as well as the anticipated subsequent phoneme – a phenomenon called *coarticulation*. Additional variability is introduced when speakers adjust their speech to the current situation and the needs of their listeners. The resulting speech signal reflects these moving articulators and sound sources in a complex and rapidly varying signal. Figure 3.2 shows a speech spectrogram of a short phrase.

Advances in speech recognition accuracy and performance are the result of thousands of person-years of scientific and engineering effort. Hence state-of-the-art recognizers include many carefully optimized, highly engineered components. From about 1990 to 2010, most state-of-the-art systems were fairly similar, and showed gradual, incremental improvements. In the following we will describe the fundamental components of a "canonical" voice recognition system, and briefly mention some recent developments.

The problem solved by canonical speech recognizers is typically characterized by Bayes' rule:

$$W^* = \arg\ \max_{\widetilde{W}}(P(\widetilde{W}|\overline{O})) \tag{3.1}$$

That is, the goal of speech recognition is to find the highest probability sequence of words, W^*, given the set of acoustic observations, $\overline{O}$. Using Bayes' rule, we get:

$$P(\widetilde{W}|\overline{O}) = \frac{P(\overline{O}|\widetilde{W})P(\widetilde{W})}{P(\overline{O})} \tag{3.2}$$

Note that $P(\overline{O})$ does not depend on the word sequence $\widetilde{W}$, so we want to find:

$$W^* = \arg\ \max_{\widetilde{W}}(P(\overline{O}|\widetilde{W})P(\widetilde{W})) \tag{3.3}$$

We evaluate $P(\overline{O}|\widetilde{W})$ using an *acoustic model* (AM), and $P(\widetilde{W})$ using a *language model* (LM).

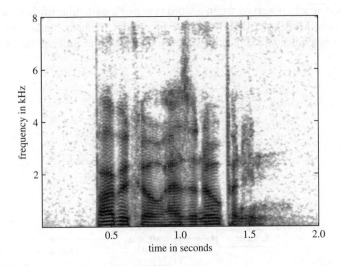

Figure 3.2 A speech spectrogram of the phrase "Barbacco has an opening". The X-axis shows time, the Y-axis, frequency. Darkness shows the amount of energy in a frequency region.

Thus, the goal of most speech recognizers is to find the sequence of words with the highest joint probability of producing the acoustic observation, given the structure of the language.

It turns out that the system diagram for a canonical speech recognition system maps well onto this equation, as shown in Figure 3.3.

The evaluation of acoustic probabilities is handled by the acoustic front-end and an acoustic model; the evaluation of probabilities of word sequences is handled by a language model; and the code which finds the best scoring word sequence is called the search component. Although these modules are logically separate, their application in speech recognition is highly interdependent.

3.2.2 Acoustic Model and Front-end

Front-end: Incoming speech is digitized and transformed to a sequence of vectors which capture the overall spectrum of the input by an "acoustic front-end". For years, the standard front-end was based on using a vector of mel-frequency cepstral coefficients (MFCC) to represent each "frame" of speech (of about 25 msec) [1]. This representation was chosen to represent the whole spectral envelope of a frame, but to suppress harmonics of the fundamental frequency. In recent years, other representations have become popular [2] (see below).

Acoustic model: In a canonical system, speech is modeled as a sequence of words, and words as sequences of phonemes. However, as mentioned above, the acoustic expression of a phoneme is very dependent on the sounds and words around them, as a result of coarticulation. Although the context dependency can span several phonemes or syllables, many systems approximate phonemes using "triphones", i.e., phonemes conditioned by their left and right phonetic context. Thus, a sequence of words is represented as a sequence of triphones. There are many possible triphones (e.g. $\approx 50^3$), and many of them occur rarely. Therefore the standard technique is to cluster them, using decision trees, [3], then create models for the clusters rather than the individual triphones.

The acoustic features observed when a word contains a particular triphone are modeled as a Hidden Markov Model (HMM), [4] – see Figure 3.4. Hidden Markov models are simple finite state machines (FSMs), with states, transitions, and probabilities associated with

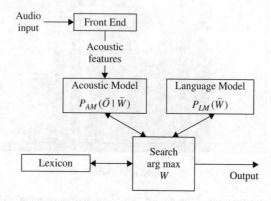

Figure 3.3 Components of a canonical speech recognition system.

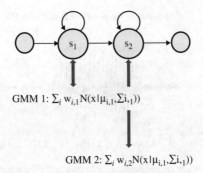

GMM 1: $\Sigma_i w_{i,1} N(x|\mu_{i,1},\Sigma i,_1))$

GMM 2: $\Sigma_i w_{i,2} N(x|\mu_{i,1},\Sigma i,_1))$

Figure 3.4 A simple HMM, consisting of states, probability distributions, and PDFs. The PDFs are Gaussian mixture models, which evaluate the probability of an input frame, given an HMM state.

the transitions. Also, each state is associated with a probability density function (PDF) over possible front-end vectors.

The probability density function is usually represented as a Gaussian Mixture Model (GMM). GMMs are well studied, easily trained PDFs which can approximate arbitrary PDF shapes well. A GMM is a weighted sum of Gaussians; each Gaussian can be written as:

$$N\left(x|\mu, \sum\right) = \frac{1}{\sqrt{\det\left(\sum\right)}\sqrt{(2\pi)^n}} * \exp\left(-\frac{1}{2}(x-\mu)^T \sum^{-1}(x-\mu)\right) \quad (3.4)$$

Where x is an input vector, μ is a vector of means, and Σ is the covariance matrix. x and μ are vectors of length n, and Σ is a square matrix of dimension n^2 and each GMM is a simple weighted sum of Gaussians, i.e.,

$$GMM\left(x|w_{1,n}, \mu_{1,n}, \sum_{1,n}\right) = \sum_i \left(w_i N\left(x|\mu_i, \sum_i\right)\right) \quad (3.5)$$

3.2.3 Aligning Speech to HMMs

In the speech stream, phonemes have different durations, so it's necessary to find an alignment between input frames and states of the HMM. That is, given input speech frames, $\overline{O}$ and a sequence of HMM states $\tilde{H}$, an alignment A maps the frames monotonically into the states of the HMMs. So the system needs to find the optimal (i.e. highest probability) alignment A between the frames (f) and the HMM states.

$$P_{AM}(\overline{O}|\tilde{H}) \cong \max_A \prod_f (P(O_f|H_{A(f)}) \quad (3.6)$$

This is often done using a form of the Viterbi algorithm [5].

For each hypothesized word sequence, the system looks up the phonemes that make up the pronunciation of each word from the lexicon, and then looks up the triphone for each phoneme in context, using decision trees. Then, given the sequence of triphones, the system looks up the sequence of HMM states. The acoustic probability of that hypothesis is the probability of the optimal alignment of these states with the input. An example of such an alignment is shown in Figure 3.5.

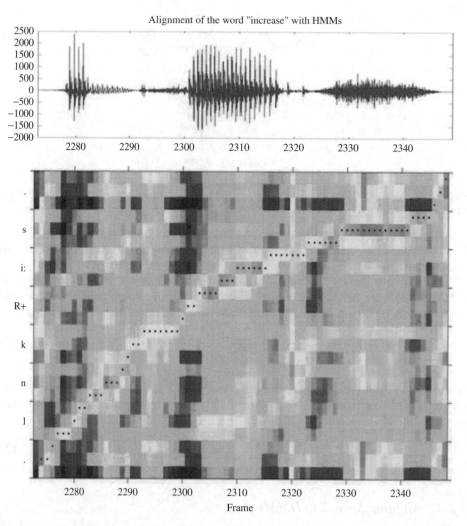

Figure 3.5   Viterbi alignment of a speech signal (x axis); against a sequence of HMMs (y axis); lighter areas indicate higher probabilities as assessed by the HMM for the given frame; dotted line shows alignment. (See color figure in color plate section).

3.2.4   Language Model

The language model computes the probability of various word sequences and helps the recognition system propose the most likely interpretation of input utterances. There are two fundamentally different types of language models used in voice recognition systems: grammar-based language models, and stochastic language models.

Grammar-based language models allow some word sequences, but not others. Often, these grammars are application-specific, and support utterances relevant to some specific task, such as making a restaurant reservation or issuing a computer command. These grammars specify the exact word sequences which the user is supposed to utter in order to instruct the system. For example, a grammar for a reservation system might recognize sequences like "find a Chinese restaurant nearby", "reservation for two at seven", or "show me the menu". The same grammar would probably not recognize sequences like "pepperoni pizza", "economic analysis of the restaurant business", or "colorless green ideas sleep furiously".

The set of word-sequences recognized by a grammar is described by a formal grammar, e.g. a finite-state machine, or a context-free grammar. Often, these grammars are written in formalism like Speech Recognition Grammar Specification (SRGS) (see [6]). While it is easy to construct simple example grammars, writing a grammar that covers all the ways a user might like to specify an input can be difficult. Thus, one might say "nearby Chinese restaurant", "please find a nearby Chinese restaurant", "I'd like to eat Chinese", or "Where can I find some dim sum?" All these sentences mean the same thing (to a restaurant ordering app), but writing a grammar that can cover all the choices can be onerous because of users' creative word choices.

Stochastic language models (originally used for free-text dictation) estimate the probability of any word sequence (but some are much more likely than others.) Thus, "Chinese restaurant" would have a reasonable probability; "restaurant Chinese" would have a somewhat smaller probability; and "nearby restaurant Chinese a find" would have a much lower probability. Attempts to write grammar-based language models to cover all the ways a user could generate English text have not been successful; so for general dictation applications, stochastic language models have been preferred. It turns out that it is much easier to make a robust, application-specific language model by using a stochastic LM, followed by NLU processing even for quite specific applications.

The job of stochastic language modeling is to calculate an approximation to $P(\widetilde{W})$, which is, by definition:

$$P(\widetilde{W}) = \prod_i P(w_i|w_{i-1}, w_{i-2}, w_{i-3}, \ldots, w_2, w_1) \tag{3.7}$$

A surprising development in speech recognition was that a simple approximation (the trigram approximation) works very well:

$$P(w_i|w_{i-1}, w_{i-2}, w_{i-3}, \ldots, w_2, w_1) \approx P(w_i|w_{i-1}, w_{i-2}) \tag{3.8}$$

The trigram approximation holds that the likelihood of the next word in a sentence depends only on the previous two words, (and an n-gram language model is the obvious generalization for longer spans). Scientifically and linguistically, this is inaccurate: many language phenomena span more than two words [7]! Yet it turns out that this approximation works very well in speech recognition [8].

3.2.5   Search: Solving Crosswords at 1000 Words a Second

The task of finding the jointly optimal word sequence to describe an acoustic observation is rather like solving a crossword, where the acoustic scores constrain the columns, and LM scores constrain the rows. However, there are too many possible word sequences to evaluate directly (a 100,000 word vocabulary generates 10^{50} ten-word sentences). The goal of the search component is to identify the correct hypothesis, and evaluate as few other hypotheses as possible. It does this using a number of heuristic techniques. One particularly important technique is called a beam search, which processes input frames one by one, and keeps "alive" a set of hypotheses with scores near that of the best hypothesis.

How much computation is required? For a large vocabulary task, on a typical frame, the search component might have very roughly 1000 active hypotheses. Updating the scores for an active hypothesis involves extending the alignment by computing acoustic model (GMM) scores for the current HMM state of that hypothesis, and the next state, and then updating the alignment. If the hypothesis is at the end of a word, (about 20 word-end hypotheses per frame), then the system also needs to look up LM scores for potential next words (about 100 new words per word ending). Thus, we need about 2000 GMM, 1000 alignment computations, and 2000 LM lookups, per frame. At a typical frame rate of 100 Hz, we are computing about 200 k GMMs, 200 k LM lookups, and 100 k alignment updates per second.

3.2.6   Training Acoustic and Language Models

The HMMs used in acoustic models are created from large datasets by an elaborate training process. Speech data is transcribed, and provided to a training algorithm which utilizes a maximum likelihood objective function. This algorithm estimates acoustic model parameters so as to maximize the likelihood of observing the training data, given the transcriptions. The heart of this process bootstraps an initial approximate acoustic model into an improved version, by aligning the training speech against the transcriptions and re-training the HMMs. This process is repeated many times to produce Gaussian mixture models which score the training data as high likelihood.

However, the goal of speech recognition is not to represent the most likely sequence of acoustic states, but rather to give correct word sequence hypotheses higher probabilities than incorrect hypotheses. Hence, various forms of discriminative training have been developed, which adjust the acoustic models so as to decrease various measures related to the recognition error rate [9–11].

The resulting acoustic models routinely include thousands of states, hundreds of thousands of mixture model components, and millions of parameters. Canonical systems used "supervised" training, i.e. used both speech and the associated transcriptions for training. As speech data sets have grown, there has been substantial effort to find training schemes using un-transcribed or "lightly labeled" data.

Stochastic language models are trained from large text datasets containing billions of words. Large text databases are collected from the World Wide Web, from specialized text databases, and from deployed voice recognition applications. The underlying training algorithm is much simpler than that used in acoustic training (it is basically a form of counting), but finding good data, weighting data carefully, and handling unobserved word sequences require considerable engineering skill. The resulting language models often include hundreds of thousands to billions of n-grams, and have billions of parameters.

3.2.7   Adapting Acoustic and Language Models for Speaker Dependent Recognition

People speak differently. The words they pick, and the way they say words is influenced by their anatomy, accent, education, and the intentional style of speaking (e.g., dictating a formal document vs. an informal SMS message).

The resulting differences in pronunciation may confuse a speaker-independent voice recognition system, which may not have encountered training exemplars with these particular combinations of characteristics. In contrast, a speaker dependent system which models a single speaker might achieve a higher accuracy than a speaker-independent system. However, users are not inclined to record thousands of hours of speech in order to train a voice recognition system, and it is thus highly desirable to produce speaker-dependent acoustic and language models by adapting speaker-independent models utilizing limited data from a particular user.

There are many kinds of adaptation for acoustic models. Early work often used MAP (*maximum a posteriori*) training, which modifies the means and variances of the GMMs used by the HMMs. MAP adaptation is generally "data hungry", since it needs to utilize training examples for most of the GMMs used in the system. More data-efficient alternatives modify GMM parameters for whole classes of triphones (e.g. MLLR, maximum likelihood linear regression [12]). Transformations can be applied either to the models or to the input features. While "canonical" adaptation is "supervised" (i.e. uses speech data with transcriptions), some forms of adaptation now are unsupervised, using the input speech data and recognition hypotheses without a person checking the correctness of the transcriptions.

Language models are also adapted for users or tasks. Adaptation can modify either single parameters (i.e. adjust the counts which model data for a particular n-gram, analogous to MAP acoustic adaptation), or can effectively adapt clusters of parameters (analogous to MLLR). For instance, when building a language model for a new domain, one can use interpolation weights to combine n-gram statistics from different corpora.

3.2.8   Alternatives to the "Canonical" System

In the previous paragraphs, we have outlined the basics of a voice recognition system. Many alternatives have been proposed; we mention a few prominent alternatives in Table 3.1.

3.2.9   Performance

Speech recognition accuracy has steadily increased in performance during the past several decades. When early dictation systems were introduced in the late 1980s, a few early adopters eagerly embraced them and used them successfully. Many others found the error rate too high and decided that speech recognition "wasn't really ready yet". Speech recognition performance had a bit of a breakthrough in popular consciousness in 2010, when David Pogue, the New York times technology reporter, reported error rates on general dictation of less than 1% [22]. Most speakers do not yet show nearly this level of performance, but performance has steadily increased each year through a combination of better algorithms, more computation, and utilization of ever-larger training corpora. In fact, on one specialized task, recognizing overlapping speech from multiple speakers, a voice recognition system was able to perform better than human listeners [23].

Over the past decade, the authors' experience has been that the average word error rate on a large vocabulary dictation task has decreased by approximately 18% per year. This has meant that each year, the proportion of the user population that experiences acceptable performance without any system training has increased steadily. This progress has allowed us to take on

Table 3.1 Alternative voice recognition options

Technique	Explanation	Reference
Combining AM and LM scores		
Weighting AM vs. LM	Instead of the formula in equation II, $$W^* = \arg \max_{\widetilde{W}}(P\overline{O}|\widetilde{W})^\alpha P(\widetilde{W}))$$ with $\alpha < 1$ is used because the AM observations across frames are not really independent.	[8]
Alternative front ends		
PLP	Perceptual linear predictive analysis: find a linear predictive model for the input speech, using a perceptual weighting.	[2]
RASTA	"Relative spectra": a front end designed to filter out slow variations in channel and background noise characteristics.	[13]
Delta and double delta	Given a frame based front-end, extend the vector to include differences and second order differences between frames around the frame in question.	[14]
Dimensionality reduction	Heteroscedastic linear discriminant analysis (HLDA): given MFCC or other speech features, along with delta and double delta features, or "stacked frames" (i.e. a range of frames), using a linear transformation to decrease the dimension of the front end feature.	[15]
Acoustic modeling		
Approximations to full covariance	Using the full covariance matrixes in HMM requires substantial computation and training data. Earlier versions of HMM used simpler versions of covariance matrixes such as "single variance", diagonal covariance.	[16]
Vocal Tract Length Normalization	VTLN is a speaker-dependent technique which modifies acoustic features in order to take into account differences in the length of speakers' vocal tracts.	[17]
Discriminative training		
MMIE	Maximum Mutual Information Estimation: a training method which adjusts parameters of an acoustic model so as to maximize the probability of a correct word sequence vs. all other word sequences.	[9]
MCE	Minimum classification error: a training method which adjusts parameters of an acoustic model to minimize the number of words incorrectly recognized.	[10]
MPE	Minimum phone error: a training method which adjusts parameters of an acoustic model to minimize "phone errors", i.e., the number of phonemes incorrectly recognized.	[11]
LM		
Back-off	In a classic n-gram model, it is very important to predict not only the probability of observed n-grams, but also n-grams that do not appear in the training corpus.	[18]

Table 3.1   (*continued*)

Technique	Explanation	Reference
Exponential models	Exponential models (also known as maximum entropy models) estimate the probability of word sequences by multiplying many different probability estimates and other functions and weigh these estimates in the log domain. They can include longer range features than n-gram models.	[19]
Neural Net LMs	Neural network language models extend exponential models by allowing the indicator functions to be determined automatically via non-linear transformations of input data.	[20]
System organization		
FSTs	In order to reduce redundant computation, it is desirable to represent the phonetic decision trees which map phonemes to triphones, the phoneme sequences that make up words, and grammars as finite state machines (called weighted finite state transducers, WFSTs), then combine them into a large FSM and optimize this FSM. Compiling all the information into a more uniform data structure helps with efficiency, but can have problems with dynamic grammars or vocabulary lists.	[21]

challenging applications such as voice search, but also more challenging environments, such as in-car voice control. Finally, improving accuracy has meant that voice recognition has now become a viable front end for sophisticated natural language processing, giving rise to a whole new class of interfaces.

3.3   Deep Neural Networks for Voice Recognition

The pattern of steady improvements of "canonical" voice recognition systems has been disrupted in the last few years by the introduction of deep neural nets (DNNs), which are a form of artificial neural networks (ANN). ANNs are computational models, inspired by the brain, that are capable of machine learning and pattern recognition. They may be viewed as systems of interconnected "neurons" that can compute values from inputs by feeding information through the network.

Like other machine learning methods, neural networks have been used to solve a wide variety of tasks that are difficult to solve using ordinary rule-based programming, including computer vision and voice recognition.

In the field of voice recognition, ANNs were popular during late 1980s and early 1990s. These early, relatively simple ANN models did not significantly outperform the successful combination of HMMs with acoustic models based on GMMs. Researchers achieved some success using artificial neural networks with a single layer of nonlinear hidden units to predict HMM states from windows of acoustic coefficients [24].

At that time, however, neither the hardware nor the learning algorithms were adequate for training neural networks with many hidden layers on large amounts of data, and the performance benefits of using neural networks with a single hidden layer and context-independent

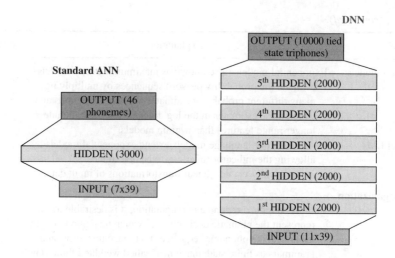

Figure 3.6   A standard ANN used for ASR in the 1990s vs. a DNN used today.

phonemes as output were not sufficient to seriously challenge GMMs. As a result, the main practical contribution of neural networks at that time was to provide extra features in tandem with GMMs, or "bottleneck" systems that used ANN to extract additional features for GMMs. ANN were used with some success in voice recognition systems and commercial products only in a limited number of cases [25].

Up until a few years ago, most state of the art speech recognition systems were thus based on HMMs that used GMMs to model the HMM emission distributions. It was not until recently that new research demonstrated that hybrid acoustic models utilizing more complex DNNs, trained in a manner that was less likely to get "stuck" in a local optimum, could drastically improve performance on a small-scale phone recognition task [26]. These results were later extended to a large vocabulary voice search task [27, 28]. Since then, several groups have achieved dramatic gains due to the use of deep neural network acoustic models on large vocabulary continuous speech recognition (LVCSR) tasks [27]. Following this trend, systems using DNNs are quickly becoming the new state-of-the-art technique in voice recognition.

In practice, DNNs used for voice recognition are multi-layer perceptron neural networks with up to 5–9 layers of 1000–2000 units each. While the ANNs used in the 1990s output the context independent phonemes, DNNs use a very large number of tied state triphones (like GMMs). A comparison between the two models is shown in Figure 3.6.

DNNs are usually pre-trained with the Restricted Boltzmann Machine algorithm and fine-tuned with standard back-propagation. The segmentations are usually produced by existing GMM-HMM systems. The DNN training scheme, shown in Figure 3.7, consists of a number of distinct phases.

At run-time, a DNN is a standard feed-forward neural network with many layers of sigmoidal units and a top-most layer of softmax units. It can be executed efficiently both on conventional and parallel hardware.

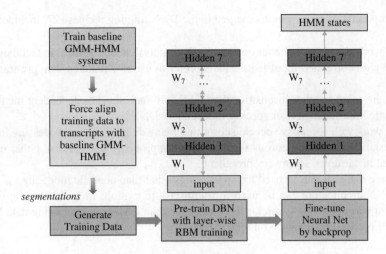

Figure 3.7   DNN training.

DNN can be used for ASR in two ways:

1. Using DNN to extract features for GMM (bottleneck features). This can be done by inserting a bottleneck layer in the DNN and using the activation of the units in that layer as features for GMM.
2. Using DNN outputs (tied triphones probabilities) directly in the decoder (DNN-HMM hybrid model).

The first method enables quick improvements to existing GMM based ASR systems, with error rate reductions of 10–15%, but the second method yields larger improvements, usually resulting in an error reduction of 20–30% compared to state-of-the-art GMM systems.

Three main factors were responsible for the recent resurgence of neural networks as high-quality acoustic models:

1. The use of deeper networks makes them more powerful, hence deep neural networks (DNN) instead of shallow neural networks.
2. Initializing the weights appropriately and using much faster hardware makes it possible to train deep neural networks effectively: DNN are pre-trained with the Restricted Boltzmann Machine algorithm and fine-tuned with standard back-propagation; GPUs are used to speed-up the training.
3. Using a larger number of context-dependent output units instead of context-independent phonemes. A very large output layer that accommodates the large number of HMM tied triphone states greatly improves DNN performances. Importantly, this choice keeps the decoding algorithm largely unchanged.

Other important findings which emerged in the DNN training recipes [27] include:

1. DNNs work significantly better on filter-bank outputs than on MFCCs. In fact they are able to deal with correlated input features and prefer to use raw features than pre-transformed features.
2. DNNs are more speaker-insensitive than GMM. In fact, speaker-dependent methods provide little improvement over speaker-independent DNNs.
3. DNNs work well for noisy speech, subsuming many de-noising pre-processing methods.
4. Using standard logistic neurons is reasonable but probably not optimal. Other units, like rectified linear units, seem very promising.
5. The same methods can be used for applications other than acoustic modeling.
6. The DNN architecture can be used for multi-task (e.g. multi-lingual) learning in several different ways and DNNs are far more effective than GMMs at leveraging data from one task to improve performance on related tasks.

3.4   Hardware Optimization

The algorithms described in the previous sections require quite substantial computational resources. Chip-makers increasingly recognize the importance of speech-based interfaces, so they are developing specialized processor architectures optimized for voice and NLU processing, as well as other input sensors.

Modern users complement their desktop computers and televisions with mobile devices, (laptops, tablets, smartphones, GPS devices), where battery life is often a limiting factor. These devices have become more complex, combining multiple functionalities into one single form factor, as vendors participate in an "arms race" to provide the next "must-have" bestselling device. Although users expect increased functionality, they have not changed their expectations for battery life: a laptop computer needs several hours of battery life; a smart phone should last a whole day without recharging. But the battery is part of the device, impacting both weight and size.

3.4.1   Lower Power Wake-up Computation

This has led to a need to reduce power consumption on mobile devices. Software can disable temporarily unused capabilities (Bluetooth, Wi-Fi, Camera, GPS, microphones) and quickly re-enable them when needed. Devices may even put themselves into various low-power modes, with the system able to respond to fewer and fewer actions. Think of Energy Star compliant TVs and other devices, but with more than just three (On-Off-Standby) states. How, then, does the system "wake up"? A physical action by the user, such as pressing a power key on the device, is the most common means today.

However, devices today have a variety of sensors that can also be used for this purpose. An infrared sensor can detect a signal from a remote control. A light sensor can trigger when the device is pulled out of a pocket. A motion sensor can detect movement. A camera can detect people. A microphone voice wakeup can detect voice activity or a particular phrase.

This is accomplished through low-power, digital signal processing (DSP)-based "wake-up-word" recognition, which allows users to speak to devices without having first to turn them on, further reducing the number of steps separating user intent and desired outcome. For instance, the Intel-inspired Ultrabook is integrating these capabilities, responding to "Hello Dragon" by waking up and listening to a users' command or taking dictation.

The security aspect starts to loom large here. A TV responds to its known signals, regardless of the origin of the signal. Anyone who has the remote control is able to operate it – or, indeed, anyone with a compatible remote control. While a living room typically has at most one TV in it, a meeting room may have 20 people, all with mobile phones. For a person to attempt to wake up his or her phone and actually wake up someone else's would be most unwelcome! Hence, there needs to be an element of security through personalization. The motion sensor may only respond to certain motions. The camera sensor may only respond to certain user(s), a technology known as "Facial Recognition", and the voice wakeup may only respond to a particular phrase as spoken by a particular user – "Voice Biometrics" technology.

3.4.2 Hardware Optimization for Specific Computations

These sensors all draw power, especially if they are always active, and when the algorithms are run on the main CPU they incur a substantial drain on the battery. Having a full audio system running, with multiple microphones, acoustic echo cancellation and beam forming, draws significant amounts of power. Manufacturers are thus developing specialized hardware for these sensor tasks to reduce this power load, or are relying on DSPs, typically running at a much lower clock speed than the main CPU, say at 10 MHz, rather than 1–2 GHz.

The probability density function (pdf) associated with a single n-dimensional Gaussian Model is shown above in equation (1.4) The overall pdf of a Gaussian Mixture Model is a weighted sum of such pdfs, and some systems may have 100,000 or more of these pdfs, potentially needing to be evaluated 100 times a second. Algorithmic optimizations (only computing "likely" pdfs) and model approximations (such as assuming the covariance matrix is diagonal) are applied to reduce the computational load. The advent of SIMD (single-instruction, multiple data) hardware was a big breakthrough, as it enabled these linear algebra computations to be done for four or eight features at a time.

The most recent advance has been the use of Graphical Processing Units (GPUs). Initially GPUs were used to accelerate 3D computer graphics (particularly for games), which make extensive use of linear algebra. GPUs help with the pdfs such as those described above, but have proved particularly effective in the computation of DNNs.

As discussed above, DNNs have multiple layers of nodes, and each layer of nodes is the output from a mostly linear process applied to the layer of nodes immediately below it. Layers with 1000 nodes are common, with 5–10 layers, hence applying the DNN effectively requires the calculation of 5–10 matrix-vector multiplies, where each matrix is on the order of 1000×1000, and this occurs many times per second. Training the DNN is even more computationally expensive. Recent research has shown that training on a very small amount of data could take three months, but that using a GPU cut the time to three days, a 30-fold reduction in time [28].

3.5 Signal Enhancement Techniques for Robust Voice Recognition

In real-world speech recognition applications, the desired speech signal is typically mixed acoustically with many interfering signals, such as background noise, loudspeaker output, competing speech, or reverberation. This is especially true in situations when the microphone system is far from the user's lips – for instance, in vehicles or home applications. In the worst case, the interfering signals can even dominate the desired signal, severely degrading the performance of a voice recognizer. With speech becoming more and more important as an efficient and essential instrument for user-machine interaction, noise robustness in adverse environments is a key element of successful speech dialog systems.

3.5.1   Robust Voice Recognition

Noise robustness can be achieved by modifying the voice recognition process, or by using a dedicated speech enhancement front end. Today's systems typically use a combination of both.

State-of-the-art techniques in robust voice recognition include using noise robust features such as MFCCs or neural networks, and training the acoustic models with noisy speech data that is representative of the kinds of noise present in normal use of the application. However, due to the large diversity of acoustic environments, it is impossible to cover all possible noise situations during training. Several approaches have been developed for rapidly adapting the parameters of the acoustic models to the noise conditions which are momentarily present in the input signal. These techniques have been successfully applied – for instance to enable robust distant-talk in varying reverberant environments.

Speech enhancement algorithms can be roughly grouped into single- and multi-channel methods. Due to the specific statistical properties of the various noise sources and environments, there is no universal solution that works well for all signals and interferers. Depending on the application, the speech enhancement front-end therefore often combines different methods. Most common is the combination of single-channel noise suppression with multi-channel techniques such as noise cancellation and spatial filtering.

3.5.2   Single-channel Noise Suppression

Single-channel noise suppression techniques are commonly based on the principle of spectral weighting. In this method, the signal is initially decomposed into overlapping data blocks with a duration of about 20–30 milliseconds. Each of these blocks is then transformed to the frequency or subband domain using either a Short-Term Fourier Transform (STFT) or a suitable analysis filterbank. Next, the spectral components of the noisy speech signal are weighted by attenuation factors, which are calculated as a function of the estimated instantaneous signal-to-noise ratio (SNR) in the frequency band or subband. This function is chosen so that spectral components with a low SNR are attenuated, while those with a high SNR are not. The goal is to create a best estimate of the spectral coefficients of the clean speech signal. Given the enhanced spectral coefficients, a clean time domain signal can be synthesized and passed to the recognizer. Alternatively, the feature extraction can be performed directly on the enhanced spectral coefficients, which avoids the transformation back into the time-domain.

A large variety of linear and non-linear algorithms have been developed for calculating the spectral weighting function. These algorithms mainly differ in the underlying optimization criteria, as well as in the assumptions about the statistical characteristics of speech and noise. The most common examples for weighting functions are spectral subtraction, the Wiener filter, and the minimum mean-square error (MMSE) estimator [29]. The single-channel noise suppression scheme is illustrated by the generalized block diagram in figure 3.8. Figure 3.9

shows a spectrogram of the noisy phrase 'Barbacco has an opening' and the spectrogram of the enhanced signal after applying the depicted spectral weighting coefficients.

Single-channel noise suppression algorithms work well for stationary background noises like fan noise from an air-conditioning system, a PC fan, or driving noise in a vehicle, but they are not very suitable for non-stationary interferers like speech and music. In a single-channel system, the background noise can mostly be tracked in speech pauses only, due to the typically high frequency overlap of speech and interference in the noisy speech signal. This limits the single-channel noise reduction schemes mainly to slowly time-varying background noises which do not significantly change during speech activity.

Several optimizations have been proposed to overcome this restriction. These include model-based approaches which utilize an explicit clean speech model, or typical spatio-temporal features of speech and specific interferers in order to segregate speech from non-stationary noise. Efficient methods have been developed to reduce fan, wind-buffets in convertibles [30], and impulsive road or babble-noise.

Another drawback of single-channel noise suppression is the inherent speech distortion of the spectral weighting techniques, which significantly worsens at lower signal-to-noise ratios. Due to the SNR-dependent attenuation of the approach, more and more components of the desired speech signal are suppressed as the background noise increases. This increasing speech distortion degrades recognizer performance.

3.5.3 Multi-channel Noise Suppression

Unlike single-channel noise suppression, multi-channel approaches can provide low speech distortion and a high effectiveness against non-stationary interferers. Drawbacks are an increased computational complexity, and that additional microphones or input channels are required. Multi-channel approaches can be grouped into noise-cancellation techniques which use a separate noise reference channel, and spatial filtering techniques such as beamforming (discussed below).

3.5.4 Noise Cancellation

Adaptive noise cancellation [31] can be applied if a correlated noise reference is available. That means that the noise signals in the primary channel (i.e. the microphone) and in the reference channel are linear transforms of a single noise source. An adaptive filter is used to identify

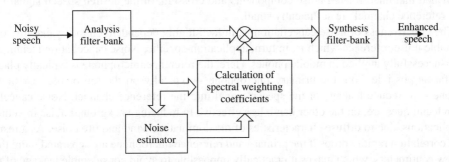

Figure 3.8   Block diagram of a single-channel noise suppression scheme based on spectral weighting.

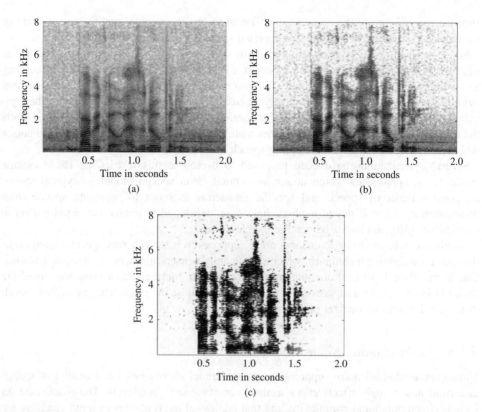

Figure 3.9 Time-frequency analysis of the noisy (a) and enhanced speech signal (b); time-frequency plot of the attenuation factor applied to the noisy speech signal (c).

the transfer function that maps the reference signal to the noise component in the primary signal. By filtering the reference signal with the transfer function, an estimate for the noise component in the primary channel is calculated. The noise estimate is afterwards subtracted from the primary signal to get an enhanced speech signal. The principle of adaptive noise cancellation is illustrated in figure 3.10. As signal and noise components superpose linearly at the microphone, the subtraction of signal components does not lead to any speech distortion, provided that uncorrelated noise components and crosstalk of the desired speech signal into the reference channel are sufficiently small.

The effectiveness of noise cancellation is, in fact, highly dependent on the availability of a suitable noise reference which is, in turn, application-specific. Noise cancellation techniques are successfully applied in mobile phones. Here, the reference microphone is typically placed as far as possible from the primary microphone – typically on the top or the rear of the phone – to reduce leakage of the speech signal into the reference channel. Noise cancellation techniques are, on the other hand, less effective in reducing background noise in vehicle applications, due to diffuse characteristics of the dominating wind and tire noise. As a result, the correlation rapidly drops if the primary and reference microphone are separated more than a few centimeters, which makes it practically impossible to avoid considerable leakage of the speech signal into the reference channel.

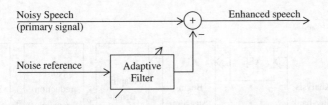

Figure 3.10   Basic structure of a noise canceller.

3.5.5   Acoustic Echo Cancellation

A classic application for noise cancellation is the removal of interfering loudspeaker signals, which is typically referred to as Acoustic Echo Cancellation (AEC) [32]. This method was originally developed to remove the echo of the far end subscriber in hands-free telephone conversations. In voice recognition, AEC is used to remove the prompt output of the speech dialog system or the stereo entertainment signal from a TV or mobile application.

Analogously to the noise canceller described above, the electric loudspeaker reference is filtered by an adaptive filter to get an estimate of the loudspeaker component in the microphone signal. The adaptive filter has to model the room transfer function, including the loudspeaker and microphone transfer functions. As the acoustic environment can change quickly – for instance due to movements of persons in the room – fast tracking capabilities of the adaptive filter are of major importance to remove loudspeaker components effectively.

Due to its robustness and simplicity, the so-called Normalized Least Mean Square (NLMS) algorithm [33] is widely used to adjust the filter coefficients of the adaptive filter. A drawback of the algorithm is that it converges slowly if the interference has a high spectral dynamic, as in the case of speech or music. For this reason, the NLMS is often implemented in the frequency or subband domain. As the spectral dynamic in the individual frequency subbands is much lower than over the complete frequency range, the tracking behavior is significantly improved. Another advantage of working in the subband domain is that AEC and spectral weighting techniques like noise reduction can be efficiently combined.

3.5.6   Beamforming

When speech is captured with an array of microphones, the resulting multi-channel signal also contains spatial information about the sound sources. This facilitates spatial filtering techniques such as *beamforming*, which extract the signal from a desired direction while attenuating noise and reverberation from other directions. The application of adaptive filtering techniques [34] allows adjusting the spatial characteristics of the beamformer to the actual sound-field, thus enabling effective suppression, even of moving sound sources. The directionality of such an adaptive beamformer, however, depends on the number of microphones, which are often limited in practical devices to 2–3 due to cost reasons.

To improve the directionality the adaptive beamformer can be combined with a so called spatial post-filter [35]. The spatial post-filter is based on the spectral weighting techniques also applied for noise reduction, but uses a spatial noise estimate of the adaptive beamformer. Although spatial filtering can significantly reduce interfering noise or competing speakers, it

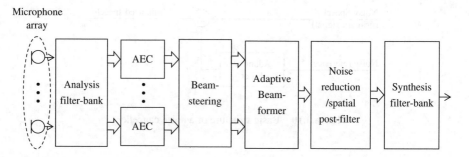

Figure 3.11   Combined speech enhancement front-end for far-talk operation of a TV.

can also be harmful if the speaker is not at the specified location. This makes it imperative to have a robust speaker localization system, especially in the mobile context, where the direction of arrival changes if the user moves or tilts a device.

A simple approach is to perform the localization acoustically, by simply selecting the direction in which the sound level is strongest [36]. This works reasonably well if there is a single speaker at a relatively close distance to the device, such as when using a tablet. A more challenging scenario arises in the context of smart TVs, or smart home appliances in general, where there may be several competing speakers at a larger distance. Such conditions prevent acoustic source localization from working reliably. In this situation, it may be preferable to track the users visually with the help of a camera and to focus on the speaker that is, for instance, facing the device. An alternative is to use gestures to indicate who the device should listen to.

As mentioned above, the speech enhancement front-end often combines several techniques to cope effectively with complex acoustic environments. Figure 3.11 shows an example of a speech enhancement front end enabling far-talk operation of a TV. In this system, acoustic echo cancellation is applied to remove the multi-channel entertainment signal, while beam-forming and noise reduction are used to suppress interfering sources such as family members and background noises.

3.6   Voice Biometrics

3.6.1   Introduction

Many voice-driven applications on a mobile device need to verify a user's identity. This is sometimes for security purposes (e.g. allowing a user to make a financial transaction) or to make sure that a spoken command was issued by the device's owner.

Voice biometrics recognizes a person based on a sample of their voice. The primary use of voice biometrics in commercial applications is speaker verification. Here, a claimed identity is validated by comparing the characteristics of voice samples obtained during a registration phase and validation phase. Voice biometrics is also used for speaker identification where a voice sample is matched to one of a set of registered users. Finally, in cases where a recording may include voice data from multiple people, such as during a conversation between an agent and customer, "speaker diarization" extracts the voice data corresponding to each speaker. All of these technologies can play a role in human-machine interaction and, particularly, when security is a requirement.

Voice biometrics will be a key component of mobile user interfaces. Traditional security methods have involved tedious measures based on personal identification numbers, passwords, tokens, etc., which are particularly awkward when interacting with a mobile device. Voice biometrics provides a much more natural and convenient method of validating the identity of a user. This has numerous applications, including everyday activities such as accessing email and waking up a mobile device. For "seamless wakeup", not only must the phrase be correct, but it must be spoken by the owner of the device. This can preserve battery life and prevent unauthorized device access. Other applications include transaction validation for mobile banking and purchase authorization.

Research dedicated to developing and improving technologies for performing speaker verification, identification, and diarization has been making progress for the last 50 years. Whereas early technologies focused mainly on template-based approaches, such as Dynamic Time Warping (DTW) [37], these have evolved towards statistical models such as the GMM (discussed above in Section 1.5.2 [39]). More recent speaker recognition technologies have used the GMM as an initial step towards modeling the voice of a user, but then apply further processing in the form of Nuisance Attribute Projection (NAP) [40], Joint Factor Analysis (JFA) [41], and Total Factor Analysis (TFA) [42]. The TFA approach, which yields a compact representation of a speaker's voice known as an I-vector (or Identity vector) represents the current state of the art in voice biometrics.

3.6.2   Existing Challenges to Voice Biometrics

One of the primary challenges in voice biometrics has been to minimize the error rate increase attributed to mismatch in the recording device between registration and validation. This can occur, for example, when a person registers their voice with a mobile phone, and then validates a web transaction using their personal computer. In this case, there will be an increase in the error rate due to the mismatch in the microphone and channel used for these different recordings. This specific topic has received a great deal of attention in the research community, and has been addressed successfully with the NAP, JFA, and TFA approaches. However, new scenarios are being considered for voice biometrics that will warrant further research. Another challenge is "voice aging". This refers to the degradation in verification accuracy as the time elapse between registration and validation increases [43]. Model adaptation is one potential solution to this problem, where the model created during registration is adapted with data from validation sessions. Of course, this is applicable only if the user accesses the system on a regular basis.

Another challenge to voice biometrics technology is to maintain acceptable levels of accuracy with minimal audio data. This is a common requirement for commercial applications. In the case of "text-dependent" speaker verification – where the same phrase must be used for registration and validation – two to three seconds (or ten syllables) can often be enough for reliable accuracy. Some applications, however, such as using a wakeup word on a mobile device, require shorter utterances for validating a user.

Whereas leveraging the temporal information and using customized background modeling improve accuracy, this topic remains a challenge. Similarly, with "text-independent" speaker verification – where a user can speak any phrase during registration or validation – typically 30–60 seconds of speech is sufficient for reasonable accuracy. However, speaker verification and identification capabilities are often desired, with much shorter utterances, such as when issuing voice commands to a mobile device, having a brief conversation with a call-center agent, etc. The National Institute of Standards and Technology (NIST) has sponsored numerous speaker recognition evaluations that have included verification of short utterances [44], and this is still an active research area.

3.6.3   New Areas of Research in Voice Biometrics

Voice biometrics technologies have advanced significantly since their inception; however a number of areas require further investigation. Countermeasures for "spoofing" attacks (orchestrated with recording playback, voice splicing, voice transformation, and text-to-speech technology) provide new challenges. A number of such attacks have been recently covered in an international speech conference [45]. Ongoing work assesses the risk of these attacks and attempts to thwart them. This can be accomplished through improved liveness detection strategies, along with algorithms for detecting synthesized speech.

Voice biometrics represents an up-and-coming area for interactive voice systems. Whereas speech recognition, natural language understanding, and text-to-speech have had more deployment history, voice biometrics technology is rapidly growing in the commercial and Government sectors. Voice biometrics provides a convenient means of validating an identity or locating a user among an enrolled population, which can reduce identity theft, fraudulent account access, and security threats. The recent algorithmic advances in the voice biometrics field, as described in this section, increase the number of use-cases and will facilitate adoption of this technology.

3.7   Speech Synthesis

Many mobile applications not only recognize and act on a user's spoken input, but also present spoken information to the user, via text-to-speech synthesis (TTS). TTS has a rich history [46], and many elements have become standardized. As shown in Figure 3.12, TTS has two components, front-end (FE) and back-end (BE) processing. Front-end processing derives information from an analysis of the text. Back-end processing renders this information into audio in two stages:

- First, it searches an indexed knowledge base of pre-analyzed speech data, and finds the indexed data which most closely matches the information provided by the front-end (unit selection).
- Second, this information is used by a speech synthesizer to generate synthetic speech.

The pre-analyzed data may be stored as encoded speech or as a set of parameters used to drive a model of speech production or both.

In Figure 3.12, the front-end is subdivided into two constituent parts: text pre-processing and text analysis. Text pre-processing is needed in "real world" applications, where a TTS system is expected to interpret a wide variety of data formats and content, ranging from short, highly stylized dialogue prompts to long, structurally complex prose. Text pre-processing is application-specific, e.g. the pre-processing required to read out customer and product information extracted from a database will differ substantially from an application designed to read

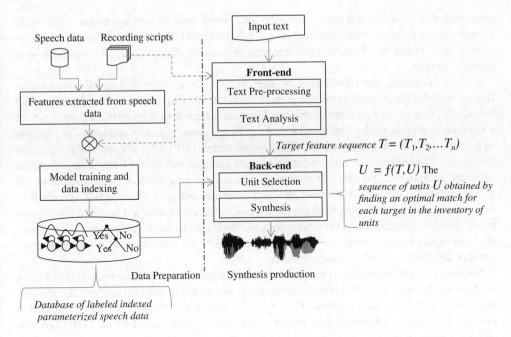

Figure 3.12   Speech synthesis architecture.

back news stories taken from an RSS feed. Also, a document may contain mark-up designed to aid visualization in a browser or on a page, e.g., titles, sub-headings. A pre-processor must re-interpret this information in a way which leads to rendered speech expressing the document's structure.

Text analysis falls into four processing activities: tokenization and normalization, syntactic analysis, prosody prediction and grapheme-to-phoneme conversion. Tokenization aids in the appropriate interpretation of orthography. For example, a telephone number is readily recognizable when written, and has a regular prosodic structure when spoken. During tokenization graphemes are grouped into tokens, where a token is defined as a sequence of characters belonging to a defined class. A digit is an example of a simple token, while a telephone number would be considered a complex token. Tokenization is particularly challenging in writing systems such as Chinese, where sentences are written as sequences of characters without white-space between words.

Text normalization is the process of converting orthography into an expanded standardized representation (e.g. $5.00 would be expanded into "five dollars") and is a precursor to further syntactic analysis. Syntactic analysis typically includes part of speech and robust syntactic structure determination. Together, these processes aid in the selection of phonetic pronunciations and in the prediction of prosodic structure [47].

Prosody may be defined as the rhythm, stress and intonation of speech, and is fundamental to the communication of a speaker's intentions (e.g., questions, statements, imperatives) and emotional state [48]. In tone languages, there is also a relationship between word meaning and specific intonation patterns. The prosody prediction component expresses the prosodic realization of the underlying meaning and structure encoded within the text using symbolic information (e.g., stress patterns, intonation and breath groups) and sometimes parametric information (e.g., pitch, amplitude and duration trajectories). Parametric information may be quantized and used as a feature in the selection process, or directly within a parametric synthesizer, or both.

In most languages, the relationship between graphemes (i.e., letters) and the representation of sounds (the phonemes) is complex. In order to simplify the selection of the correct sounds, TTS systems first convert the grapheme sequence into a phonetic sequence which more closely represents the sounds to be spoken. TTS systems typically employ a combination of large pronunciation lexica and grapheme-to-phoneme (G2P) rules to convert the input into a sequence of phonemes. A pronunciation lexicon contains hundreds of thousands of entries (typically morphemes, but also full-form words) each consisting of phonetic representations of pronunciations of the word, but sometimes also other information such as part-of-speech. Pronunciations may be taken directly from the lexicon, or derived through a morphological parse of the word in combination with lexical lookup. No lexicon can be complete, as new words are continually being introduced into a language. G2Ps use phonological rules to generate pronunciations for out-of-vocabulary words.

The final stage in generating a phonemic sequence is post lexical processing, where common continuous speech production affects such liaison, assimilation, deletion and vowel reduction are applied to the phoneme sequence [49]. Speaker-specific transforms may also be applied to convert the canonical pronunciation stored in the lexicon or produced by the G2P rules into idiomatic pronunciations.

As previously stated, the back-end consists of two stages: unit selection and synthesis. There are two widely adopted forms of synthesis the most widely adopted of which is concatenative synthesis, whereby the selected sound fragments indexed by the units are optionally joined. Signal processing such as Pitch Synchronous Overlap and Add (PSOLA) may be used as an aid to smoothing the joins and to offer greater prosodic control at the cost of some signal degradation [47]. Parametric synthesis – for example, the widely used "HMM synthesis" approach – uses frames of spectrum and excitation parameters to drive a parametric speech synthesizer [50].

Table 3.2 highlights the differences between concatenation and parametric methods. As can be seen, concatenation offers the best fidelity at the cost of flexibility and size, while parametric synthesis offers considerable flexibility at a much smaller size but at the cost of fidelity. As a result, parametric solutions are typically deployed in embedded applications where memory is a limiting factor.

Unit selection [51, 52] attempts to find an optimal sequence of units U from the pre-generated database which describes a target sequence T of features produced by the front-end for an analyzed sentence (Figure 3.12). Two heuristically derived cost functions are used to constrain the search and selection. These are unit costs (how closely unit features in the database match those of an element in the target sequence) and join costs (how well adjacent units match). Typically dynamic programming is used to construct a globally optimal sequence of units which minimizes the unit and join costs.

$$U = \underset{u}{\operatorname{argmin}} \sum_{n=1}^{N} Unit(T_n, U_n) + \sum_{n=1}^{N-1} Join(U_n, U_{n+1})$$

Table 3.2 Differences between concatenation and parametric methods

Category	Concatenation synthesis	Parametric synthesis
Speech quality	Uneven quality, highly natural at best. Typically offers good segmental quality, but may suffer from poor prosody.	Consistent speech quality, but with a synthetic "processed" characteristic.
Corpus-size	Quality is critically dependent upon the size of the sound database	Works well when trained on a small amount of data
Signal manipulation	Minimal to none	Signal manipulation by default. Suitable for speaker and style adaptation.
Basic Unit topology	Waveforms	Speech parameters
System footprint	Simple coding of the speech inventory leads to a large system footprint	Heavy modeling of the speech signal results in a small system footprint. Systems are resilient to reduction in system footprint.
Generation quality	Quality is variable depending upon the length of continuous speech selected from the unit inventory. For example, limit domain systems, which tend to return long stretches of stored speech during selection, typically produce very natural synthesis.	Smooth and stable, more predictable behavior with respect to previously unseen contexts.
Corpus-quality	Need accurately labeled data	Tolerant towards labeling mistakes

In HMM selection, the target sequence T is used to construct an HMM from the concatenation of context clustered triphone HMMs. An optimal sequence of parameter vectors can be derived to maximize the following:

$$O = \arg \max_{o} P(O|\lambda, N)$$

Where O is the parameter vector sequence to be optimized, λ is an HMM and N is the length of the sequence. In contrast to unit selection methods, which determine optimality based on local unit costs and join costs, statistical methods attempt to construct an optimal sequence which avoids abrupt step changes between states through a consideration of 2nd order features [50]. While still not widely deployed, there is also an emerging trend to hybridize these two approaches [53]. Hybrid methods use the state sequence to simultaneously generate candidate parametric and unit sequences. The decision on which method to select is made at each state, based on phonological rules for the language and an appreciation of the modeling power of the parametric solution.

There are two fundamental challenges to generating natural synthetic speech. The first challenge is representation, which is the ability of the FE to identify and robustly extract features which closely correlate with characteristics observed in spoken language, and the companion of this, which is the ability to identify and robustly label speech data with the same features. A database of speech indexed with too few features will lead to poor unit discrimination, while an FE which can only generate a subset of the indexed features will lead to units in the database never being considered for training or selection. In other words, the expressive power of the FE must match the expressive power of the indexing.

The second challenge is *sparsity*, that is sufficient sound examples must exist to adequately represent the expressive power of the features produce by the FE. In concatenation synthesis, sparsity means that the system is forced to select a poorly matching sound, simply because it cannot find a good approximation. In HMM synthesis, sparsity results in poorly trained models. The audible effects of sparsity increase as the styles of speech become increasingly expressive. Sparsity can, to some extent, be mitigated through more powerful speech synthesis models, which are capable of generating synthetic sounds from high-level features. Recently, techniques such as CAT (Cluster Adaptive Training, [54]) and DNN (Deep Neural Networks [Zen et al., 2013 55]) are being used to make the best use of the available training data by avoiding fragmentation that compounds the effects of sparsity.

As suggested by Table 3.2, the commercial success of concatenation is due to the fact that high fidelity synthesis is possible, provided care is taken to control the recording style and to ensuring sufficient sound coverage in key application domains, when constructing the speech unit database. Surprisingly good results can be achieved with relatively simple FE analysis and simple BE synthesis. However, technologically, such approaches are likely to be an evolutionary cul-de-sac. While many of the traditional markets are well served by these systems, they are expensive and time-consuming to produce.

The growing commercial desire for highly expressive personalized agents is driving a trend towards trainable systems, both in the FE, where statistical classifiers are replacing rule-based analysis methods, and in the BE, where statistical selection and hybridized parametric systems are promising flexibility combined with fidelity [53]. The desire to synthesize complex texts such as news and Wikipedia entries more naturally is forcing developers to consider increasingly complex text analysis through the inclusion of semantic and pragmatics knowledge in the FE, and consequently, increasingly complex statistical mappings between these abstract concepts and their acoustic realization in the BE [54].

3.8   Natural Language Understanding

As we have suggested, speech is a particularly effective way of interacting with a mobile assistant. The user's spoken utterances constitute commands for specific system actions and requests for the system to retrieve relevant information. A user utterance is first passed to an automatic speech recognition module that converts it to text. The recognized text is then passed to a natural language understanding (NLU) module that extracts meaning from the utterance. In real-world situations, the recognized text may contain errors, and thus it is common practice to output a "top-N" list of alternative hypotheses or a results lattice to make it possible for the NLU to explore alternative candidate answers. Accurate meaning extraction is necessary for the system to perform the correct action or retrieve the desired information.

The complexity of the NLU module depends on the capabilities that the system offers to the user, and on the linguistic variation expected in the user's utterance. Many current spoken dialog systems are limited in the range of tasks that they can perform, and require relatively limited and predictable inputs from the user in order to carry out those tasks. For example, a restaurant reservation system needs to fill a certain number of slots (restaurant, time, number of people) in a standard action-template in order to make a reservation. Likewise, a television interface might only need a specification of a particular movie or show to determine which channel it can be viewed on and to put it up on the screen.

Some systems have a high degree of *system initiative*. They ask fully specified questions to fill in the necessary slots, and they expect answers to only the questions that were posed. For example, for the question, "who would you like to call?" a database lookup can fill in the phone number slot if the recognized text matches a known name. Or, for the question, "on what date is your flight?" a regular expression can be used to match the various ways a user could express a date.

3.8.1 Mixed Initiative Conversations

In [56], Walker & Whitaker observed that more natural conversations display *mixed initiative*. In human-to-human dialogs, one person may offer additional relevant information beyond what was asked, or else may ask the other person to change away from the current task. Dialog systems hoping to perform in a mixed-initiative setting must therefore prepare for utterances that do not directly provide the expected information. A restaurant system might ask "where do you want to eat?" when the user is focusing on the time and instead replies with "We want a reservation at 7." The utterance is relevant to one of the restaurant-reservation slots, but not the one that the system was expecting. This requires an NLU component that can decode and interpret more complex inputs than simple direct-answer phrases.

There are also many ways in which a user can give a direct answer to a slot-filling question. For example, the user could respond to the question "where do you want to eat?" with a variety of restaurant characteristics, as illustrated in the first column of the following table:

User utterance	Slot-value pairs
Thai Basil	Name: "Thai Basil"
An Indian restaurant	Cuisine: "Indian"
A restaurant in San Francisco	Location: "San Francisco"
I'd like to eat at that Michelin-starred	Cuisine: "Italian"
Italian restaurant in San Francisco	Rating: "Michelin-starred"
tomorrow at 8 pm	Location: "San Francisco"
	Date: "tomorrow"
	Time: "8 pm"

The system needs to fill a high-level slot that identifies a particular restaurant, but the information may be provided indirectly from a collection of lower-level slot-values that narrow down the set of possibilities. Note that the responses can express different slots in different orders, with variation in the natural language phrasing. The last expression shows that a single response can provide fillers for multiple slots in a fairly elaborate natural language description.

The target slot names are domain-specific; they typically correspond to column names in a back end database. Given a set of slot-value pairs, application logic can retrieve results from the back-end database.

The task of the NLU module is to map from utterances in the first column to their slot-value interpretations in the second column. The NLU module must handle the variations in phrasing and ordering, if it is to find the meaning with a high level of accuracy. A simple strategy for NLU is to match the utterance against a template pattern that instantiates the values for the slots:

Query	Pattern template
Thai Basil	[name]
a Chinese restaurant	a [cuisine] restaurant
a Chinese restaurant in San Francisco	a [cuisine] restaurant in [location]

In this simple approach, every phrase variation would require its own template. A more sophisticated pattern matcher could use regular expressions, context-free grammars, or rules in more expressive linguistic formalisms, so that a few rules could cover a large number of variations. In any case, these approaches need to resolve ambiguities that arise in the matching.

The templates or rules give the context in which specific entities or key phrases are likely to occur. The Named Entity Recognition (NER) task is often a separate processing step that picks out the substrings that identify particular entities of interest (the restaurant name and cuisine-type in this example). Machine learning approaches like the one in [57] are typically used for named entity detection. These techniques have been designed to handle variations in phrasing and ambiguities in the matching, but they require a large set of example utterances marked up with the correct slot-value pairs. The marked-up utterances are then converted into *IOB* notation, where each word is given one of the following kinds of tags:

Tag type	Description
I	Inside a slot
O	Outside a slot
B	Begins a slot

The *I* and *B* tags have a slot name associated with them. An example of an IOB-tagged utterance is:

IOB tags	O	B Cuisine	O	O	B Location	I Location
Utterance	An	Italian	Restaurant	In	San	Francisco

The IOB-tagged utterances comprise the *training data* for the machine learning algorithm. At this point, the task can be regarded as a sequence classification problem. A common approach for sequence classification is to predict each label in the sequence separately. For each word, the classifier needs to combine features from the surrounding words and the previous tags to best estimate the probability of the current tag. A well-studied approach for combining evidence in a probabilistic framework is the conditional maximum entropy model described in [58]:

$$p(a_i|b_i) = \frac{1}{Z(b_i)} \prod_{j=1\,\ldots\,k} \alpha_j^{f_j(a_i,b_i)}$$

where a_i and b_i are the tag and available context (resp.) for the word at position i. The $f_j(a_i, b_i)$ denote features that encode information that is extracted from the available context, which typically consists of a few previous tags, the current word, and a few surrounding words. The α_j are the parameters of the model; they effectively weight the importance of each feature in estimating the probability. A search procedure (e.g., Viterbi) is then used to find the maximum probability tag sequence.

It is not desirable to collect training data that mentions every possible value for the slots. Furthermore, features that include explicit words do not directly generalize to similar words. For these reasons, machine-learned approaches often use external dictionaries. If a word or phrase exists as a known value in a dictionary, the model can use that knowledge as a feature. Both [57] and previously [59] use maximum entropy models to combine contextual features and features from external resources such as dictionaries. More generally, words can automatically be assigned to classes based on co-occurrence statistics as shown in [60], and features based on these word classes can improve the generalization capability of the resulting model, as shown in [61].

Recent neural net approaches such as [62] attempt to leverage an automatically derived mapping of words to a continuous vector space where similar words are expected to be "close". Features in this approach can then directly make use of particular coordinates of these vector representations.

The conditional random field (CRF), described in [63] is another model for sequence classification that produces a single probability for the entire sequence of labels, rather than one label at a time. See [64] for an example of CRFs applied to named entity detection.

3.8.2   Limitations of Slot and Filler Technology

A mobile assistant can complete many tasks successfully, based only on values that the NER component identifies for the slots of an action template. Regarding the slot values as a set of individual constraints on elements in a back-end database, systems often take their conjunction (e.g. "Cuisine: Italian" and "Location: San Francisco") as a constraint on the appropriate entries ("Zingari's", "Barbacco") to extract from the back-end. This rudimentary form of natural language understanding is not enough if the user is allowed to interact with more flexible utterances and more general constraints.

Consider the contrast between "an Italian restaurant *with* live music" and "an Italian restaurant *without* live music". These mention the same attributes, but because of the preposition they describe two completely different sets of restaurants. The NLU has to recognize that the prepositions express different relations, that "without" has to be interpreted as a negative *constraint* on the fillers that can fill the Name slot and not a specific set of restaurants. Prepositions like "with" and "without", and other so-called function words, are often ignored as stop-words in traditional information retrieval or search systems, but the NLU of a mobile assistant must pay careful attention to them.

Natural languages also encode meaning by virtue of the way that the particular words of an utterance are ordered. The description "an Italian restaurant with a good wine list" does not pick out the same establishments as "a good restaurant with an Italian wine list", although presumably there will be many restaurants that satisfy both descriptions. Here the NLU must translate the order of words into a particular set of grammatical relations or dependencies, taking into account the fact that adjectives in English typically modify the noun that they immediately precede. These relations are made explicit in the following dependency diagrams.

Figure 3.13 represents the output of a *dependency parser*, a stage of NLU processing that typically operates on the result of a name-entity recognizer. A dependency parser detects the meaningful relationships between words, in this case showing that "Italian" is a modifier of "restaurant", that the preposition "with" introduces another restriction on the restaurant, and that "good" modifies "wine list".

Dependency parsers also detect relationships within full clauses, identifying the action of an event and the participants and their specific roles. The subject and object labels shown in the Figure 3.14 restrict the search for movies to those in which Harry is saved by Ron and not the other way around. The grammatical patterns that code dependencies can be quite complicated and can overlap in many ways. This is a potential source of alternative interpretations in addition to the ambiguity of named-entity recognition, as shown in Figure 3.15.

According to the grammatical patterns of English, the "after" prepositional phrase can be a modifier of either "book" or "table". The interpretation in the first case would be an instruction to do the booking later in the day, after the meeting. The second and more likely interpretation in this context is that the booking should be done now for a table at the later time. A dependency parser might have a bias for one of the grammatical patterns over the other, but

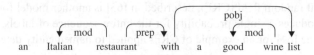

Figure 3.13   Dependency diagram 1.

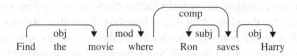

Figure 3.14   Dependency diagram 2.

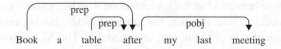

Figure 3.15   Dependency diagram 3.

the most plausible meaning might be based on information available elsewhere in the dialog system – for example, in an AI and reasoning module that can take into account a model of this particular's past behavior or knowledge of general restaurant booking conventions.

Machine learning approaches have also been defined for dependency parsing [65, 66]. As for named entity recognition, the task is formulated as a classification problem, driven by the information in a large corpus annotated with dependency relations. One technique considers all possible dependencies over the words of the sentence and picks the maximum spanning tree, the collection of dependencies that scores best against the training data [67]. Other techniques process a sentence incrementally from left to right, estimating at each point which of a small set of actions will best match the training data [68]. The actions can introduce a new dependency for the next word or save the next word on a stack for a later decision.

There are also parsers that produce dependency structures or their equivalent by means of large-scale manually written grammars [69–71]. These are based on general theories of language and typically produce representations with much more linguistic detail. Also, they do not require the construction of expensive annotated corpora and are therefore not restricted by the phenomena that such a corpus exemplifies. However, they may consume more computational resources than statistically trained parsers, and they require more linguistic expertise for their development and maintenance. These considerations may dictate which parsing module is most effective in a particular mobile assistant configuration.

Dependency structures bring out the key grammatical relations that hold among the words of a sentence. But further processing is necessary for the system to understand the concepts that the words denote and to translate them into appropriate system actions. Many words have multiple and unrelated senses that the NLU component must be able to distinguish. It is often the case that only one meaning is possible, given the tasks that the mobile assistant can perform. The word "book" is ambiguous even as a verb (make a reservation vs. put someone in jail), but the first sense is the only plausible one for a reservation-making assistant. It takes a little more work to disambiguate the word "play":

- Who played Serena Williams?
- Who played James Bond?

The same word has the "compete-with" sense in the first question and the "act-as" meaning in the second. Which sense is selected depends on the type of the object. The compete-with sense applies if the object is an athlete, the act-as is selected if the object denotes a theatrical character. Disambiguation thus depends on named entity recognition (to find the names), reference resolution (to identity the entities that the names refer to), and parsing (to assign the object grammatical relation). In addition, disambiguation depends on ontological reasoning: the back-end knowledge component knows that Serena Williams is a tennis player, tennis players are classified as athletes, and that category matches the object-type condition only of the compete-with sense.

The reasoning needed for disambiguation is not tied to just looking up the categories of names. Category information for definite and indefinite descriptions is also needed for disambiguation, as in:

- Who played the winner of the French Open?
- Who played a Stradivarius?

The first case needs information about what sorts of entities are winners of sporting events, namely athletes; the second case requires information about what a Stradivarius is, namely a stringed instrument. This information can then be fed into a module that can traverse the ontology to reason, for example, that stringed instruments are objects that make sounds and thus match the object-type condition for play in the sense of making music.

These examples fit within RDFS [72], a small ontology language used in conjunction with RDF, the "resource description framework" [73] for representing simple information about entities in the Semantic Web. RDFS permits representation of types of object (as in Serena Williams is a tennis player and a person), generalization relationships between these types (as in a tennis player is an athlete), and typing for relationships (as in a winner is a person).

More complex chains of reasoning can also be required for disambiguation, including combining information about multiple entities or descriptions. In these more complex cases, a more powerful ontology language may be required, such as the W3C OWL Web Ontology Language [74]. OWL extends RDFS with the ability to define categories (such as defining a man as a male person) and providing local typing (such as the child of a person being a person). Ontological reasoners are a special case of the more general knowledge representation and reasoning capabilities that may be needed not only to resolve more subtle cases of ambiguity, but also for the planning and inferences that underlie more flexible dialog interactions (see section 3.10.6). These may require the ability to perform more intricate logical deductions (e.g., in first-order predicate logic) at a much higher computational cost than relatively simple ontology-based reasoning.

Certain words in a user input take their meaning from the context in which they are uttered. This is the case for pronouns and other descriptions that refer to entities mentioned earlier in the conversation. If the system suggests a particular restaurant that satisfies all of the requirements that the user has previously specified, the user may further ask "Does it have a good wine list?" The system then must recognize (through a process called *anaphora resolution* – see Mitkov, [75]) that the intended referent of the pronoun "it" is the just-suggested restaurant. The user may even ask "What about the wine list?", a question that does not contain an explicit pronoun but yet is still interpreted as a reference to the wine list of the suggested restaurant. The linkage of the definite description ("*the* wine list") to the particular restaurant depends on information from an ontology that lists the parts and properties of typical restaurants.

There are other words and expressions that do not refer to entities previously mentioned in the conversation. Rather, they point directly to objects or aspects of the situation in which the dialog takes place. Demonstrative pronouns (this, that, those ...) and other so-called *indexical* expressions (now, yesterday, here, there ...) fall into this class. If the dialog takes place while the user is driving in a car, the user may point to a particular restaurant and ask, "Does that restaurant have a good wine list?" In this case the dialog system must recognize that the user is making a pointing gesture, identify the target of that gesture as the indicated restaurant, and offer that information to the NLU component so that it can assign the proper referent to the "that" of the user's utterance.

Other aspects of the dialog context (e.g., current location and time) must similarly be taken into account if the user asks "Are there any good restaurants near here?" or "What movies are starting an hour from now?" These examples show that a dialog system must be able to manage and correlate information of different modalities coming in from different channels. One tool for handling and synchronizing multi-modal information is EMMA, the W3C recommendation for an extensible multi-modal annotation mark-up language [76].

3.9 Multi-turn Dialog Management

An NLU module of the form discussed above can suffice for a *single-turn* dialog system, in which the user interaction is complete after a single utterance. However, in *multi-turn* dialog systems, the NLU must interpret a user utterance in the context of the questions, statements, and actions of the system, as well as the previous user utterances. This requires that the system needs to be able to recognize and track the user's intent throughout the dialog.

It is useful to divide the space of user intents into dialog intents and domain intents [77]. Dialog intents signal the start of a sub-dialogue to clarify, correct, or initiate a new topic and they are domain-independent. Domain intents signal the user's desire to inform the system or demand a specific system's action. As argued by Young (1993) [77], both types of intents need to be modeled and tracked through the course of a complex, multi-turn dialog.

An approach to intent tracking is referred to as *dialog state tracking* [78]. Each user's utterance is first processed by the NLU module to find (via classification) the dialog intent (inform/ask/correct) and the domain intent (play/record movies, reserve tables), and to extract the slot-value pairs from the utterance. Extracted information (which includes probabilities to model uncertainty) from the current utterance is fed to a dynamic model (such as a Dynamic Bayesian Network) as an observation. Belief updating via Bayesian update is then employed to help remove or reduce uncertainty based on the system's belief state prior to the current utterance.

System	Where would you like to eat?
User	An Italian restaurant in San Francisco.
System	I found several Italian restaurants in San Francisco. Here they are ...
User	Actually I prefer a Chinese restaurant tonight at 7.
System	I found several Chinese restaurants in San Francisco with tables available at 7 pm tonight. Here they are ...

In this example, in order to interpret the last user utterance correctly, a dialog state tracker classifies the utterance as a correction intent and hence overwrites the cuisine type mentioned in the previous utterance. As a result, the system is able to combine the cuisine, date, and time slots from the last user utterance with the location from the first user utterance. Such a system architecture is attractive, as it is robust in dealing with the uncertainty and ambiguity inherent in the output of a speech recognition/NLU pipeline.

While tracking dialog intents is necessary for dealing with the natural flow of spoken dialog, recognizing the domain intents is also needed so that the system is aware of the user's ultimate goal and can take actions to fulfill it. The user's domain intents are often complex and organized in a hierarchical manner similar to an AI plan [79]. Therefore, various hierarchical structures, ranging from and-or task networks [79] to probabilistic hierarchical HMMs [80], have been proposed to model complex intents. Robust probabilistic modeling of complex intents, while moving beyond slot-value pairs, requires even more expressive representations that can combine both probabilistic and logical constructs. Such a hybrid modeling approach is an active current research area in AI [81].

Based on the state of the dialog, the system must adjust its expectations and find a suitable response. *Dialog managers* like RavenClaw [82] are used to guide the control flow so that the system prompts for the information that is necessary for the system to complete its tasks. Dialog managers in mixed-initiative settings must use the NLU module to detect task changes at unexpected times in the conversation. Complex dialogs also require an error recovery strategy.

Thus, natural language understanding in dialog needs to happen in close collaboration with a dialog management strategy. As we have observed, utterance complexity can range from simple words or phrases that exactly match a known list, all the way to open-ended utterances that give extra information or ask to switch tasks at unexpected times. Accurate NLU modules handle linguistic variation using a combination of training data and manually specified linguistic resources, including dictionaries, grammars, and ontologies. One of the challenges for the NLU module is to interpret properly utterances that are only fragments of full sentences, simple words or phrases. The utterance "9 am" would be understood as filling the departure-time slot in a flight-reservation dialog if the system had just asked "When do you want to leave?", but the arrival-time slot for the question "When do you want to arrive?" The dialog manager holds the state of the dialog and can provide the dialog-context information that may simplify the task of interpreting fragmentary inputs.

An early, simple proposal for communicating contextual information to the NLU component is for the dialog manager to predict a set of linguistic environments that will help the NLU to "make sense" of the user's next utterance [83]. If the system has asked "When do you want to leave?" then the dialog manager can provide the declarative prefix "I want to leave at ... " to concatenate to the front of whatever the user says. If the user's reply is "9 am", the result after concatenation is a full, interpretable sentence to which a complete meaning can be assigned, according to the normal patterns of the language. In a mixed initiative setting, the user is not constrained to give a direct or minimal answer to the system's question, and so the dialog manager can provide a set of alternative prefixes in anticipation of a range of user utterances:

- [I want to leave] "at 9 am"
- [I want to leave on] "Tuesday at 9 am"

The claim of this approach is that there are a small set of linguistic patterns that provide the environment for a natural and meaningful user response to a query; it would be very odd and confusing, even to a human, if the user responded to this question with "Boston" instead of a time or date. Of course, the user may choose not to address the question at all and to provide other information about the trip or even to shift to another task. The natural utterance in that case will be a full sentence, and the dialog manager can anticipate that with the empty linguistic environment:

- [] "I want a flight to Boston"

This is one way in which the dialog manager and the NLU component can be partners in determining the meaning of a user's next utterance. The dialog manager can feed expectations, based on the current state of the dialog, forward to the NLU in a way that may simplify the overall system, while also leading to more appropriate dialog behaviors.

The output of the NLU module provides the information that the dialog manager needs to determine the user's intents and desires (e.g. search for a restaurant in the area, watch a movie, book a flight, or simply get information on the first president of the United States). The dialog manager also takes into account the system's capabilities (e.g., access to a local TV listing, ability to control Netflix, or access to real time traffic information and ability to give driving directions), the user's behaviors and preferences, and the interaction history.

If the user's intents and desires can be satisfied, then the system simply performs the appropriate domain action. Otherwise, its task is to figure out "what to say next" according to a *dialog strategy* [84] in order to get more information from the user and eventually satisfy the user's needs. Once the question of "what to say" is answered, the natural language generation (NLG) module takes over and answers the question of "how to say it (i.e., decide the best way to communicate it to the user).

Even though the dialog manager is such an essential component of a spoken dialog system, the research and implementation community differs on its definition and necessary functionalities. However, it is agreed that a dialog manager should minimally cover two fundamental aspects of an interaction system, namely to keep track of the dialog state and to decide the next course of action.

There are many different ways that these two functionalities can be implemented. Most commercial systems and some research systems largely rely on some form of Finite State Machine (FSM) [84]. The FSM approach requires every change in the dialog to be explicitly represented as a transition between two states in the network, and assumes that the user inputs can be restricted or directed by the system prompts. This means that the dialog manager is not very flexible and cannot handle unexpected situations. It is not practical for a more complex system to adopt such an approach, because one would have to specify fully all of the possible choices in every turn of the dialog. Furthermore, such an approach makes it almost impossible to have any degree of mixed initiative.

These deficiencies give rise to the Functional Model approach [85–88]. This is essentially an extension to the traditional FSM that allows the finite state machine to assume any topology, to invoke any arbitrary action at each state, and to assume any arbitrarily complex pre-condition on the transitions. These extensions make it possible to allow for over-specified user utterances, which is a form of mixed-initiative. In contrast, the information State Update approach [89, 90, 82] uses frames or tree structures as control mechanisms, while also allowing for unexpected user utterances. However, the conversations that are handled by any of these systems are usually of the slot-filling type. The system only asks users questions when a parameter of the required task is not specified.

In order to handle more complex tasks that involve collaborative problem solving, intelligent assistants, and tutorial dialogs, the dialog system is often implemented with planning technologies [91, 92]. More recently, statistical systems using machine learning approaches, more specifically Reinforcement Learning (RL) approaches, have become more prevalent in the research community. These approaches model a dialog strategy as a sequential decision-making process called Partially Observable Markov Decision Process (POMDP). Frampton and Lemon (2009) [93] give a comprehensive overview of the research efforts aimed at employing RL in spoken dialog systems.

These approaches give the developers a precise formal model for data-driven optimization instead of depending on strategies by experts with expertise and intuition. They also provide possibilities for generalization to unseen states and adaptation to unknown contexts. However, they have been criticized for the large amount of data needed to train for the dialog strategies and the need for more principled techniques to construct the state space, reward function and objective function used for strategy optimization. There is also a lack of understanding of how the learned policy from such a system can be intuitively understood by humans and modified if needed. Furthermore, the complexity of solving a POMDP often limits the richness of the dialog system representation.

More recent studies in the area begin to address these issues, such as using hierarchical RL to reduce the size of the state space [94]. Another strategy is to learn a simulated environment from a small collection of "Wizard of Oz" data, making it possible to employ RL techniques without extensive human-human dialog data [95].

Recent years have seen the commercial deployment of spoken dialog systems that are used by large populations on a day-to-day basis, particularly because of their distribution on mobile devices. However, these systems still lack many important functionalities. They can perform certain predefined tasks very well, based on slot-filling conversations, but they often revert either to generic web search or domain-specific search for particular services (restaurants, phone contacts, movies).

There are virtually no collaborative planning or problem-solving capabilities to extend system behavior to handle difficult or unpredicted inputs, or inputs that require clarification and refinement through a sequence of flexible, multi-turn interactions with the user that take into account more of the history and context of the dialog. The dialog research community is experimenting with new technologies and new system configurations that will support much more natural and effective conversational agents in a broader range of domains and settings. These systems will appear in the not too far distant future, as new techniques for integrating machine learning, artificial intelligence and reasoning, user-interface design, and natural language understanding are developed.

3.10 Planning and Reasoning

In this section, we will examine some of the deeper levels of processing that will play an increasingly important role in future dialog systems. These include deeper semantic analysis, discourse representation, pragmatics, and knowledge representation and reasoning. We will begin by highlighting some of the technical challenges in the context of a running example, followed by a brief overview of the relevant state-of-the-art.

3.10.1 Technical Challenges

Consider the following mock dialog with an automated Virtual Assistant (VA) of the future:

1. **Bob>** Book a table at Zingari's after my last meeting and let Tom and Brian know to meet me there.
2. **VA>** Sorry, but there aren't any tables open until 9 pm. Would you like me to find you another Italian restaurant in the area at about 6:30 pm?
3. **Bob>** Can you find a table at a restaurant with a good wine list?
4. **VA>** Barbacco has an opening. It's in the Financial District but the travel time is about the same.
5. **Bob>** Ok. That sounds good.

Utterance (1) is ambiguous in terms of the time of the requested reservation: should it take place after the meeting or should it be performed now? Resolution of this ambiguity requires background commonsense knowledge that needs to be represented by the system: reservations should be done as early as possible, otherwise the space in a restaurant might fill up. There are default assumptions that are also necessary to correctly interpret this exchange: the referenced "last meeting" is today's last meeting and not tomorrow's.

The reasoning here involves an appeal to general principles of communication [96]. People tend to be as informative as needed; if the person wanted the dinner tomorrow, he would have said so. However, this is again no more than a defeasible assumption, as it need not always hold; imagine the dialog having been prefaced with some discussion about what the person had planned for tomorrow. Similarly, the interpretation of the prepositional phrase, "after my last meeting" must be handled in a similar fashion, as tomorrow or the day after, for example, would also satisfy that constraint.

Of course, the above inferences regarding the time of the planned dinner are only approximate; a scheduler will need more exact information. The system could explicitly ask for the time, but a truly helpful assistant should make an effort to "fill in the blanks" as much as possible; here, the system should attempt to create some reasonable expectation regarding the best time for the dinner. Toward that end, the VA can try to base its reasoning on knowledge of past behavior; it may know that Bob works until 5 pm and that he typically spends 30 minutes on email before leaving work. Such information would reside in a *user model* of typical behavior that, as we shall see, should likely also include information about the user's desires or preferences. In addition, the system should factor in travel time based on possible locations of whatever restaurant is chosen. Finally, the identity of Tom and Brian must be ascertained. This is information that, again, would be kept in a user model of friends and phone numbers.

It is worth highlighting the important general principle that this utterance illustrates – the dialog system must be able to take a variety of contextual factors into account as it engages with the user. These include elements of the conversational history, as we have indicated before, but also many non-conversational aspects of the user and the situation in which the dialog takes place. Appropriate system behavior in different settings may depend, for example, on the time of day, the user's physical location, current or anticipated traffic conditions, music that the user has recently listened to or movies that have been recently watched. The dialog system must receive and interpret signals from a variety of sensors and must maintain a history not only of the conversation, but also of previous events and actions.

In utterance (2), we see that the initial search for a restaurant satisfying the expressed and implied constraints fails. An uninformative response such as "No" or "I could not find a restaurant for you" would hardly be helpful. A pragmatically more informative response is provided here in terms of a gloss of the reasons for the failure, and then an attempt to propose alternatives. Alternatives can be sought by relaxing some of the less important constraints; in this case, the specification of the type of restaurant and the time of the dinner are relaxed. This activity should be the responsibility of a dialog management module that guides the system and user to reach agreement on an executable set of constraints, while accessing the user model as much as possible to capture the relative importance of the constraints.

In utterance (3), we have what is known in the literature as an *indirect speech act* [97–99], which is discussed in more detail in section 3.13.3. If we take the utterance literally, it can be answered with a simple yes or no. However, neither would be very satisfying; the utterance, in fact, is a disguised request to perform an action. The action implicitly referred to is the dinner reservation. With respect to management of the dialog, notice that the user himself has implicitly responded to the question posed in (2). This again falls out of general principles of communication having to do with conciseness of expression. Since the user does not disagree, there is an implicit confirmation and, in the process, a new constraint is exposed by the user – the availability of a good wine list. At some point, these constraints must be collected to drive a search. This requires that the interpretation of utterances should be properly constrained by the relevant previous utterances. In (3), it means that the request should be interpreted as a request to find an *Italian* restaurant with a good wine list. At this point, a number of relevant databases and websites can be brought to bear on the search.

As the dialog progresses to (4), the VA informs the user that it has dropped one of the earlier constraints ("restaurant in the area") in preference to others, such as "the same travel time", "Italian restaurant", and "tonight", through the same sort of process employed during the processing of utterance (2). Bob then confirms the proposal in (5) and the dialog ends. The VA can now go to an appropriate online website to make the reservation and send the requested emails to Tom and Brian. However, the duties of a good assistant are not yet complete. It must be *persistent and proactive,* monitoring the plan for unexpected events such as delays and responding in a helpful manner.

Given these technical challenges, what follows is an overview of some of the most common approaches found in the literature to address these problems.

3.10.2   Semantic Analysis and Discourse Representation

Much of the dialog that involves virtual assistance is about actions that need to be taken in the world. From the perspective of semantic and discourse levels of analysis, one approach is to *reify* events and map the syntactic structure into a first order logical representation in which constant symbols stand for particular events (say, the assassination of Kennedy) that can have properties (e.g., a "killing") and relations to other events. Consider the fragment, "Can you find me a restaurant", of utterance (3). The following formula represents such a translation:

$$\exists e1 \exists e2 \exists x (surface_{request(e1,e2)} \wedge agent(e1, Bob) \wedge agent(e2, find)$$

$$\wedge \ object(e2, x) \wedge restaurant(x))$$

This can be glossed as saying that the variable $e1$ represents a surface request (see below) whose agent is Bob and the request is in terms of $e2$, an event of type "find" performed by agent VA(a virtual assistant is also sometimes called a Personal Assistant, PA). The object of the "finding" event is x, which is a restaurant. The advantage of such a representation is that one can conjoin additional properties of an event in a declarative fashion: for example, one can add additional constraints, such as that the restaurant be Italian: italian (x). One problem, however, is that from a logical perspective the addition of such a constraint must be done by re-writing the entire formula, as the constraint must occur within the scope of the existential quantifiers.

One solution to this problem has been proposed by Discourse Representation Theory (DRT), according to which a dynamic model of the evolving discourse is maintained in structures that can be augmented with new information. The figure below shows such a Discourse Representation Structure (DRS) for the complete utterance (3). The "box" illustrated in Figure 3.16 lists a set of reference markers corresponding to the variables that we have mentioned: $e1, e2, x,$

```
e1,e2,x,y
Surface_request(e1,e2)
Agent(e1,Bob),
Agent(e2,PA)
Find(e2),                    ∃e1∃e2∃x∃y.surface_request(e1,e2)∧agent(e1,Bob)
Restaurant(x),               ∧agent(e2,PA)∧find(e2)∧restaurant(x)∧object(e2,x)
Object(e2,x)                 ∧food(x,Italian)∧open(x,t)∧has(x,y)∧wine(y)∧good(y)
Food(x,Italian),
Open(x,t),
has(x,y),
Wine(y)
Good(y)
```

Figure 3.16   Discourse Representation Structure and statement in first order logic for "Can you find me an Italian restaurant with a good wine list?"

and y, followed by a set of conditions involving those variables. Such structures can be augmented with new information and then, if desired, translated into a first order logic form for more general reasoning (as shown to the right).

3.10.3   Pragmatics

There are many pragmatic issues that arose during the analysis of our target user-VA interaction. One particularly succinct expression of a set of general maxims that conversational agents should adhere to was proposed by the philosopher Grice [96]. These maxims state that a speaker should:

1. Be truthful.
2. Be as informative (but not more so) as necessary.
3. Be relevant.
4. Be clear, brief and, in general, avoid ambiguity.

How these maxims are to be captured computationally is, of course, a major challenge. They also represent only desirable default behavior; agents can violate them and, in the process, communicate something by doing so [100]. These maxims also reflect the *efficiency* of language – the fact that one can communicate much more than what is explicitly said. We saw an example of this in that utterance (3) needed to be interpreted relative to the prevailing context, in order to correctly ascribe to the user a request to find an Italian restaurant with a good wine list (and at the desired time and date).

The proper treatment of speech acts is also a central topic in pragmatics. Speech acts are utterances that are best interpreted as actions that can change the world somehow (specifically, the beliefs and intentions of other agents) as opposed to having a truth value based relative to a state of affairs. In the case of our VA, speech acts must be translated into *intentions* or commitments on the part of the VA to do something for the user. Often, the intention is not explicit but must be inferred. A simple example of the need for such inferences can be found in the above exchange; it probably escaped the reader that nowhere in the dialog does the user specify that he or she wants to eat dinner at the restaurant. This might appear to be a trivial observation, but if the VA decided to make a reservation at a restaurant which, for example, only served meals during the day or drinks at night, that choice would not satisfy the user's desires.

In the initial DRS for utterance (3), we have that the utterance was interpreted as a request to find a restaurant but, through a process of *intention recognition*, this is transformed to the structure shown in Figure 3.17. This represents a request to reserve the restaurant, the actual intent of the utterance. The plan recognition process works backwards from the find action and reasons that if a user wants to find a restaurant, it is probably because he wants to go to the restaurant, and if he wants to go to the restaurant a plausible precondition is having a reservation. In the helpful role that the VA plays, that translates to a task for the system to reserve the restaurant for the user.

In additional to logic-based approaches to plan recognition [91, 92], probabilistic methods have been developed which can explicitly deal with the uncertainty regarding the system's knowledge of the (inner) mental state of the user [80].

3.10.4   Dialog Management as Collaboration

Many approaches have been put forward for managing a dialog and providing for assistance. Most approaches are grounded in the observation that dialogs commonly involve some associated task. A dialog, then, is a collaborative activity between two agents: a give-and-take in which the agents are participating in a task and, in the process, exchange information that allows them to jointly accomplish the task. The major approaches might be divided in the following way:

- Master-slave sorts of approaches, in which the interaction is tracked and managed relative to a set of pre-defined task hierarchies (which we will call *recipes*), sometimes with the speech acts represented explicitly.
- Plan-based approaches, which model the collaboration as a joint planning process in which a plan is a rich structure in which a recipe might not be defined *a priori* but constructed at run time.
- Learning approaches that attempt to acquire prototypical dialog interactions.

We will expand a bit on plan-based approaches, as they provide an attractive approach for also modeling the collaborative support that a VA needs to provide.

Plan-based models to dialog understanding architect an agent – be it a human, a computer system or a team – in terms of the beliefs, desires and intentions of an agent. Without going into

---

> e1,e2,x,y
>
> Request(e1,e2)
> Agent(e1,Bob), Agent(e2,PA)
> **Reserve(e2)**, Restaurant(x), Object(e2,x)
> Food(x,Italian), Open(x,t)
> Has(x,y), wine(y), Good(y)

Figure 3.17   Intent structure for "Can you find me an Italian restaurant with a good wine list?"

philosophical details, it is enough to think of beliefs as capturing the information that a user has in a particular situation (for example, the availability of a table at a particular restaurant or the route to a particular location). Of course, an agent's beliefs can be wrong, and one of the responsibilities of the VA is to try to detect any mis-information that the user might have (e.g., he might incorrectly think that a particular restaurant is in another part of town). In a dialog setting, the agent's beliefs might reference those objects that are most "salient" or currently under discussion (e.g., "*the* Italian restaurant" that I am going to is "Zingari").

Desires reflect the user's preferences (e.g., the user might prefer Italian over Chinese food; or the user might prefer to travel by freeway whenever possible). Intentions capture what the agent is *committed to*. For example, the system might be committed to making sure that the user gets to the restaurant in time for the reservation. It will therefore persistently monitor the status of the user's progress toward satisfying that intention.

One important role of a virtual assistant is to assist a user in elaborating the high-level intentions or *plans* that the user has communicated to the VA, using elements from a library of task recipes. Once a plan is elaborated upon, the resulting set of potential options (if there is more than one) are examined, and the one that has higher payoff (or utility, for example, *get the user to the restaurant on time*) is chosen for action.

3.10.5 Planning and Re-planning

Planning makes use of recipes that, as we said, encodes alternative means for performing actions. In our mock dialog, suppose that are three ways of making a reservation: online by going to Opentable, going to the restaurant website, or calling the restaurant directly. The user and the system would then jointly try to fill in the details to those recipes or compose them in some way. Each agent brings with it complementary capabilities and responsibilities. The dialog, then, captures the contributions that each makes or proposes to make toward advancing that joint goal.

Recipes include logical constraints on constituent nodes. They can be either deep structures defined *a priori* and instantiated for a particular goal, or they can be shallow decompositions which can be composed during the planning process. Figure 3.18 illustrates examples of the latter type in the context of our example. Building a complex structure out of these at planning time allows for more flexibility in accommodating unexpected contingencies.

3.10.6 Knowledge Representation and Reasoning

The system's knowledge about, for example, types of restaurants, travel times, and beverages and menus for restaurants are stored in a knowledge representation. First order logic, a very

Figure 3.18 Recipe structure.

expressive knowledge representation, is one of the many options. For tractability, alternatives include ontology-based description logics that have been explored in connection with the Semantic Web. These are particularly effective for many examples of word-sense disambiguation, as mentioned in section 3.8.2. In some cases, specialized knowledge representations might be needed for temporal relations, default knowledge, and constraints, and these may require specialized reasoners.

Throughout, we have been referring to a user model of preferences. Preferences are directly related to the agent's desires. Preferences can be offered by a user, elicited by the VA, or learned by observation. They can be expressed either qualitatively, in terms of an ordering on states of the world, or numerically in terms of utilities. For example, we might have a simple user model that states that a user likes Peet's coffee. However, more interesting are statements which express comparative preferences, as in preferring Peet's over Starbucks. These are often called *ceteris paribus* preferences, as they are meant to capture generalizations, all other things being equal. Exceptions need to be dealt with, however. For example, one might also prefer espresso over American drip coffee, and from that one might want to be able to conclude that one would prefer Starbucks over a café (such as Peet's) with a broken espresso machine, that being the exceptional condition. As these preferences are acquired from the user, they must be checked for consistency, as the user might accidentally state circular or inconsistent preferences. If so, the system should engage the user in a dialog to correct such problems.

3.10.7   Monitoring

Let us continue the above scenario as follows.

1. [VA notices that it is: 5:30 pm and Bob has not left the office, cannot get to the restaurant on time. MA phones Bob, via TTS]
2. **VA>** Bob, you're running late. Should I change the reservation?
3. **Bob>** Yes, I'll leave in about 30 minutes.
4. [VA replans: Barbacco is not available later. Based on stored preferences, VA reserves at another similar restaurant.]
5. **VA>** Barbacco could not extend your reservation, so I reserved a table for you at Capannina. They have a good wine list. I'll also let Tom and Brian know. Is that OK?
6. **Bob>** Yes. Thanks.
7. [VA sends text messages to Tom and Brian, sets up another monitor]

At the end of the original scenario, the VA sets up a monitor (a conditional intention) to make sure that the plan involving the choice of restaurant, reservation and dinner attendance can be completed. The intention is relative to the VA's beliefs. The conditional intention should be checked at every time step to see if it is triggered by some changing circumstances. In this case, if the VA comes to believe that Bob has not left his meeting by the expected time, then it should form the intention to replan the reservation activity.

In the remainder of the scenario, that is exactly what happens: the VA notices that Bob has not left the meeting. It checks the restaurant and finds that the reservation cannot be extended and so, as part of helpful collaborative support, it looks for alternative restaurants based on its knowledge of Bob's requirements for dinner and his preferences: the area of the restaurant, the type of food and the wine list. It locates a substitute, Capannina, which, although not in the same area, has the same travel time. It goes ahead and drops the location constraint, goes ahead and books the restaurant and lets Brian and Tom know about the changes.

3.10.8 Suggested Readings

This section has provided only a cursory overview of the relevant areas of research. The interested reader can refer to the following for supplementary information.

1. A good overview of commonsense reasoning, first order logic, techniques such as reification and alternative event representations can be found in the text of Davis (1990) [101]. In addition, the proceedings of the bi-annual conference on Logical Formalizations of Commonsense Reasoning will provide the reader with the most current work in this area [102].
2. A Good presentation of DRT can be found in Kamp and Reyle (1993) [103] and Gamut (1991) [104]. A related approach is Segmented Discourse Representation Theory, SDRT [105]. A good overview of the field of pragmatics can be found in Levinson (1983) [106], including detailed discussions on Grice's Maxim's of conversation [96]. Details on speech acts can be found in [97, 99]. The proceedings of the annual conferences on plan recognition provide a good overview of alternative approaches.
3. Good introductions to the field of knowledge representation can be found in introductory texts [107, 101]; more recent research in this area, including default reasoning, can be found in the proceedings of the annual KR conferences [108]. An overview of description logics can be found in [109]. Utility theory as related to the representation of preferences is also discussed in [107]. A number of tools have been developed for representing preferences [110].
4. Seminal papers on plan-based approaches to dialog processing include [111, 112]. The information-based approach to dialog processing is presented in [113]. Recent work in dialog processing can be found in [114]

3.11 Question Answering

General questions are bound to arise in interactions with the kind of virtual assistant described earlier. For example, a user might well ask "Are there any restaurants that serve vegetarian food near here?" as a prelude to deciding where to make a reservation. Answering the question need not involve the system in taking any task actions beyond finding the relevant information, but these Question Answering (QA) interactions can be an important part of formulating or refining the task.

Question answering has a long history as a stand-alone task, from natural language interfaces to databases [115], through supporting intelligence analysts [116], to the recent success of IBM's Watson/DeepQA system playing Jeopardy! against human champions [117]. But, as mentioned, question answering also plays a natural role, embedded as part of broader spoken language dialog applications. For example, a user might want to ask about the start time of the movie they plan to watch before deciding on a time for a dinner reservation; or the system might decide for itself that it needs to get this information, in order to check that the user's choice of time is compatible with their other stated plans. Successfully answering a question requires all of the NL components (recognition, parsing, meaning, and reasoning) for the key steps:

- Question analysis: what is the questioner looking for?
- Locate information relevant to the answer.
- Determine the answer and the evidence for the answer.
- Present the information to the questioner.

All of these steps have to be able to situate the meaning and answers in terms of the questioner, their preferences and their world. "Will there be time to get to the movie?" requires knowing the questioner's schedule.

3.11.1   Question Analysis

Analysis of a question may involve not only determining what the question is asking, but also why the question is being asked. "Are there any restaurants that serve vegetarian food near here?" is technically a question with a yes or no answer, but a useful answer would provide some information about what restaurants are available. "Useful answer" depends on the asker's intent (if all they are trying to do is confirm previous information, then "yes" is the appropriate answer, not a complete description of the information). Determining the asker's intent depends on the conversation, the domain, and knowledge of the world (in this case, it is necessary to know the asker's location).

Question analysis typically determines the key terms in the question and their relations. The key terms (entities) are usually the nouns. Relations can be the main predicate (which largely signals the asker's intent) or constraints on the answer. For example, for "Are there any restaurants that serve vegetarian food near here?" the intent is a list of restaurants that meet the constraints of serving vegetarian food and are near the asker's current location; "restaurants", "vegetarian food" and "here" are the entities.

Common simple questions can be handled with patterns ("What time is <event>?"), but are not robust to variations and less common linguistic patterns. Parsing (see Section 3.8.2) is a common approach but subtle constructions of language can be hard to parse accurately, and thus parsing is typically complemented by statistical entity- and intent-detection (see Section 3.8.1). Dictionaries of known entities (such as lists of movie actors, medication names, book titles, political figures, …) can also be effective to locate common types of entities, particularly for specific domains.

3.11.2   Find Relevant Information

Once the intent(s) and the entities have been determined, we locate information that might be relevant to the intent. For the example, for "Are there any restaurants that serve vegetarian food near here?" we could look in databases of restaurants, look in a business directory based on the asker's location, or do a general Internet search. From those results, we can compile a list of restaurants that meet the constraints ("have vegetarian dishes on the menu" and "near current location").

Much information exists only in structured form (tables or databases, such as historical player information for baseball teams, while other kinds of information exists only in unstructured form (documents in natural language, such as movie plot summaries). Accessing information in structured sources requires precise question analysis (the asker may not know how the database designer created the field names, so the question analysis needs to map the semantics of the question to the database fields). Accessing unstructured information with search (either on the Internet or in a specific set of source documents) requires less precision, but leads to many more potential items that could be answers, making it harder to select the correct answer.

3.11.3 Answers and Evidence

Most work on QA has focused on "factoid" answers, where the answer to a question is suc-cinctly encapsulated in a single place in a document (e.g. What is the capital of Peru? "Lima is the Peruvian capital.") There has been less work on non-factoid answers [118], especially ones that have to be put together from evidence coming from more than one place (e.g., text passages from different parts of the same document, text passages from different documents, combinations of structured data and raw text).

The evidence supporting both factoid and non-factoid answers may differ substantially from the language of the user's question, which has led to recent interest in acquiring paraphrase information to drive textual inference [119]. Non-factoid answers arise either from inherently complex questions (Why is the sky blue, but often red in the morning and evening?), or from qualified answers to simple questions (When can I reserve a table at Café Rouge: if it's a table for 2, at 7 pm; a table for 4 at 7:30; at table for more than 4 at 8:30; or, Café Rouge is closed today).

3.11.4 Presenting the Answer

Having assembled the evidence needed to provide an answer, the system must find a way of presenting it to the user. This involves strategic decisions about how much of the evidence to present, and tactical decisions about the best form of presenting it [120]. These decisions are informed by the answer medium (spoken, text on a mobile display). However, for QA embed-ded in dialog applications, the background goals (question under discussion) have a significant impact on strategic decisions. At the tactical level for generating NL, possibly dispersed text passages must be stitched together in a coherent and natural way, while NL answers from structured data must either be generated from the data, or textual search carried out to find text snippets to present or support the answer.

Since machine learning systems can potentially benefit and adapt from getting feedback on their actions, capturing user reactions to system answers is desirable. However, this is not likely to succeed if the feedback has to be given in an unnatural and obtrusive way (e.g., thumbs up/down on answers). Instead, more subtle clues to success or failure need to be detected (e.g., the user repeating /reformulating questions, abandoning tasks, the number of steps required to complete a task [121]).

3.12   Distributed Voice Interface Architecture

Users increasingly expect to be able to perform the types of tasks discussed in the preceding sections on a growing array of devices. While their display form factors and CPU power differ, these devices all have substantial on-board computation, displays, and are all connectable to the Web. Users also have an expectation of a consistent experience and a continuity of interactions across smartphones, tablets, Ultrabooks, cars, wearables, and TVs. For example, a user may query his or her smartphone – or smart watch or glasses – "What's the score of the Celtics game?" and, upon arrival at home, instruct the TV to "Put the game on".

To achieve such a continuity of functionality and the interaction model, the voice interface framework needs to be flexibly distributable across devices and the Cloud. This permits the optimization of computation, availability (if connectivity fails), and latency, as well as making the user's preferences and interaction history available across these devices. In order to resolve the command "Put the game on", the TV-based interface would access the Cloud-based user profile and conversation history, adapt it during use, and update it back on the server after the session ends, making it available for access from the next device with which the user interacts.

3.12.1   Distributed User Interfaces

Devices like phones or TVs often form hubs for other mobile devices, which may have less computational capacity, but which may provide a more effective and direct interface to the user, supplementing or replacing the UI of the hub device. For example, a phone may be connected to smart glasses, a connected watch, and a wireless headset, distributing the task of channeling information between the user and the device making more effective, natural, and pleasant use of the peripherals.

As discussed earlier in section 3.8.2, the various modalities may overlap and complement one another. Consider for instance a graphical representation on a map of all towns named 'Springfield', displayed in response to the user entering a navigation destination. Pointing to the correct target, she specifies "this one", with gesture recognition identifying the position of the target, and the spoken command conveying the nature of the directive and the time at which the finger position is to be measured.

As the manufacturers of these devices vie for predominance of their particular platforms, users are dividing their attention and interact with multiple devices every day. This has exacerbated the need for a 'portable' experience – a continuity of the functionality, the interaction model, the user's preferences and interaction history across these devices.

A pan-device user profile will soon become essential. This stored profile will contain general facts learned about the user, such as a preference for a specific music style or news category, information relevant in the short term such as the status of an ongoing airline reservation and booking dialogue, and also information relevant to voice recognition, such as acoustic models. Devices will use voice biometrics to recognize users, access a server-based user profile, adapt it during use, and update it back on the server after the session ends, making it available for access from the next device with which the user interacts.

A user might search for restaurants on his smartphone, which would then be stored in his or her user profile. Later, on a desktop PC, with a larger screen available, he or she might choose one venue. Getting into his or her car, this location would be retrieved by the onboard navigation from the user profile and, as the recent dialogue context is also available via the profile, a simple command like 'drive to the restaurant' is sufficient to instruct the navigation system.

In this model, the human-machine interface design no longer focuses on the interaction between the human and a particular machine. The machine's role is turning into a secondary one: it becomes one of many instantiations of a common, multi-device strategy for directly linking the user to data and services.

A common construct for this type of abstracted interface is the 'virtual assistant', which mediates between the user and information and services. Such an assistant has access to a server-based user profile which aggregates relevant information from all devices favored by the user. The manifestation of the assistant may itself be personalized, so that it exhibits the same TTS voice, speaking style and visual guise, to ensure the cross-device dialog forms a continuous, personalized experience. The user thus no longer focuses on interacting with hardware, but on being connected to the desired information, either directly or through the virtual assistant. The interaction model shifts from the use of a human-machine interface to a personalized human-*service* interface.

Just as the UI can be distributed across multiple interaction sessions on multiple devices, the services may also be spread out across multiple devices and sources. Again, the hardware moves to the background as the UI groups functionality and data by domains, not by device.

Consider, for example, an automotive music app that has access to music on the car's hard drive juke box, an SD memory card, a connected phone, and internet services. A user experience designer may elect to make content from all these sources available in one homogeneous set, so a request "play some jazz" yields matching music from all the sources.

3.12.2 *Distributed Speech and Language Technology*

For speech and language UIs, the location of the processing is largely determined by the following considerations:

- Capacity of the platform: CPU, memory, and power profile.
- Degree of connectivity: the speed, reliability and bandwidth of the connection, and additional costs for connectivity, e.g., data plan limits.
- The type and size of models needed for voice recognition and understanding of the application's domains. For example, are there 100,000 city names that need to be recognized in a variety of contexts, or only a few hundred names in a user's contact list?

The following device classes exemplify the spectrum of different platforms:

- PCs: ample CPU and memory, continuous power supply. Typically connected to the internet. Native domains: commands to operate software and PC, text dictation.
- Mobile phones and tablets: limited CPU and memory, battery-driven. Often connected to the internet, connection may be more costly and not stable (e.g., when coverage is lost).
- Cars: limited CPU and memory, continuous power supply. Often connected to the internet, connection may be more costly and not stable.
- TVs: limited CPU and memory, continuous power supply. Often connected to the internet, not every user might connect the TV.
- Cloud servers – vast CPU and memory resources feeding multiple interactions at the same time. Connected to the internet and other large data sources.

Increased connectivity has given rise to hybrid architectures that blur the lines between traditional embedded versus server-based setups and also lead to a number of functions and domains that are expected on all personal devices – for instance, information search, media playback, and dictation.

When considering how to allocate tasks in a distributed architecture, the mantra used to be *do the processing where the data is* – which, with growing connectivity bandwidth, is no longer a requirement but still a good guideline. UI consistency is also an essential aspect; say the natural language or dialogue components run entirely on a remote server. If a data connection is lost, it is easy for the user to understand that connected data services like web search are interrupted, but it may not be clear that the capacity to converse in a natural language dialogue is also no longer available.

In a speech and language UI, the embedded, "on-board" speech recognition typically handles speech commands to operate the given device. This may be using a grammar-based command-and-control type recognizer or a small statistical language model (SLM) enabling natural language processing. Contemporary mobile platforms, however, reach their limits when attempting to recognize speech with SLMs filled with large slot lists with tens of thousands of city names. This task would thus be carried out by a server-based recognizer. In many cases, it is advisable to perform recognition both on the embedded platform and on the server at the same time, compare the confidences of the results, and then select the best one, to avoid latencies one would incur if only triggering the server speech recognition after a low-confidence onboard outcome.

Other tasks that may reside on the onboard recognizer are wake-up-word detection, possibly in combination with voice biometrics – to activate a device and authenticate the user, respectively, speech activity and end point detection to segment the audio, and speech recognition feature extraction, so that only the features rather the full audio need to be transferred to the recognition server.

The user profile is beneficial for storing personal preferences relevant for dialogue decisions, speaker characteristics and vernacular for acoustic and language modeling. It can also hold biometrics information so that the user's identity can be verified to authorize use of a service or access of data. This is most powerful if the profile is available to the user from any device, but it should also be functional if the device loses connectivity, e.g., when a car drives through a tunnel. This is solved by a master user profile in the Cloud with synchronized copies on local devices, or by making the phone the hub for the profile, as it accompanies the user at most times.

Storing such a profile on a server also offers the advantage that the *group* of these profiles forms an entity of its own containing a vast amount of information, and allows deriving statistics of the user population or parts thereof. A news service may be interested in finding the trending news search keywords found in all logs connected to the group of profiles. A music web store could query the group of profiles for the artists most commonly requested by male listeners aged 18–25 in the state of California.

Often, profiles of different users are connected to each other, say, because users A and B are in each other's email address book or linked via a social network. If this information is stored in the user profile, the group of connected profiles spans a large graph that, for instance, could allow a user's virtual assistant to conduct queries such as, "do I have any friends or friends of friends that are in the town I'm currently visiting?" or "what music are my friends listening to?" Even more so than with server-based recognition and logging, when storing data in user profiles, privacy and data security are a key matter for designing and operating the server infrastructure.

Lastly, on the output side, the TTS (text-to-speech) and language generation usually run on the user's device, unless a high-quality voice requires more memory than available locally, or if an entire app is hosted and a server solution is more convenient to develop and maintain.

3.13   Conclusion

Interacting with voice-driven NLU interfaces on a broad variety of devices spanning smartphones, tablets, TVs, cars and kiosks is becoming a routine part of daily life. These interfaces allow people to use complicated functions on their devices in a simpler and more natural way. Rather than issuing a series of atomic commands, a user can increasingly express his or her overall intent in natural language, with the system determining the underlying steps. Such natural language interactions are becoming practical in an increasing number of environments: on the street; in an automobile; in the living room; and on new devices.

All of this new functionality begs the question of how to best incorporate natural language understanding into today's visual interfaces. There is a broad range of possible approaches, including the "virtual assistant" category, which had a banner year in 2013, with Apple's Siri, Samsung's S-Voice, Dragon Assistant, Google Now, and nearly 60 other offerings on the market.

The assistant can be viewed as a separate entity, one which is primarily conversational and with a personality. It interprets a user's input and mediates between the user, the native user interface of the device, and a variety of on- and off-board applications. In some cases, the assistant even renders retrieved information in a re-formatted fashion within its own UI – and in this sense acts as a lens and filter on the Web.

An alternative design, which may be termed "ambient NLU", retains the look-and-feel of native device and application interfaces, but "embeds" context-sensitive NLU. By speaking to this interface, the user can retrieve information as well as open and control familiar-looking applications. The system engages in conversation when needed to complete multi-turn transactions or resolve ambiguity. Rather than being front-and-center, the NLU aims to be unobtrusive, efficient and open. To the extent possible, it accomplishes tasks based on a single utterance and never limits from where the user can get information. In contrast to the assistant that aims to help the user address the shortcomings of the existing UI, ambient NLU holds the promise of becoming an inherent part of an improved UI.

In either case, voice and language understanding are now viewed as a new fundamental building block – a new dimension added to the traditional visual UI that permits access and manipulation of assets that may not be visible, located either on the device or in the cloud. Over the next few years, we will see active exploration of the best way to make use of these new dimensions, and a rapid revision of the current "shrunken desktop" metaphor as designers experiment with new ways of structuring the experience.

This voice revolution has been driven by ongoing, incremental improvements in many component technologies, and the improvements have accelerated in the last few years. There has been a continuous growth in the number of situations where the system "just" works. This performance increase has been due to progress on many complementary fronts, including:

- voice recognition technology, particularly DNNs;
- signal acquisition enhancements;
- improved TTS and voice biometrics modeling;
- meaning extraction combining structural methods with machine learning;
- conversational dialog, probabilistic plan recognition, knowledge representation and reasoning;
- question answering.

Many factors have contributed to this progress:

- Increases in available computational power, including special purpose computation devices.
- Increases in size of available training corpora.
- Improvements in statistical modeling.
- Thousands of person years of engineering effort.

Despite this progress, many challenges remain. Or, to put it more positively: we can expect further improvements over the next years. Creating conversational agents with a deep understanding of human language is a both a challenge and a promise.

Acknowledgements

The authors would like to thank Dario Albesano, Markus Buck, Joev Dubach, Nils Lenke, Franco Mana, Paul Vozila, and Puming Zhan for their contributions to this chapter.

References

1. Davis, S., Mermelstein, P. (1980). Comparison of parametric representations for monosyllabic word recognition in continuously spoken sentences. *IEEE Transactions on Acoustics, Speech and Signal Processing* **28**(4), 357–366.
2. Hermansky, H. (1990). Perceptual linear predictive (PLP) analysis of speech. *The Journal of the Acoustical Society of America* **87**, 1738.
3. Bahl, L., Bakis, R., Bellegarda, J., Brown, P., Burshtein, D., Das, S., De Souza, P., Gopalakrishnan, P., Jelinek, F., Kanevsky, D. (1989). Large vocabulary natural language continuous speech recognition. *International Conference on Acoustics, Speech, and Signal Processing, 1989 (ICASSP-89)*.
4. Rabiner, L. (1989). A tutorial on hidden Markov models and selected applications in speech recognition. *Proceedings of the IEEE* **77**(2), 257–286.
5. Ney, H., Ortmanns, S. (2000). Progress in dynamic programming search for LVCSR. *Proceedings of the IEEE* **88**(8), 1224–1240.
6. Hunt, A., McGlashan, S. (2004). *Speech recognition grammar specification version 1.0.* W3C Recommendation. http://www.w3.org/TR/speech-grammar/.
7. Chomsky, N. (2002). *Syntactic structures.* Mouton de Gruyter.
8. Jelinek, F. (1997). *Statistical methods for speech recognition.*: MIT press.
9. Bahl, L., Brown, P., De Souza, P., Mercer, R. (1986). Maximum mutual information estimation of hidden Markov model parameters for speech recognition. *IEEE International Conference on Acoustics, Speech, and Signal Processing, ICASSP '86*.

10. McDermott, E., Hazen, T.J., Le Roux, J., Nakamura, A., Katagiri, S. (2007). Discriminative training for large-vocabulary speech recognition using minimum classification error. *IEEE Transactions on Audio, Speech, and Language Processing* **15**(1), 203–223.

11. Povey, D., Woodland, P.C. (2002). *Minimum Phone Error and I-Smoothing for Improved Discrimative Training.* International Conference on Acoustics, Speech, and Signal Processing (ICASSP).

12. Leggetter, C.J., Woodland, P.C. (1995). Maximum likelihood linear regression for speaker adaptation of continuous density hidden Markov models. *Computer Speech & Language* **9**(2), 171–185.

13. Hermansky, H., Morgan, N. (1994). RASTA processing of speech. *IEEE Transactions on Speech and Audio Processing* **2**(4), 578–589.

14. Furui, S. (1986). Speaker-independent isolated word recognition based on emphasized spectral dynamics. *IEEE International Conference on Acoustics, Speech, and Signal Processing, ICASSP'86.*

15. Kumar, N., Andreou, A.G. (1998). Heteroscedastic discriminant analysis and reduced rank HMMs for improved speech recognition. *Speech Communication* **26**(4), 283–297.

16. Sim, K., Gales, M. (2004). Basis superposition precision matrix modelling for large vocabulary continuous speech recognition. *Proceedings of IEEE International Conference on Acoustics, Speech, and Signal Processing, 2004. (ICASSP'04).*

17. Lee, L., Rose, R.A. (1998). Frequency warping approach to speaker normalization. *IEEE Transactions on Speech and Audio Processing* **6**(1), 49–60.

18. Kneser, R., Ney, H. (1995). Improved backing-off for M-gram language modeling. *1995 International Conference on Acoustics, Speech, and Signal Processing, ICASSP-95.*

19. Chen, S.F. (2009). Performance prediction for exponential language models. *Proceedings of Human Language Technologies: The 2009 Annual Conference of the North American Chapter of the Association for Computational Linguistics.* Association for Computational Linguistics.

20. Kuo, H.-K., Arisoy, E., Emami, A., Vozila, P. (2012). *Large Scale Hierarchical Neural Network Language Models. INTERSPEECH.*

21. Pereira, F.C., Riley, M.D. (1997). 15 Speech Recognition by Composition of Weighted Finite Automata. *Finite-state language processing*, 431.

22. Pogue, D. (2010). TechnoFiles: Talk to the machine. *Scientific American Magazine* **303**(6), 40–40.

23. Hershey, J.R., Rennie, S.J., Olsen, P.A., Kristjansson, T.T. (2010). Super-human multi-talker speech recognition: A graphical modeling approach. *Computer Speech & Language* **24**(1), 45–66.

24. Bourlard, H.A., Morgan, N. (1994). *Connectionist speech recognition: a hybrid approach.* Vol. 247. Springer.

25. Gemello, R., Albesano, D., Mana, F. (1997). *Continuous speech recognition with neural networks and stationary-transitional acoustic units.* International Conference on Neural Networks.

26. Mohamed, A., Dahl, G.E., Hinton, G. (2012). Acoustic Modeling Using Deep Belief Networks. *IEEE Transactions on Audio, Speech, and Language Processing* **20**(1), 14–22.

27. Hinton, G., Li, D., Dong, Y., Dahl, G.E., Mohamed, A., Jaitly, N., Senior, A., Vanhoucke, V., Nguyen, P., Sainath, T.N., Kingsbury, B. (2012). Deep Neural Networks for Acoustic Modeling in Speech Recognition: The Shared Views of Four Research Groups. *Signal Processing Magazine* **29**(6), 82–97.

28. Dahl, G.E., Dong, Y., Li, D., Acero, A. (2012). Context-Dependent Pre-Trained Deep Neural Networks for Large-Vocabulary Speech Recognition. *IEEE Transactions on Audio, Speech, and Language Processing* **20**(1), 30–42.

29. Loizou, P.C. (2013). *Speech enhancement: theory and practice.* CRC press.

30. Hofmann, C., Wolff, T., Buck, M., Haulick, T., Kellermann, W.A. (2012). Morphological Approach to Single-Channel Wind-Noise Suppression. *Proceedings of International Workshop on Acoustic Signal Enhancement (IWAENC 2012).*

31. Widrow, B., Stearns, S.D. (1985). *Adaptive signal processing.* Vol. 15. IET.

32. Breining, C., Dreiscitel, P., Hansler, E., Mader, A., Nitsch, B., Puder, H., Schertler, T., Schmidt, G., Tilp, J. (1999). Acoustic echo control. An application of very-high-order adaptive filters. *Signal Processing Magazine* **16**(4), 42–69.

33. Haykin, S.S. (2005). *Adaptive Filter Theory, 4/e.* Pearson Education India.

34. Griffiths, L.J., Jim, C.W. (1982). An alternative approach to linearly constrained adaptive beamforming. *IEEE Transactions on Antennas and Propagation* **30**(1), 27–34.

35. Wolf, T., Buck, M. (2010). A generalized view on microphone array postfilters. *Proc. International Workshop on Acoustic Signal Enhancement, Tel Aviv, Israel.*

36. DiBiase, J.H., Silverman, H.F., Brandstein, M.S. (2001). Robust localization in reverberant rooms. In: *Microphone Arrays.* Springer. 157–180.

37. Furui, S. (1981). Cepstral analysis technique for automatic speaker verification. *IEEE Transactions on Acoustics, Speech and Signal Processing* **29**(2), 254–272.

38. Furui, S. (1981). Comparison of speaker recognition methods using statistical features and dynamic features. *IEEE Transactions on Acoustics, Speech and Signal Processing* **29**(3), 342–350.

39. Reynolds, D.A., Quatieri, T.F., Dunn, R.B. (2000). Speaker verification using adapted Gaussian mixture models. *Digital signal processing* **10**(1), 19–41.

40. Solomonoff, A., Campbell, W.M., Boardman, I. (2005). Advances In Channel Compensation For SVM Speaker Recognition. *Proceedings of IEEE International Conference on Acoustics, Speech, and Signal Processing, 2005 (ICASSP '05)*.

41. Kenny, P., Boulianne, G., Ouellet, P., Dumouchel, P. (2007). Joint factor analysis versus eigenchannels in speaker recognition. *IEEE Transactions on Audio, Speech, and Language Processing* **15**(4), 1435–1447.

42. Dehak, N., Kenny, P.J., Dehak, R., Ouellet, P., Dumouchel, P. (2011). Front-end factor analysis for speaker verification. *IEEE Transactions on Audio, Speech, and Language Processing* **19**(4), 788–798.

43. Mistretta, W., Farrell, K. (1998). Model adaptation methods for speaker verification. *Proceedings of the 1998 IEEE International Conference on Acoustics, Speech and Signal Processing*.

44. Speaker Recognition Evaluation (2013). http://www.itl.nist.gov/iad/mig/tests/spk/.

45. Evans, N., Yamagishi, J., Kinnunen, T. (2013). Spoofing and Countermeasures for Speaker Verification: a Need for Standard Corpora, Protocols and Metrics. *SLTC Newsletter*.

46. Klatt, D.H. (1987). Review of text-to-speech conversion for English. *Journal of the Acoustical Society of America* **82**(3), 737–793.

47. Taylor, P. (2009). *Text-to-speech synthesis*. Cambridge University Press.

48. Ladd, D.R. (2008). *Intonational phonology*. Cambridge University Press.

49. Ladefoged, P., Johnstone, K. (2011). *A course in phonetics*. CengageBrain.com.

50. Yoshimura, T., Tokuda, K., Masuko, T., Kobayashi, T., Kitamura, T. (1999). *Simultaneous modeling of spectrum, pitch and duration in HMM-based speech synthesis*.

51. Hunt, A.J., Black, A.W. (1996). Unit selection in a concatenative speech synthesis system using a large speech database. *Proceedings of 1996 IEEE International Conference on Acoustics, Speech, and Signal Processing, ICASSP-96*.

52. Donovan, R.E. (1996). *Trainable speech synthesis*, PhD Thesis, University of Cambridge.

53. Pollet, V., Breen, A. (2008). Synthesis by generation and concatenation of multiform segments. *INTERSPEECH*.

54. Chen, L., Gales, M.J., Wan, V., Latorre, J., Akamine, M. (2012). Exploring Rich Expressive Information from Audiobook Data Using Cluster Adaptive Training. *INTERSPEECH*.

55. Zen, H., Senior, A., Schuster, M. (2013). *Statistical parametric speech synthesis using deep neural networks*. *International Conference on Acoustics, Speech, and Signal Processing, ICASSP-13*. Vancouver.

56. Walker, M., Whittaker, S. (1990). Mixed initiative in dialogue: An investigation into discourse segmentation. *Proceedings of the 28th annual meeting on Association for Computational Linguistics*.

57. Florian, R., Hassan, H., Ittycheriah, A., Jing, H., Kambhatla, N., Luo, X., Nicolov, N., Roukos, S., Zhang, T. (2004). A Statistical Model for Multilingual Entity Detection and Tracking. *HLT-NAACL*.

58. Berger, A.L., Pietra, V.J.D., Pietra, S.A.D. (1996). A maximum entropy approach to natural language processing. *Computational Linguistics* **22**(1), 39–71.

59. Borthwick, A., Sterling, J., Agichtein, E., Grishman, R. (1998). Exploiting diverse knowledge sources via maximum entropy in named entity recognition. *Proc. of the Sixth Workshop on Very Large Corpora*.

60. Brown F, deSouza V, Mercer RL, Pietra VJD, Lai JC. (1992). Class-based n-gram models of natural language. *Computational Linguistics* **18**, 467–479.

61. Miller, S., Guinness, J., Zamanian, A. (2004). Name tagging with word clusters and discriminative training. *HLT-NAACL* 337–342.

62. Collobert, R., Weston, J., Bottou, L., Karlen, M., Kavukcuoglu, K., Kuksa, P. (2011). Natural language processing (almost) from scratch. *The Journal of Machine Learning Research* **12**, 2493–2537.

63. Lafferty J, McCallum A, Pereira FC. (2001). *Conditional random fields: Probabilistic models for segmenting and labeling sequence data*.

64. Finkel, J.R., Grenager, T., Manning, C. (2005). *Incorporating non-local information into information extraction systems by Gibbs sampling*. *Proceedings of the 43rd Annual Meeting on Association for Computational Linguistics*.

65. Ballesteros, M., Nivre, J. (2013). Going to the roots of dependency parsing. *Computational Linguistics* **39**(1), 5–13.

66. Kübler, S., McDonald, R., Nivre, J. (2009). *Dependency parsing*. Morgan & Claypool Publishers.

67. McDonald, R., Pereira, F., Ribarov, K., Hajič, J. (2005). Non-projective dependency parsing using spanning tree algorithms. *Proceedings of the conference on Human Language Technology and Empirical Methods in Natural Language Processing.*

68. Nivre, J. (2008). Algorithms for deterministic incremental dependency parsing. *Computational Linguistics* **34**(4), 513–553.

69. Riezler, S., King, T.H., Kaplan, R.M., Crouch, R., Maxwell III,, J.T., Johnson, M. (2002). Parsing the Wall Street Journal using a Lexical-Functional Grammar and discriminative estimation techniques. *Proceedings of the 40th Annual Meeting on Association for Computational Linguistics.*

70. Flickinger, D. (2000). On building a more efficient grammar by exploiting types. *Natural Language Engineering* **6**(1), 15–28.

71. Callmeier, U. (2002). *Preprocessing and encoding techniques in PET. Collaborative language engineering. A case study in efficient grammar-based processing.* Stanford, CA: CSLI Publications.

72. Brinkley, D., Guha, R. (2004). *RDF vocabulary description language 1.0: RDF schema.* W3C Recommendation. Available at http://www. w3. org/TR/PR-rdf-schema.

73. Manola, F., Miller, E., McBride, B. (2004). *RDF primer.* W3C recommendation; 10, 1–107.

74. Hitzler, P., Krötzsch, M., Parsia, B., Patel-Schneider, P.F., Rudolph, S. (2009). *OWL 2 Web Ontology Language primer.* W3C recommendation **27**, 1–123. http://www.w3.org/TR/owl2-primer/

75. Mitkov, R. (2002). *Anaphora resolution.* Vol. 134. Longman, London.

76. Johnson, M. (2009). *EMMA: Extensible MultiModal Annotation markup language.* http://www.w3.org /TR/emma/.

77. Young, S.R. (1993). *Dialog Structure and Plan Recognition in Spontaneous Spoken Dialog.* DTIC Document.

78. Williams, J.D. (2013). *The Dialog State Tracking Challenge.* SIGdial 2013. http://www.sigdial.org/workshops /sigdial2013/proceedings/index.html.

79. Ferguson, G., Allen, J.F. (1993). Generic plan recognition for dialogue systems. *Proceedings of the workshop on Human Language Technology.* Association for Computational Linguistics.

80. Bui, H.H. (2003). A general model for online probabilistic plan recognition. *IJCAI.*

81. Domingos, P., Lowd, D. (2009). Markov logic: An interface layer for artificial intelligence. *Synthesis Lectures on Artificial Intelligence and Machine Learning* **3**(1), 1–155.

82. Bohus, D., Rudnicky, A.I. (2009). The RavenClaw dialog management framework: Architecture and systems. *Computer Speech & Language* **23**(3), 332–361.

83. Bobrow, D.G., Kaplan, R.M., Kay, M., Norman, D.A., Thompson, H., Winograd, T. (1977). GUS, a frame-driven dialog system. *Artificial intelligence* **8**(2), 155–173.

84. Pieraccini, R., Huerta, J. (2005). Where do we go from here? Research and commercial spoken dialog systems. *6th SIGdial Workshop on Discourse and Dialogue.*

85. Pieraccini, R., Levin, E., Eckert, W. (1997). AMICA: the AT&T mixed initiative conversational architecture. *Eurospeech.*

86. Pieraccini, R., Caskey, S., Dayanidhi, K., Carpenter, B., Phillips, M. (2001). ETUDE, a recursive dialog manager with embedded user interface patterns. *IEEE Workshop on Automatic Speech Recognition and Understanding, 2001 (ASRU'01).*

87. Carpenter, B., Caskey, S., Dayanidhi, K., Drouin, C., Pieraccini, R. (2002). *A Portable, Server-Side Dialog Framework for VoiceXML.* Proc. Of ICSLP 2002. Denver, CO.

88. Seneff, S., Polifroni, J. (2000). Dialogue management in the Mercury flight reservation system. *Proceedings of the 2000 ANLP/NAACL Workshop on Conversational systems – Volume 3.* Association for Computational Linguistics.

89. Larsson, S., Traum, D.R. (2000). Information state and dialogue management in the TRINDI dialogue move engine toolkit. *Natural language engineering* **6**(3–4), 323–340.

90. Lemon, O., Bracy, A., Gruenstein, A., Peters, S. (2001). The WITAS multi-modal dialogue system I. *INTERSPEECH.*

91. Rich, C., Sidner, C.L. (1998). COLLAGEN: A collaboration manager for software interface agents. *User Modeling and User-Adapted Interaction* **8**(3–4), 315–350.

92. Blaylock, N., Allen, J. (2005). A collaborative problem-solving model of dialogue. *6th SIGdial Workshop on Discourse and Dialogue.*

93. Frampton, M., Lemon, O. (2009). Recent research advances in Reinforcement Learning in Spoken Dialogue Systems. *Knowledge Eng. Review* **24**(4), 375–408.

94. Lemon, O., Liu, X., Shapiro, D., Tollander, C. (2006). Hierarchical Reinforcement Learning of Dialogue Policies in a development environment for dialogue systems: REALL-DUDE. *BRANDIAL'06, Proceedings of the 10th Workshop on the Semantics and Pragmatics of Dialogue.*

95. Rieser, V., Lemon, O. (2011). *Reinforcement learning for adaptive dialogue systems.* Springer.

96. Grice, H.P. (1975). Logic and conversation. *Syntax and Semantics*, Vol. 3: Speech Acts, 41–58.

97. Searle, J.R. (1969). *Speech acts: An essay in the philosophy of language.* Vol. 626. Cambridge University Press.

98. Austin, J. (1962). *How to do things with words (William James Lectures).* Oxford University Press.

99. Cohen, P.R., Perrault, C.R. (1979). Elements of a plan-based theory of speech acts. *Cognitive science* **3**(3), 177–212.

100. Lenke, N. (1993). Regelverletzungen zu kommunikativen Zwecken. *KODIKAS,/ CODE* **16**, 71–82.

101. Davis, E. (1990). *Representations of commonsense knowledge.* Morgan Kaufmann Publishers Inc.

102. Commonsense Reasoning (2013). *Commonsense Reasoning ~ Home*; http://www.commonsensereasoning.org/.

103. Kamp, H., Reyle, U. (1993). *From discourse to logic: Introduction to model theoretic semantics of natural language, formal logic and discourse representation theory.* Springer.

104. Gamut, L. (1991). *Logic, Language and Meaning, volume II, Intentional Logic and Logical Grammar.* University of Chicago Press, Chicago, IL.

105. Lascarides, A., Asher, N. (2007). Segmented discourse representation theory: Dynamic semantics with discourse structure. *Computing meaning*, 87–124. Springer.

106. Levinson, S.C. (1983). *Pragmatics (Cambridge textbooks in linguistics).*

107. Russell, S.J., Norvig, P., Canny, J.F., Malik, J.M., Edwards, D.D. (1995). *Artificial intelligence: a modern approach.* Vol. 74. Prentice Hall, Englewood Cliffs.

108. KR, Inc. (2013). *Principles of Knowledge Representation and Reasoning.* http://www.kr.org/.

109. Baader, F. (2003). *The description logic handbook: theory, implementation, and applications.* Cambridge university press.

110. Boutilier, C., Brafman, R.I., Domshlak, C., Hoos, H.H., Poole, D. (2004). CP-nets: A tool for representing and reasoning with conditional ceteris paribus preference statements. *J. Artif. Intell. Res.(JAIR)* **21**, 135–191.

111. Grosz, B.J., Sidner, C.L. (1986). Attention, intentions, and the structure of discourse. *Computational linguistics* **12**(3), 175–204.

112. Allen, J. (1987). *Natural language understanding.* Vol. 2. Benjamin/Cummings Menlo Park, CA.

113. Traum, D.R., Larsson, S. (2003). The information state approach to dialogue management. *Current and new directions in discourse and dialogue*, 325–353. Springer.

114. SIGdial: *Special Interest Group on Discourse and Dialog* (2013). http://www.sigdial.

115. Woods, W.A., Kaplan, R.M., Nash-Webber, B., Center, M.S. (1972). *The lunar sciences natural language information system: Final report.* Bolt Beranek and Newman.

116. AQUAINT (2013). *Advanced Question Answering for Intelligence.* http://www-nlpir.nist.gov/projects/aquaint/

117. Ferrucci, D.A. (2012). Introduction to This is Watson. *IBM Journal of Research and Development* **56**(3.4), 1:1–1:15.

118. Surdeanu, M., Ciaramita, M., Zaragoza, H. (2011). Learning to rank answers to non-factoid questions from web collections. *Computational Linguistics* **37**(2), 351–383.

119. De Marneffe, M.–C., Rafferty, A.N., Manning, C.D. (2008). Finding Contradictions in Text. *ACL.*

120. Demberg, V., Winterboer, A., Moore, J.D. (2011). A strategy for information presentation in spoken dialog systems. *Computational Linguistics* **37**(3), 489–539.

121. Diekema, A.R., Yilmazel, O., Liddy, E.D. (2004). Evaluation of restricted domain question-answering systems. *Proceedings of the ACL2004 Workshop on Question Answering in Restricted Domain.*

Further reading

Allen, J., Kautz, H., Pelavin, R., Tenenberg, J. (1991). *Reasoning about plans.* Morgan Kaufmann San Mateo, CA.

Allen, J.F. (2003). *Natural language processing.*

Chen, C.H. (1976). *Pattern Recognition and Artificial Intelligence: Proceedings of the Joint Workshop on Pattern Recognition and Artificial Intelligence, Held at Hyannis, Massachusetts, June 1–3, 1976.* Acad. Press.

Dayanidhi, B.C.S.C.K., Pieraccini, C.D.R. (2002). *A portable, server-side dialog framework for VoiceXML.*

Graham, S., McKeown, D., Kiuhara, S., Harris, K.R. (2012). A meta-analysis of writing instruction for students in the elementary grades. *Journal of Educational Psychology* **104**, 879–896.

Kautz, H.A. (1991). A formal theory of plan recognition and its implementation. In *Reasoning about plans*. Morgan Kaufmann Publishers Inc.

Lascarides, A. (2003). *Logics of conversation*. Cambridge University Press.

Perrault, C.R., Allen, J.F. (1980). A plan-based analysis of indirect speech acts. *Computational Linguistics* **6**(3–4), 167–182.

Roche, E., Schabes, Y. (1997). *Finite-state language processing*. The MIT Press.

Schlegel, K., Grandjean, D., i Scherer, K.R. (2012). Emotion recognition: Unidimensional ability or a set of modality- and emotion-specific skills. *Personality and Individual Differences* **53**(1), 16–21.

4

Visual Sensing and Gesture Interactions

Achintya K. Bhowmik
Intel Corporation, USA

4.1 Introduction

Vision plays the dominant role in our interactions with the physical world. While all our perceptual sensing and processing capabilities, such as touch, speech, hearing, smell, and taste, are important contributors as we the humans understand and navigate in our daily lives, the visual perception process involving the sensing and understanding of the world around us based on the optical information received and analysed by the human visual system is the most important and utilized. There is a good reason why we have evolved to predominantly use the daytime for activities and rest after the sun goes down. Fittingly, significant portions of our brain are dedicated to processing and comprehending visual information.

Visual displays are already the primary human interfaces to most of the electronic devices that we use in our daily lives for computing, communications, entertainment, etc., presenting the system output and responses in the form of visual information to the user. Manipulating and interacting with the visual content on the display remains an intense field of research and development, generally referred to in the published literature as human-computer interaction or human-machine interface.

As reviewed in Chapter 1, the early commercially successful means of human interaction with displays and systems included indirect methods using external devices such as television remote control and the computer mouse. With the widespread adoption of touch screen-equipped displays and touch-optimized software applications and user interfaces in recent years, displays are fast becoming two-way interaction devices that can also receive direct human inputs. However, touch-based systems are inherently two-dimensional input interfaces, limiting human interaction with the content on the display to the surface of the device. In contrast, we visually perceive and interact in the three-dimensional world, aided by a 3D visual perception system consisting of binocular imaging and inference schemes.

Interactive Displays: Natural Human-Interface Technologies, First Edition. Edited by Achintya K. Bhowmik.
© 2015 John Wiley & Sons, Ltd. Published 2015 by John Wiley & Sons, Ltd.

The addition of such advanced vision capabilities promises to expand the scope of interactive displays and systems significantly. Equipped with human-like visual sensing and inference technologies, such displays and systems would "see" and "understand" human actions in the 3D space in front of the visual display, and make the human-interaction experiences more lifelike, natural, intuitive, and immersive.

The functional block diagram of an interactive display, modified to focus only on the vision-based human interfaces and interactions, is shown below in Figure 4.1. The process begins with capturing of the user actions via real-time image acquisition. The image sensing subsystem converts the light rays originating from the scene into electrical signals representing 2D or 3D visual information and sends it to the computing subsystem. Software algorithms that are specially designed to extract meanings from the image sequences lead to the recognition of user actions, such as gestures with hands, facial expressions, or eye gazes. This intelligence is then supplied to the application layer as user inputs, which executes various processing functions on the computing hardware accordingly to generate system responses. Finally, the display subsystem produces visual output of the application in the form of rays of light that are reconstructed to form images with the eyes and the visual perception system of the user.

In this chapter, we review the fundamentals and advances in vision-based interaction technologies and applications, with a focus on visual sensing and processing technologies, as well as algorithmic approaches for implementing system intelligence to enable automatic inference and understanding of user actions. In the next section, we discuss image acquisition methods, covering both 2D and 3D imaging technologies. 3D sensing techniques are covered in details in the next three chapters. Following the overview of image acquisition technologies, we review

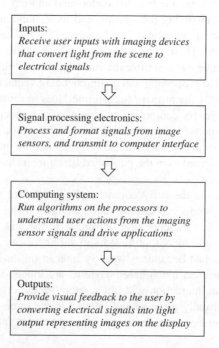

Figure 4.1   Functional block diagram of the system architecture of an interactive display focusing on vision-based sensing and interactions.

gesture interaction techniques, including gesture modelling, analysis, and recognition methods with applications in interactive displays. Finally, we also provide a review of the advances in the development of techniques for automatic understanding of facial expressions.

4.2   Imaging Technologies: 2D and 3D

Image sensing devices, primarily digital cameras that capture two-dimensional images, are increasingly becoming ubiquitous and being built into devices of all form-factors along with the image displays. Cameras are now integral elements of most mobile devices, such as phones, tablets, and laptop computers, and are increasingly appearing in all-in-one desktop computers, modern flat-panel television sets, etc. Despite the widespread adoption of imaging devices in all these systems, their applications have mostly been limited to capturing digital media, such as pictures and videos for printing or viewing on display devices, and video conferencing applications, rather than vision-based user interactions.

Traditional image sensing and acquisition devices convert the visual information in the 3D scene into a 2D array of numbers, where points in the original 3D space in the real world are mapped to discrete 2D points on the picture plane (pixels) and digital values are assigned to indicate the brightness levels corresponding to their primary colors (pixel values). This process of generating 2D images from the visual information in the 3D world can be mathematically described by a homogenous matrix formalism involving the perspective projection technique:

$$[x'] = [C][x] \tag{4.1}$$

where [x] represents the point in the 3D world, [x'] represents the transformed point on the 2D image, and [C] is the camera transformation matrix consisting of matrices corresponding to camera rotation and translation and the perspective projection matrix [1].

However, as a result of this transformation process, the 3D information cannot be truthfully recovered from the 2D images that are captured, since the pixels in the images only preserve partial information about the original 3D space. Reconstruction of 3D surfaces from single intensity images is a widely researched subject, and continues to make significant progress [2, 3]. Implementation of real-time interaction applications based on the single 2D image sensing devices, however, remains limited in scope and computationally intensive.

The human visual system consists of a binocular imaging scheme, and it is capable of depth perception. This allows us to navigate and interact in the 3D environment with ease. Similarly, rich human-computer interface tasks with complex interaction schemes are better accomplished using 3D image sensing devices which also capture depth or range information in addition to the color values for a pixel. Interactive applications utilizing real-time 3D imaging are starting to become popular, especially in the gaming and entertainment console systems in the living rooms and 3D user interfaces on personal computers [4, 5]. While there are various ways of capturing 3D visual information, three of the most prominent methods are projected structured light, stereo-3D imaging, and time-of-flight range imaging techniques [6]. Chapters 5, 6, and 7 provide in-depth reviews of these 3D sensing technologies.

In the case of structured light-based 3D sensing methods, a patterned or "structured" beam of light, typically infrared, is projected onto the object or scene of interest. The image of the light pattern deformed due to the shape of the object or scene is then captured using an image sensor. Finally, the depth map and 3D geometric shape of the object or scene are determined using this distortion of the projected optical pattern. This is conceptually illustrated in Figure 4.2 [7]. In Chapter 5, Zhang, *et al.*, cover both the fundamental principles and state-of-the-art developments in structured light 3D imaging techniques and applications.

The stereo imaging-based 3D computer vision techniques attempt to mimic the human-visual system, in which two calibrated imaging devices, laterally displaced from each other, capture synchronized images of the scene. Subsequently, the depth for the points, mapped to the corresponding image pixels, is extracted from the binocular disparity. The basic principles behind this technique are illustrated in Figure 4.3, where C_1 and C_2 are the two camera centers with focal length f, forming images of a point in the 3D world, P, at positions A and B in their respective image planes.

In this simple case, where the cameras are parallel and calibrated, it can be shown that the distance of the object, perpendicular to the baseline connecting the two camera centers, is inversely proportional to the binocular disparity:

$$depth = f \times L/\Delta \tag{4.2}$$

Algorithms for determining binocular disparity and depth information from stereo images have been widely researched and further advances continue to be made [8]. In Chapter 6, Lazaros provides an in-depth review of the developments in stereo-imaging systems and algorithms.

The time-of-flight 3D imaging method measures the distance of the object points, hence the depth map, by illuminating an object or scene with a beam of modulated infrared light and determining the time it takes for the light to travel back to an imaging device after being reflected from the object or scene, typically using a phase-shift measurement technique [9]. The system typically comprises a full-field range imaging capability, including amplitude-modulated illumination source and an image sensor array.

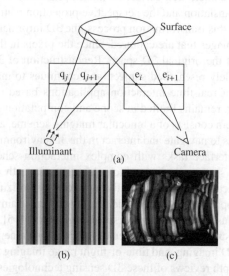

Figure 4.2 Principles of a projected structured light 3D image capture method. (a) An illumination pattern is projected onto the scene and the reflected image is captured by a camera. The depth of a point is determined from its relative displacement in the pattern and the image. (b) An illustrative example of a projected stripe pattern. In practical applications, infrared light is typically used with more complex patterns. (c) The captured image of the stripe pattern reflected from the 3D object. Source: Zhang, Curless and Seitz, 2002. Reproduced with permission from IEEE. (See color figure in color plate section).

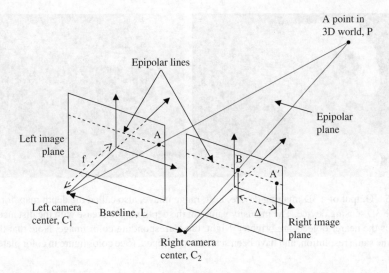

Figure 4.3 Basics of stereo-3D imaging method, illustrated with the simple case of aligned and calibrated camera pair with optical centers at C_1 and C_2, respectively, separated by the baseline distance of L. The point, P, in the 3D world is imaged at points A and B on the left and the right cameras, respectively. A′ on the right image plane corresponds to the point A on the left image plane. The distance between B and A′ on the epipolar line is called the binocular disparity, Δ, which can be shown to be inversely proportional to the distance, or depth, of the point P from the baseline.

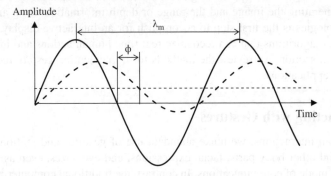

Figure 4.4 Principles of 3D imaging using the time-of-flight range measurement techniques. The solid sinusoidal curve is the amplitude-modulated infrared light that is emitted onto the scene by a source, and the dashed curve is the reflected signal that is detected by an imaging device. Note that the reflected signal is attenuated and phase-shifted by an angle ϕ relative to the emitted signal, and includes a background signal that is assumed to be constant. The distance or the depth map is determined using the phase shift and the modulation wavelength.

Figure 4.4 conceptually illustrates the method for converting the phase shifts of the reflected optical signal to the distance of the point. The reflected signal, shown in the dashed curve, is phase-shifted by ϕ relative to the original emitted signal. It is also attenuated in strength and the detector picks up some background signal as well, which is assumed to be constant. With this configuration, it can be shown that the distance of the object that reflected the signal,

$$d = (\lambda_{\mathrm{m}}/2) \times (\phi/2\pi),\tag{4.3}$$

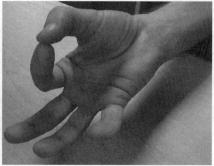

Figure 4.5   Output of a 3D imaging device. Left, range image, also called the depth map, for a hand in front of the 3D sensing device. The intensity values in the depth map decrease with the distance from the sensor, thus the nearer objects are brighter. Right, the corresponding color image. Note that the images are not of the same resolution, and have been scaled and cropped. (See color figure in color plate section).

where λ_m is the modulation wavelength of the optical signal. In Chapter 7, Nieuwenhove reviews the principles of time-of-flight depth-imaging methods and system design considerations for applications in interactive displays and systems.

Generally, the output of a 3D imaging device is a range image, also called depth map, often along with a corresponding color image of the scene. An example of this is shown in Figure 4.5, where the depth values are scaled to display as an 8-bit image, such that the points nearer to the sensing device appear brighter. As shown in the functional block diagram in Figure 4.1, generating the image and the range or depth information using any of these 3D sensing technologies is the first step to accomplish for an interactive display. The next step is to implement algorithms that can recognize real-time human actions and inputs using this data. In the next section, we review the methods for implementing gesture recognition tasks for interactive applications.

4.3   Interacting with Gestures

For inter-human interactions, we make abundant use of gestures and motions with fingers, hands, head and other body parts, facial expressions, and eye gazes, even as we use speech as the primary mode of communications. In contrast, the traditional computer input interfaces based on mouse, keyboard, or even touch screens allow limited interactive experiences. Thus, there has been significant interest and efforts made in adding gesture recognition capabilities in human-computer interactions using computer vision technologies, with a goal towards making the user experiences more natural and efficient.

The general objective of implementing a gesture recognition system is to make the computer automatically understand human actions, instructions, and expressions, by recognizing the gestures and motions articulated by the user in front of an interactive display.

Early implementations of systems developed to recognize human gestures in the 3D space were based on sensor-rich body-worn devices such as hand gloves. This approach gained popularity among the human-computer interaction researchers, following the commercial introduction of a data glove that was designed to provide hand gesture, position, and orientation information [10]. The data glove, connected to a host computer, allowed the wearer to drive a 3D hand model on the interactive display for real-time manipulation of objects in a 3D environment. A number of review articles provide in-depth surveys of the glove-based research and developments [11, 12].

While the glove-based approaches demonstrated the efficacy and versatility of 3D interactions, broad consumer adoption would require marker-less implementations that do not require wearing such tracking devices on the body. The recent developments of sophisticated and yet cost-effective and small form-factor imaging devices, advanced pattern recognition algorithms, and powerful computational resources, make computer-vision based natural gesture recognition schemes feasible.

In recent years, much research has been devoted to human gesture recognition using computer vision techniques, modelling and statistical methods based on real-time capture and analysis of 2D and 3D image sequences, and a vast number articles have been published detailing research results as well as broad reviews and surveys [13–16]. Broadly, the algorithmic approaches can be divided into two major categories: 3D model-based techniques that use shape or skeletal models of human hands and body, and view-based techniques that use 2D intensity image sequences or low-level features extracted from the visual images of the hand or other body parts. As each of these approaches has its strengths and limitations, the optimum solutions are likely to be hybrid methods that utilize relevant aspects of each technique. Figure 4.6 shows the typical algorithmic steps and flow in the vision-based gesture recognition processes.

The process starts with real-time acquisition of visual images of the user, providing inputs with gestures in the 3D space. The first step after receiving the visual images as inputs is to localize and segment the target object, such as the hand for recognizing gestures or the face for recognizing expressions. The traditional 2D camera-based systems use color or motion cues for image segmentation, which are prone to errors due to varying and unpredictable background color and ambient lighting conditions. The use of 3D-sensing cameras, such as those reviewed in the previous section, provides an additional important cue, such as the proximity of the hand to the imaging device for initial detection and tracking, and it allows segmenting based on depth.

As an example, Van den Bergh and Van Gool experimentally demonstrated the advantages of using a depth camera for hand segmentation in real-time gesture recognition applications over the color-based approaches [35]. As shown in Figure 4.7, the use of color probability approach does not segment the hand from the face when the hand overlaps the face. However, when a depth-based thresholding is added to remove the face, a clean segmentation is achieved.

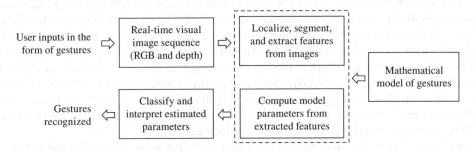

Figure 4.6 Block diagram showing the algorithmic steps and flow of vision-based gesture recognition processes.

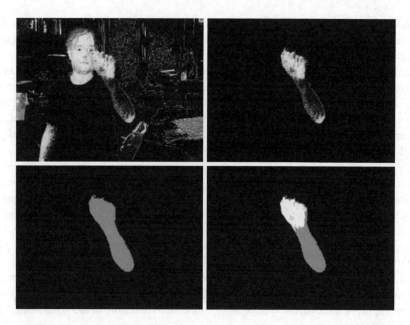

Figure 4.7   An example of using depth and color images for hand segmentation. Top left, skin color probability determined from the RGB image. Bottom left, depth image of the hand after threshold-ing to remove the background including the face. Top right, skin color probability limited only to foreground pixels determined from thresholded depth image. Bottom right, combined result showing segmented hand. Source: Van den Bergh & Van Gool 2011. Reproduced with permission from IEEE.

After the target objects are identified and segmented, specific features from the images are extracted, such as contours, edges, or special features such as fingertips, face, and limb sil-houettes, etc. Generally, a mathematical model is developed for the target gestures, including both the temporal and spatial attributes of the gestures, modelled with a set of parameters. Subsequent to the feature detection and extraction process, the parameters of this model are computed, using the features extracted from the images. Finally, the gestures performed by the user are recognized by classifying and interpreting the model parameters that were estimated at the analysis step.

Early work on 3D model-based approaches generally focused on finding a set of kinematic parameters for the 3D hand or body model, such that the 2D projected geometries of the model would closely match the corresponding edge-based images [18]. Simply stated, the parameters of the 3D model are varied until the appearance of the articulated model resembles the captured image. The models that have been used can be categorized as either volumetric or skeletal in their construction. The volumetric models typically represent the human hands or body as a collection of adjoined cylinders, with varying diameters and lengths. In such cases, the matching exercise has the goal of determining the parameters of the cylinders that bring the 3D model into correspondence with the recorded images.

In contrast, the skeletal models consist of parameters based on joint angles and segment lengths. In both cases, use of constraints based on physiology to restrict the degrees of freedom for motions that are consistent with human anatomy helps to bound the analysis space. As an

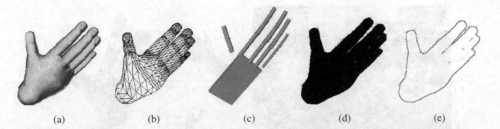

(a)   (b)   (c)   (d)   (e)

Figure 4.8   Different hand models to represent the same hand pose: (a) 3D textured volumetric model, (b) wireframe 3D model, (c) 3D skeletal model, (d) 2D silhouette, (e) 2D contour or edges. Source: Pavlovic, Sharma, and Huang 1997. Reproduced with permission from IEEE.

example, Figure 4.8 shows the representation of the same human hand pose using different models [13].

Recently, Stenger *et al.* used a Kalman filter-based method to estimate the pose of the human hand that minimizes the geometric errors between the 2D projections of the 3D hand model and the edges extracted from the hand images [19]. 3D model-based approaches have typically been computationally heavy, though their versatility in human-computer interactions have long been recognized [13].

With the advent of small-footprint and low-cost depth-sensing cameras, more efficient and robust methods are being developed. For example, Melax *et al.* recently reported a computationally-efficient method that fits a 3D model of the hand into the depth image or 3D point cloud acquired by a 3D imaging device, on a frame-by-frame basis, and imposes constraints based on physiology for tracking of the hand and the individual fingers, despite occasional occlusions [20]. As shown in Figure 4.9, this method enables versatile manipulation of objects in the 3D space visually represented in an interactive display system that is equipped with a real-time 3D sensing device. Similarly, robust tracking of human body poses using real-time range data captured by depth cameras has also been reported [21].

While the view-based, also referred to as appearance-based, approaches have been reported to be computationally lighter than the model-based approaches, they have typically lacked the generality that can be achieved with the 3D model-based techniques, and thus so far have found relatively limited applicability in human-computer interaction schemes. However, promising developments are increasingly being reported in the recent years.

Generally, this approach involves comparing the visual images or features with a predefined template representing a set of gestures. Much of the early work focused on the relatively simpler cases of static hand pose recognition using algorithms adapted from general object recognition schemes. However, it is well recognized that the implementation of natural human-computer interactions requires the recognition of dynamic gesture inputs to understand the intent of human actions rather than just static poses. Various statistical modelling based approaches, along with image processing and pattern recognition techniques, have been reported for automatic recognition of human poses, gestures, and dynamic actions, including principal component analysis using sets of training data, hidden Markov models, Kalman filtering, particle filtering, conditional density propagation or "condensation" algorithms, finite state machine techniques, etc. [22–29].

Incorporating gesture-based interactions in practical applications requires careful consideration of human-factors aspects of physical interaction to ensure a comfortable and intuitive user

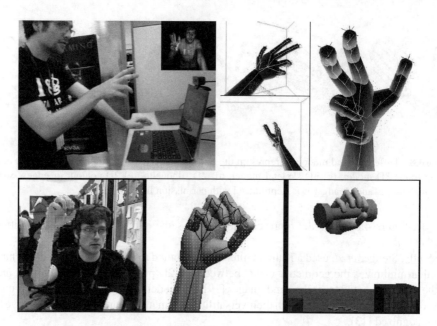

Figure 4.9 An articulated 3D model based hand skeletal tracking technique that enables fine-grain versatile manipulation of objects in the 3D space. A 3D model of the hand is fit to the depth map or point cloud acquired by a depth sensing device, taking into account physiological constraints.

experience. Clearly, the meaning and implication of gestures are subjective, different people may use different types of gestures to convey the same intention, and even the same person may use varying gestures at different times and context. Several researchers have reported detailed analysis of human motion behavior [30, 31].

An efficient approach to gesture recognition research has been to first understand and model human motion behavior, then proceed to developing the algorithms for recognition of user actions based on gesture motions. This is the underlying principle behind the hidden Markov model-based approaches, where the human behavior is treated as a large set of mental or intentional states that are represented by individual control characteristics and statistical probabilities of interstate transition [22–25]. In simple words, what we will do at the next moment is a smooth transition from our action at the current moment, and this transition can be represented by a statistical probability, since there would generally be a number of future actions to pick from that could follow the current action. So, the task of understanding the intended human interactions with gestures consists of recognizing the current hand and finger poses and a prediction of the next likely set of actions.

As an example, Pentland's work builds on the model that the basic elements of cortical processing can be described by Kalman filters that can be linked together to model the larger sets of behavior [25]. Using this hypothesis, they described the human motion behavior as a set of dynamic models represented by Kalman filters that are linked by a Markov chain of probabilistic transitions, and demonstrated the power of this technique in predicting the behavior of automobile drivers from their initial preparatory movements. The same approach would be applicable in general human-computer interfaces and interactions. This is conceptually

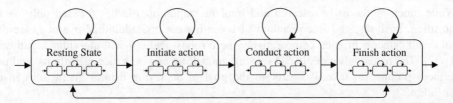

Figure 4.10 A simplified conceptual representation of a human action chain by a Markov dynamic model. The major states are linked by inter-state transition probabilities. Each state also consists of inter-linked sub-states. A smooth gesture action recognition system would consist of interpretation of the current pose and prediction, along with the anticipation of the next set of actions.

illustrated in Figure 4.10, where a specific human action chain is described by a Markov dynamic model consisting of distinguishable states that are linked by probabilistic transitions. Generally, each state also consists of a number of sub-states, indicative of the complexity of mathematical models underlying human behavior. A combination of the recognition of user actions, coupled with successful predictions and anticipations of subsequent user motions, would make the interaction experience fluid by making the systems more responsive to human gesture inputs.

In addition to the development of sensing technologies and recognition algorithms, human computer interaction researchers also continue to study and categorize classes of gesture interfaces for interactive displays and systems [32, 33]. As part of human discourse, we often perform non-specific gestures that are intended to accompany spoken language, such as waving of hands or fingers in space while speaking. By nature, this kind of gestures are not strictly defined, often unintentional, and vary widely between people and circumstances. We also often make deliberate gestures that are intended to imply specific communications or instructions, either independently articulated or reinforcing spoken discourses.

It is the second category of gesture articulations that is more relevant to the realm of implementing human-computer interactions. After reviewing the existing body of work on comprehending human gestures with the goal of advancing human computer interactions towards natural inter-human discourses and communications, Quek *et al.* [32] broadly classified gesture implementation into two predominant clusters: semaphoric and manipulative gestures. The semaphoric approach employs a stylized dictionary of static or dynamic poses or gesture sets, where the system is designed to record and identify the closest match of the performed poses or gestures from this predefined library. Practically, this approach can only provide the application of a small subset of the rich variety of daily interactions that humans deploy in real-life interactions with the physical world.

Manipulative gestures, on the other hand, involve controlling the virtual entities on an interactive display with hand or body gestures, where the movements of the displayed objects are correlated with the real-time user actions. While the manipulative gesture-based approaches are more versatile than the semaphoric approaches, Quek and Wexelblat [32, 33] also point to the shortcomings of current implementations when compared with our conversational and natural interactions. Implementations that are limited to and optimized for specific systems and applications perform reasonably well, but the goal remains to develop versatile methods that would make natural and intuitive lifelike gesture interactions possible between the humans and the content on an interactive display.

While much of the early research and implementation approaches focused only on the acquisition, analyses, and interpretation of color images, the additional use of range data from depth cameras provide comparatively more robust and efficient algorithmic pathways [34–37]. The recent advances in 3D sensing, modelling and inference algorithms, and user interfaces promise to take us nearer to the goal of achieving natural interactions in the near future.

Besides making gestures with hands, fingers, and other limbs, we also make abundant use of eye gazes, facial gestures and expressions in our interactions with fellow human beings. In Chapter 8, Drewes provides an extensive review of eye gaze tracking technologies, algorithms, and applications utilizing gaze-based interactions.

Detection and recognition of human faces and facial expressions are a widely researched area in the field of computer vision as well [38–45]. In Chapter 10, Poh *et al.* reviews face recognition technologies as part of multimodal biometrics. In addition to face detection and recognition, the importance of facial expressions in multimodal human communications cannot be overstated, as the emotions conveyed via facial gestures can amplify, or literally change the meaning of, the discourses conveyed via speech or hand gestures. It is not surprising that we always attempt to look at the person who is talking to us in order to properly understand his or her intent.

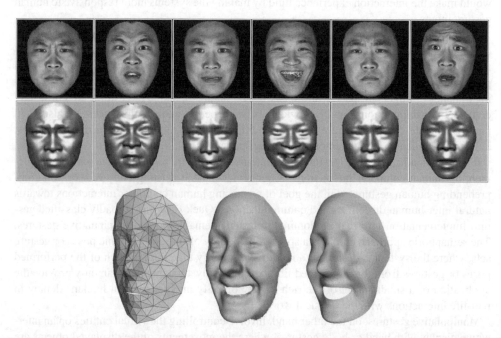

Figure 4.11   Top, an example of 3D facial range models developed by Wang *et al.*, showing six expressions: anger, disgust, fear, happiness, sadness, and surprise (from left to right). Upper row shows the textured models and the row below shows the corresponding shaded models. Source: Wang, Yin, Wei, and Sun 2006. Reproduced with permission from IEEE. Bottom, 3D face model approach used by Mpiperis *et al.* in face and expression recognition. Left, an initial 3D mesh model. Middle, 3D surface of a human face captured using 3D imaging technique. Right, the 3D mesh model fitted to the 3D surface. Source: Mpiperis, Malassiotis, Strintzis 2008. Reproduced with permission from IEEE. (See color figure in color plate section).

Early work on developing algorithms for automatic interpretation of human facial expressions were based on traditional 2D images and intensity-based analyses, such as those detailed in a number of comprehensive review articles that have been published in the recent years [40, 41]. Recently, advances are also being made that utilize 3D models and point cloud data generated by 3D sensing devices [42–45]. For example, Wang *et al.* reported a method to extract primitive 3D facial expression features and then use these features to classify expressions followed by a person-independent expression recognition method, as illustrated in the top of Figure 4.11 [42].

As another example, Mpiperis *et al.* reported a 3D-model based approach that achieves both identity-invariant facial expression recognition and expression-invariant face recognition [43]. As shown in the bottom of Figure 4.11, this approach fits a deformable facial surface mesh model to the cloud of 3D points collected from the human face acquired by a 3D sensor. The initial mesh model is designed to be neutral, and it conforms to the exhibited expression, once fitted to the surface generated by the 3D point cloud of the face under study. The facial expression is then recognized by a mathematical procedure to find correspondence within a previously determined set. While further algorithmic developments continue to make progress in several research labs around the world, the incorporation of facial expression understanding, along with hand and body gesture recognition techniques, is set to significantly enhance vision-based human interface schemes for the future interactive displays and systems.

4.4 Summary

Science fiction pioneers have long envisioned computers that would understand and respond to human actions and instructions in natural settings in the 3D environment. As an example, the acclaimed 2002 American science fiction movie *Minority Report*, directed by Steven Spielberg, depicts a future in the year 2054, where the computer has a spatial operating interface that allows the user to interact with the multimedia content on the display with gestures of the hands in the 3D space in front of it. While this was considered prescient foretelling of the future at the time, scientists and engineers have already developed technologies and systems that are helping to realize this dream several decades sooner. In fact, the real-world implementations are already more elegant, as the advances in computer vision technologies allow 3D gesture interactions with the computers without having to use any special gloves or other body-worn devices, as envisioned in the movie.

In this chapter, we have reviewed the developments in vision-based 3D sensing and interaction technologies. While the early work on computer vision was largely based on the analysis of intensity images acquired by 2D cameras, the recent developments in 3D imaging techniques allow efficient and real-time acquisition of depth map and 3D point cloud. In parallel, developments in gesture recognition algorithms have also made significant progress in the recent years, on both 3D model-based approaches as well as image- or feature-based approaches. This, coupled with the research on understanding and modelling of human motion behavior, is bringing natural human-computer interactions in the 3D space in front of interactive displays close to reality. Interactive displays that can "feel" the touch of our fingers are already ubiquitous, and now the addition of advanced visual sensing and recognition technologies is allowing the development of a new class of interactive displays that can "see" and "understand" our actions in the 3D space in front of it.

References

1. Trucco, E., Verri, A. (1998). *Introductory Techniques for 3-D Computer Vision.* Prentice Hall.
2. Saxena, A., Sun, M., Ng, A. (2008). Make3D: Learning 3-D Scene Structure from a Single Still Image. *IEEE Transactions on Pattern Analysis, Machine Intelligence.*
3. Chen, T., Zhu, Z., Shamir, A., Hu, S., Cohen-Or, D. (2013). 3-Sweep: Extracting Editable Objects from a Single Photo. *ACM Transactions on Graphics* **32**(5).
4. Microsoft Corporation (2013). www.xbox.com/en-US/kinect. Retrieved Nov 16, 2013.
5. Intel Corporation (2013). www.intel.com/software/perceptual. Retrieved Nov 16, 2013.
6. Bhowmik, A. (2013). Natural, Intuitive User Interfaces with Perceptual Computing Technologies. *Inf Display* **29**, 6.
7. Zhang, L., Curless, B., Seitz, S. (2002). Rapid Shape Acquisition Using Color Structured Light, Multi-pass Dynamic Programming. *IEEE Int Symp 3D Data Proc Vis Trans* 24–36.
8. Brown, M., Burschka, D., Hager, G. (2003). Advances in Computational Stereo. *IEEE Trans Pattern Analysis, Machine Int* **25**, 8.
9. May, S., Droeschel, D., Holz, D., Fuchs, S., Malis, E., Nuchter, A., Hertzberg, J. (2009). Three-dimensional mapping with time-of-flight cameras. *Journal of Field Robotics – Three-Dimensional Mapping, Part 2*, **26**(11–12), 934–965.
10. Zimmerman, T., Lanier, J., Blanchard, C., Bryson, S., Harvill, Y. (1987). A hand gesture interface device, Proceeding of the Conference on Human Factors in Computing Systems. *Graphics Interface*, 189–192.
11. Sturman, D., Zeltzer, D. (1994). A survey of glove-based input. *IEEE Computer Graphics, Applications* **14**(1), 30–39.
12. Dipietro, L., Sabatini, A., Dario, P. (2008). A Survey of Glove-Based Systems, Their Applications, *IEEE Transactions on Systems, Man, Cybernetics, Part C: Applications, Reviews* **38**(4), 461–482.
13. Pavlovic, V., Sharma, R., Huang, T. (1997). Visual interpretation of hand gestures for human-computer interaction: A review. *IEEE Trans Pattern Analysis, Machine Intelligence* **19**(7), 677–695.
14. Derpanis, K. (2004). *A Review of Vision-Based Hand Gestures.* Internal report, Centre for Vision Research, York University, Canada.
15. Mitra, S., Acharya, T. (2007). Gesture Recognition: A Survey. *IEEE Transactions on Systems, Man,, Cybernetics – Part C: Applications, Reviews* **37**, 311–324.
16. Wu, Y., Huang, T. (1999). *Vision-Based Gesture Recognition: A Review, Proceedings of the International Gesture Workshop on Gesture-Based Communication in Human-Computer Interaction*, pp. 103–115, Springer-Verlag.
17. Garg, P., Aggarwal, N., Sofat, S. (2009). Vision Based Hand Gesture Recognition. *World Academy of Science, Engineering, Technology* **25**, 972–977.
18. Rehg, J., Kanade, T. (1994). Visual tracking of high DOF articulated structures: An application to human hand tracking. *European Conference on Computer Vision B* **35–46**.
19. Stenger, B., Mendonca, P., Cipolla, R. (2001). Model-based 3D tracking of an articulated hand. *IEEE Conference on Computer Vision, Pattern Recognition* **II**, 310–315.
20. Melax, S., Keselman, L., Orsten, S. (2013). Dynamics Based 3D Skeletal Hand Tracking. *Proceedings of the ACM SIGGRAPH Symposium on Interactive 3D Graphics, Games*, 184–184.
21. Ganapathi, V., Plagemann, C., Koller, D., Thrun, S. (2012). Real Time Human Pose Tracking from Range Data. *Lecture Notes in Computer Science* **7577**, 738–751.
22. Yamato, J., Ohya, J., Ishii, K. (1992). Recognizing human action in time-sequential images using hidden Markov model. *Proceedings of IEEE Computer Vision, Pattern Recognition*, 379–385.
23. Kobayashi, T., Haruyama, S. (1997). Partly-hidden Markov model, its application to gesture recognition. *IEEE International Conference on Acoustics, Speech, Signal Processing* **4**, 3081–3084.
24. Yang, J., Xu, Y., Chen, C.S. (1997). Human action learning via hidden Markov model. *IEEE Trans on Systems, Man,, Cybernetics Part A: Systems, Humans* **27**(1), 34–44.
25. Pentland, A., Liu, A. (1999). Modelling and Prediction of Human Behavior. *Neural Computation* **11**, 229 – 242.
26. Arulampalam, M., Maskell, S., Gordon, N., Clapp, T. (2002). A tutorial on particle filters for online nonlinear/non-Gaussian Bayesian tracking. *IEEE Transactions on Signal Processing* **50**(2), 174–188.
27. Isard, M., Blake, A. (1998). Condensation – conditional density propagation for visual tracking. *International Journal of Computer Vision* **29**(1), 5–28.
28. Bobick, A., Wilson, A. (1997). A State-based Approach to the Representation, Recognition of Gesture. *IEEE Transactions on Pattern Analysis, Machine Intelligence* **19**, 1325–1337.

29. Imagawa, K., Lu, S., Igi, S. (1998). Color-based hands tracking system for sign language recognition. *Proceedings of IEEE International Conference on Automatic Face, Gesture Recognition*, 462–467.
30. Aggarwal, J., Cai, Q. (1999). Human Motion Analysis: A Review. *Computer Vision, Image Understanding* **73**(3), 428–440.
31. Gavrila, D.M. (1999). The visual analysis of human movement: a survey. *Computer Vision, Image Understanding* **73**, 82–98.
32. Quek, F., McNeill, D., Bryll, R., Duncan, S., Ma, X., Kirbas, C., McCullough, K.E., Ansari, R. (2002). Multimodal human discourse: gesture, speech. *ACM Transactions on Computer-Human Interaction* **9**, 171–193.
33. Wexelblat, A. (1995). An approach to natural gesture in virtual environments. *ACM Transactions on Computer-Human Interaction* **2**, 179–200.
34. Ye, M., Zhang, Q., Wang, L., Zhu, J., Yang, R., Gall, J. (2013). A survey on human motion analysis from depth data, Time-of-Flight, Depth Imaging Sensors, Algorithms,, Applications. *Lecture Note in Computer Science* **8200**, 149–187.
35. Van den Bergh, M., Van Gool, L. (2011). Combining RGB, ToF cameras for real-time 3D hand gesture interaction. *IEEE Workshop on Applications of Computer Vision*, 66–72.
36. Kurakin, A., Zhang, Z., Liu, Z. (2012). A real time system for dynamic hand gesture recognition with a depth sensor. *European Signal Processing Conference,* 1975–1979.
37. Liu, X., Fuijimura, K. (2004). Hand gesture recognition using depth data. *Proceedings of the Sixth IEEE international conference on Automatic face, gesture recognition*, 529–534.
38. Yang, M., Kriegman, D., Ahuja, N. (2002). Detecting Faces in Images: A Survey. *IEEE Transactions on Pattern Analysis, Machine Intelligence* **24**(1), 34–58.
39. Tolba, A., El-Baz, A., El-Harby, A. (2006). Face Recognition: A Literature Review. *International Journal of Signal Processing* **2**(2), 88–103.
40. Fasel, B., Luttin, J. (2003). Automatic Facial Expression Analysis: a survey. *Pattern Recognition* **36**(1), 259–275.
41. Pantic, M., Rothkrantz, L. (2000). Automatic Analysis of Facial Expressions: The State of the Art. *IEEE Transactions on Pattern Analysis, Machine Intelligence* **22**, 1424–1445.
42. Wang, J., Yin, L., Wei, X., Sun, Y. (2006). 3D facial expression recognition based on primitive surface feature distribution. *IEEE Conference on Computer Vision, Pattern Recognition* **2**, 1399–1406.
43. Mpiperis, I., Malassiotis, S., Strintzis, M.G. (2008). Bilinear Models for 3-D Face, Facial Expression Recognition. *IEEE Transactions on Information Forensics, Security* **3**(3), 498–511.
44. Yin, L., Chen, X., Sun, Y., Worm, T., Reale, M. (2008). A High-Resolution 3D Dynamic Facial Expression Database. *The 8th International Conference on Automatic Face, Gesture Recognition,* 17–19.
45. Allen, B., Curless, B., Popovic, Z. (2003). The space of human body shapes: reconstruction, parameterization from range scans. *Proceedings of ACM SIGGRAPH* **22**(3), 587–594.

5

Real-Time 3D Sensing With Structured Light Techniques

Tyler Bell, Nikolaus Karpinsky and Song Zhang
Department of Mechanical Engineering, Iowa State University, Ames, Iowa

5.1   Introduction

With recent advancements of 3D computational methods, sensing dynamic 3D scenes in real time has become vital in numerous areas (e.g., manufacturing, medical sciences, computer sciences, homeland security, entertainment). Three-dimensional sensors have also started to become a common tool used to interact with 3D content on displays; Microsoft *Kinect* is a good example.

3D sensing technologies have been improving over the past few decades, with improvements accelerating in recent years. Numerous techniques have been developed, including time-of-flight, stereovision, space-time stereovision, structured light, digital holography, and digital fringe projection. Each technology has been developed to meet a certain need and works extremely well for its specialized application, but there is no single technology that can conquer the vast requirements of 3D sensing in general. Zhang (2013) [1] edited a handbook that collected the major 3D sensing technologies, which provides a source for engineers to use such that they may select the proper technology to adopt for their specific needs.

Structured light methods have the potential to be one of the most important 3D sensing technologies for both scientific research and industrial practices, mainly because of the rapid advancement of digital technologies. Real-time 3D sensing became viable in recent decades, and has emerged as a mainstream technology because the high computational demands of real-time sensing can finally be met with the substantial computational power of today's personal computers; even tablets are becoming fast enough to be used for these purposes.

Real-time 3D sensing typically refers to acquiring, processing, and redisplaying the sensed 3D data at speeds of at least 24 Hz. Though a challenging task to tackle, advanced scanning

Interactive Displays: Natural Human-Interface Technologies, First Edition. Edited by Achintya K. Bhowmik.
© 2015 John Wiley & Sons, Ltd. Published 2015 by John Wiley & Sons, Ltd.

technologies developed over the past few years, including Microsoft's *Kinect* and some of the technologies developed by us at Iowa State University, have addressed these challenges. Interestingly, almost all of these real-time 3D sensing techniques are optical methods, meaning that there is no physical touch by a sensor to the scene being captured. Yet, because they are optics-based, it is difficult for such systems to sense surfaces with certain optical properties (e.g., shiny, transparent, or purely black).

Among these structured light methods, digital fringe projection (DFP) techniques are special, in that the structured patterns vary sinusoidally and were derived from laser interferometers. DFP techniques have demonstrated to be overwhelmingly advantageous over other types of structured light techniques and have been extensively adopted in numerous disciplines [2, 3].

Real-time 3D sensing cannot advance any further without first finding applications in which it needs to be used, and improved upon. Human-computer interaction is one of the greatest applications of the real-time 3D sensing technique, since interactions with computers require a high response speed and, thus, are real-time in nature. The majority of high-resolution, real-time 3D sensing technologies can capture either black-and-white (b/w) texture or texture with directional light. This may not be desirable for applications within human-computer interaction, where the natural color of a texture may be required. This chapter will present the technology we have developed using near infrared (NIR) light for 3D sensing and the simultaneous capture of natural color texture in real time.

Typical structured light methods can reach up to 120 Hz due to speed limitations of digital video projectors [4]. Furthermore, most of the video projectors are nonlinear, making it difficult to generate high quality phase without nonlinearity calibration and correction [5]. Though numerous nonlinear calibration techniques [6–11] have been developed and have demonstrated their successes, we found that the problem is more complex than it seems to be, since the projector's nonlinear gamma actually changes over time.

To overcome the limitations of conventional DFP techniques, [12] proposed the squared binary defocusing technique. Instead of using 8-bit grayscale patterns, the binary defocusing technique only requires 1-bit binary structured patterns. Sinusoidal fringe patterns are then naturally blended together by positioning the object away from the focal plane of the projector. This technique is not affected by the nonlinearity of the projector, since only two intensity values are used. Moreover, since only 1-bit structured patterns are required, the binary defocusing technique drastically reduces the data transfer rate, making it possible to achieve faster than 120 Hz 3D shape measurement speed. Taking advantage of the digital-light-processing (DLP) *Discovery* platform, Zhang et al. (2010) [13] have successfully developed a system that could achieve tens of kiloHertz (kHz) 3D shape measurement speed. This chapter will also report the most recent advancements we have made on superfast 3D sensing.

This chapter reviews the principles of structured light technologies. It should be emphasized that most of the technologies covered in this chapter have already been published in either conference proceedings or journal articles; this chapter, by no means, is an exhaustive survey of real-time 3D shape measurements technologies. It focuses on technologies that we have explored over the past few years, and heavily relies on these previous research publications [1, 4, 13–15].

Section 5.2 reviews the structured light method and summarizes different structured patterns developed over the years. Section 5.3 discusses the calibration of structured light-based systems. Section 5.5 presents examples which illustrate how to perform 3D sensing with a structured light method. Section 5.6 sheds light on the potential applications of real-time 3D sensing for human-computer interaction. Section 5.7 discusses our recent research on superfast 3D sensing using the binary defocusing techniques while Section 5.8 summarizes this chapter.

5.2 Structured Pattern Codifications

Optical 3D sensing methods have been extensively used due to their noninvasive nature, in which the surface being captured must not be physically measured. The stereo vision technique [16] uses two cameras to capture two 2D images from different viewing angles; this is to simulate the same process as human vision. The depth information is recovered via triangulation, which can be performed by knowing the corresponding points between two cameras. Hinging on identifying corresponding pairs between two 2D images to recover depth, it is difficult for stereo-vision techniques to achieve high accuracy if an object surface does not have strong texture variations. For example, this method cannot obtain any depth information from two uniform white flat surfaces, as the texture in each appears roughly identical. The detailed discussion regarding these techniques can be found in another chapter of this book.

Structured light systems are similar in a sense but, instead of using two 2D cameras, one projector and one camera is used. The projector projects certain types of coded structured patterns to easily determine correspondence. For a structured light system, the structured pattern design is the first and one of the key factors to determine the ultimate achievable resolution, speed, and accuracy. This section reviews a few extensively used structured pattern codifications.

5.2.1 2D Pseudo-random Codifications

The structured light technique has been extensively studied and used in the field of computer vision, machine vision, and optical metrology. Structured light is similar to the stereo vision-based method mentioned above, except that a projector is used in place of one of the cameras [17]. To establish the one-to-one mapping between a camera's pixel and the projector's pixel, one natural approach is to encode the projected image in a way such that the pixels are unique across the entire image in both the x and y directions; that is to say that every pixel can be labeled by the information represented on it [17]. Methods have been developed using such techniques as generating pseudo-random patterns or by using natural speckle patterns generated by a laser source [18]. Figure 5.1 shows some of the patterns which have been used for 3D sensing.

In one such method, the pseudo-random binary array approach, an $n1$ by $n2$ array is encoded via a pseudo-random sequence to ensure that any kernel $k1$ by $k2$ on any position over the array is unique. The pseudo-random sequence to encode the $n1$ by $n2$ array is derived via the primitive polynomial modulo n^2 method that could be mathematically described as,

$$2^n - 1 = 2^{k_1 k_2} - 1 \tag{5.1}$$

$$n_1 = 2^{k_1} - 1 \tag{5.2}$$

$$n^2 = 2^n - 1/n_1 \tag{5.3}$$

A good example of a popular consumer product today that has taken a pseudo-random codification approach in its execution of computer vision is Microsoft *Kinect*. Featuring an infrared

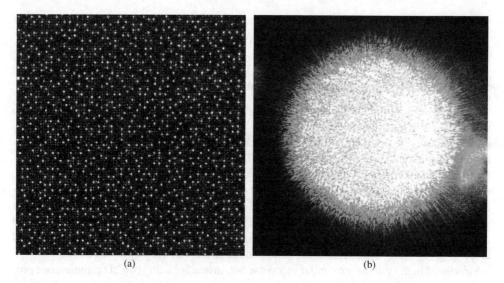

Figure 5.1 Example of pseudo-random patterns. (a) Pseudo random pattern used by Microsoft Kinect; (b) Pseudo random pattern naturally comes with a laser light. Source: Steve Jurvetson [2].

projector and an infrared camera, a beam of infrared light is projected from the projector through a diffraction grating, thus turning and focusing the light into a set of infrared dots [19]. The pseudo-random dotted pattern is then captured by the device's infrared cameras and, because the projected pattern is known by the device, a 3D scene can be constructed by triangulation. Now while the exact implementation details of this specific technology is licensed and proprietary, at a high level, the techniques are similar to other structured light approaches.

Overall, benefits of the pseudo-random codification approach are that it is easy to understand and easy to implement for 3D sensing. Some disadvantages to this technique, however, is that it may not be very tolerant to noise. Also, it is difficult for such techniques to achieve high-spatial resolution, as it is limited by the projector's resolution in both the u and v directions.

It turns out that it is not necessary to have unique correlations in 2D to perform 3D sensing if the structured light system is properly calibrated. In other words, determining (x, y, z) coordinates from a structured light system only requires one additional constraint equation besides the calibrated system constraint equations (to be discussed in detail in Section 5.3).

Therefore, the structured patterns can vary in one direction but remain constant in the other. This then eliminates the spatial resolution limits of both directions and, as a result, such techniques are extensively adopted in computer vision.

5.2.2   Binary Structured Codifications

Figure 5.2 shows the schematic diagram of a 3D sensing system using a structured light technique that contains stripes varying only in either the u or v direction, but not both.

The system includes three major units: the image acquisition unit (A), the projection unit (C), and the three-dimensional object to be measured (B). The projector shines vertical structured stripes straight onto the object's surface; the object's surface distorts these stripes from straight

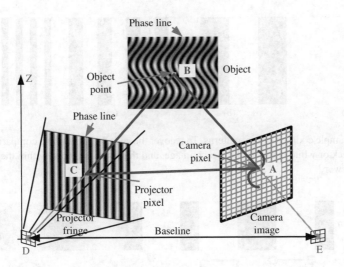

Figure 5.2 Schematic diagram of a 3D imaging system using a structured light technique. Source: Zhang 2010. Reproduuced with permission from Elsevier.

lines to curved ones if viewed from another angle. A camera then captures the distorted structured images from an angle that differs from the projection angle to form a triangle base. In such a system, correspondence is established through the structure codification information by analyzing the distortion of the known structure patterns. That is to say, the system knows which pattern is being projected and, by determining and measuring the mutations of the pattern due to the object's surface, correspondence can be established.

Binary-coded structured patterns (only 0s and 1s) are extensively used in a structured light system as they are:

1. simple: it is easy to implement since the encoding and decoding algorithms are straightforward; and
2. robust: it is robust to varying surface characteristics and noise because only two intensity levels (0 and 255) are used and expected.

In this method, every pixel of a pattern has its own unique word of 0s and 1s [17]. To determine the unique codeword for a pixel, a set of binary patterns, as illustrated by Figure 5.3, are sequentially projected and captured. One pattern for each bit of the codeword is projected and captured; to recover the unique codeword, a comparison is made at each pixel for each image, and the result is placed in the corresponding bit of the word.

Once the structured pattern is decoded and the unique codewords are known for each pixel, the correspondence can be established. Essentially, this step involves converting the pixels codeword into a projector coordinate. Once the projector coordinate for each camera coordinate is known, assuming the system is calibrated, 3D coordinates can be triangulated using,

$$[u^c, v^c, 1]^T = [P^c] [x^w, y^w, z^w, 1]^T \tag{5.4}$$

$$[u^p, v^p, 1]^T = [P^p] [x^w, y^w, z^w, 1]^T \tag{5.5}$$

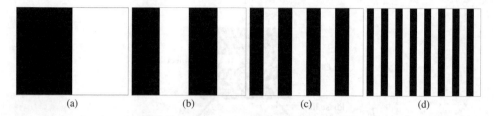

Figure 5.3   Sample codified binary patterns. To recover the code word, a simple comparison of whether the pixel is black or white is checked for each image, and the result is placed within the corresponding bit of the code word.

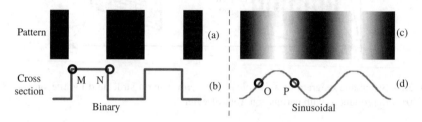

Figure 5.4   Comparison between binary and sinusoidal structured pattern. (a) Typical binary structured patterns; (b) Cross section of the binary pattern shown in (a); (c) Typical sinusoidal fringe pattern; (d) Cross-section of the sinusoidal patterns as shown in (c).

where P^c is the camera matrix, P^p is the projector matrix, (u^c, v^c) are the camera p coordinates, and (u^p, v^p) are the projector coordinates [20]. Since the codification is only in one direction, horizontal, v' is unknown, resulting in three equations and three unknowns. Solving these equations as a system of linear equations will result in the world coordinate (x^w, y^w, z^w). Section 5.3 in this chapter will discuss the calibration technique in further detail, providing more background on linear triangulation and how a structured light system can be calibrated to find matrices P^c and P^p.

Although binary codes are convenient due to their simplicity and ability to cope with noise in the system, they are not without drawbacks. The two significant drawbacks for binary codes are (1) spatial resolution and (2) large number of patterns needed for codification.

Spatial resolution for binary codes is limited by both the projector resolution and camera resolution. Figure 5.4(a) shows a binary pattern, and Figure 5.4(b) shows its corresponding cross-section. Here, black represents binary level 0, while white represents binary level 1. For a stripe between M and N in Figure 5.4(b), all points have the same intensity value, thus they cannot be differentiated. Therefore, it is difficult for binary methods to reach the projector's pixel-level spatial resolution, because the stripe width must be larger than one of the projector's pixels. Moreover, because it cannot reach pixel level correspondence, it is difficult for this technology to reach very high measurement accuracy.

The second drawback is that a large number of patterns are needed for codification, since a binary code only uses two intensity levels, binary 0 and 1; this limits the maximum number of unique codewords that can be generated to 2^n for n number of binary structured patterns. Therefore, to achieve dense 3D sensing, many binary structured patterns are required, thus making it less appealing for high-speed applications such as real-time 3D sensing.

5.2.3 N-ary Codifications

Although straightforward, robust to surface characteristics, and tolerant to noise, binary structured patterns have their drawbacks, especially when it comes to the large number of patterns needed for codification. To address this, without sacrificing spatial resolution, more intensity values can be utilized [21]. While introducing additional intensity values decreases a structured light codification's robustness, it increases its sensing speed; these trade-offs must then be weighed for the specific application that the scanner is utilized in.

Instead of using only two intensity values (0 and 255) to create unique words for every pixel in a pattern, the n-ary approach utilizes a subset of the range between these values; the most extreme case would be using all of the grayscale values (i.e., the full range 0 to 255). The codeword for each pixel can be determined via the intensity ratio calculation [22, 23]. The most basic form of this calculation assumes a spectrum of intensities, say 0 to 255, with linear "wedges" which may be placed along the vertical columns [17]. Two patterns are then projected onto the scene: one with the aforementioned wedge, and one without (constant intensity). Next, a ratio for each pixel can be derived with the values, to calculate such a ratio being present in each of the two captured patterns already projected onto the scene. In alternative methods to this approach, many wedges are projected sequentially onto the scene, while consistently increasing the wedge's period [23].

The intensity ratio method is good in terms of sensing speed, as it needs fewer patterns for codification. However, this method, as well as the aforementioned methods, are more sensitive to noise (compared to the binary structured codifications), are limited by the resolution of the projector, and are very sensitive to defocusing of the camera and the projector. To overcome the limitations of this method, phase-shifting techniques can be introduced and various structured patterns are used, such as the triangular shape [24] and the trapezoidal shape [25]. These techniques can achieve camera pixel spatial resolution and are less sensitive to defocusing; however, they are still not completely immune to the defocusing problem.

5.2.4 Continuous Sinusoidal Phase Codifications

As noted by Zhang [4], the binary, the N-ary, the triangular, and the trapezoidal patterns eventually become sinusoidal if they are blurred properly. The blurring effect often occurs when a camera captures an image out of focus and all of the sharp features of an object are blended together. Thus, sinusoidal patterns seem to be a natural choice. As shown in Figure 5.4(c) and Figure 5.4(d), the intensity varies across the image point by point. Therefore, it is feasible to reach pixel level resolution, because the intensity values between horizontal neighboring pixels are differentiable. As the spatial resolution is high, the correspondence between the projector and the camera can be determined more precisely, which allows for a higher accuracy.

Sinusoidal patterns (also known as fringe patterns) have the potential to reach pixel-level spatial resolution and, therefore, have long been studied in optical metrology. The fringe patterns used in this instance are generated by laser interference. Instead of using laser interference, where speckle noise could jeopardize the measurement quality, the digital fringe projection (DFP) technique uses a digital video projector to project computer-generated sinusoidal patterns. In principle, the DFP technique is a special kind of triangulation-based structured light method where the structured patterns vary in intensity sinusoidally. Unlike the intensity-based method, DFP techniques use the phase information to establish correspondence and are typically quite robust to surface texture variations.

The phase could be obtained using the Fourier transform profilometry (FTP) method [26] through Fourier analysis. FTP is extensively used for simple dynamic 3D shape measurement, since only a single fringe pattern is required [27]. Such a method, however, is typically very sensitive to noise and surface texture variation. To improve its robustness, Kemao (2004) [28] proposed the Windowed Fourier Transform (WFT) method. Though successful, WFT still cannot achieve high-quality 3D shape measurement for general complex 3D structures, due to the fundamental limitations of the FTP where the spatial Fourier transform is required. Modified FTP methods were developed to obtain better quality phase by using two patterns [29]. However, this is still restricted to measuring surfaces without strong texture and/or geometry variations.

To measure generic surfaces, at least three fringe patterns must be used [30]. If three or more sinusoidal structured patterns are used and their phase information is shifted, the pixel-by-pixel phase could be obtained without knowing neighboring information, thus making it immune to surface texture variations. These methods are typically referred to as phase-shifting methods.

5.2.4.1 N-step Phase-shifting Technique

Phase-shifting methods have been widely adopted in optical metrology because of the following merits [31]:

1. Dense 3D shape measurement. The phase-shifting technique permits pixel-by-pixel 3D shape measurement, making it possible to achieve camera pixel-level spatial resolution.
2. Immune to ambient light influences. Instead of utilizing intensity, the phase-shifting method analyzes phase information of the structured pattern. The ambient light influence is automatically canceled out, albeit the signal-to-noise ratio (SNR) might be sacrificed if the ambient light is too strong in comparison to the projection light.
3. Less sensitive to surface reflectivity variations. Typically, the phase-shifting method computes the phase using the arctangent function point-by-point, and the influence of surface reflectivity information is canceled out because it is constant for each pixel.
4. Permits high-speed 3D shape measurement. Since the whole measurement area can be captured and processed once, this technique, as well as other structured light techniques, can achieve high measurement speed.
5. Can achieve high measurement accuracy. Unlike other structured light techniques, the phase-shifting based method allows precise sub-pixel correspondence between the projector and the camera without any interpolation. Therefore, theoretically, it could achieve highly accurate 3D shape measurement if calibration is properly performed.

Over the years, numerous phase-shifting algorithms have been developed, including three-step, four-step, and N-step algorithms. For high-speed applications, the three-step phase-shifting algorithm is usually used, because it requires the minimum number of images to obtain phase pixel by pixel. An N-step phase-shifting algorithm with equal phase shifts can be described as:

$$I_n(x, y) = I'(x, y) + I''(x, y) \cos(\phi + 2n\pi/N) \tag{5.6}$$

where $n = 1, 2, \ldots, N$, $I'(x, y)$ is the average intensity, $I''(x, y)$ the intensity modulation, and $\phi(x, y)$ the phase to be solved for,

$$\phi(x, y) = \tan^{-1}\left[\frac{\sum_{n=1}^{N} I_n(x, y) \sin(2n\pi/N)}{\sum_{n=1}^{N} I_n(x, y) \cos(2n\pi/N)}\right] \tag{5.7}$$

This equation only provides the phase value with range $[-\pi, +\pi]$. This step is also called phase wrapping, and the phase obtained is called "wrapped phase". The wrapped phase, with modulus of 2π, can be converted to a continuous phase map, $\Phi^r(x, y)$, by adopting a spatial phase-unwrapping algorithm [32]. The spatial phase-unwrapping algorithm locates the 2π discontinuities by comparing the phase values of the neighboring pixel, and removes 2π jumps by adding or subtracting integer value of 2π.

Numerous phase unwrapping algorithms have been developed, some of which are very robust in nature [33–44]. While robust in some regards, a spatial phase unwrapping algorithm only works well for "smooth" surfaces without abrupt changes from one point to the other. Furthermore, as the unwrapped phase always refers to one point on the wrapped phase map, the phase $\Phi^r(x, y)$ obtained is called relative phase, it is difficult to uniquely correlate the phase value with the depth z value. To uniquely determine the relationship between depth z and the phase value, absolute phase is needed, which will be explained in next section [45].

5.2.4.2 Absolute Phase Recovery

The previously mentioned spatial phase unwrapping methods only recover the relative phase for each pixel, which cannot be used to measure objects with step heights greater than π or discontinuous patches. To recover absolute phase, at least one point for each continuous patch is required to have a known phase value. If sensing speed is vital, such information could be transferred by encoding the fringe patterns with a marker [46] or by projecting an additional pattern [47]. To obtain absolute phase pixel by pixel, more images are typically required, and a temporal phase unwrapping algorithm is usually adopted. Instead of looking at the phase values of neighboring pixels, a temporal phase unwrapping algorithm uses clues from other phase values in the same camera pixel.

Researchers have developed many temporal phase unwrapping methods, including two- [48] or multi-frequency [49] phase-shifting and the gray-coding plus phase-shifting [50] methods. For the gray-coding plus phase-shifting method to obtain absolute phase, a sequence of designed binary coding patterns uniquely defines the location of each 2π phase jump to create a fringe order, $k(x, y)$, so that the phase can be recovered pixel by pixel by referring to the binary coded patterns. In brief, a unique codeword $k(x, y)$, similar to the binary structured light method, is assigned to each 2π phase-change period for phase unwrapping. Each codeword is formed by a sequence of binary structured patterns. Once $k(x, y)$ is determined, the phase can be unwrapped pixel by pixel without looking at neighboring phase values. That is, the absolute phase can be obtained by:

$$\Phi(x, y) = \phi(x, y) + k(x, y) \times 2\pi \tag{5.8}$$

As aforementioned, the phase obtained from a single-frequency method is within the range of $[-\pi, \pi)$. When a fringe pattern contains more than one stripe, the phase must be unwrapped to obtain the continuous phase map. This means that if another set of wider fringe patterns (a single fringe stripe can cover the whole image) is used to obtain a phase map without 2π discontinuities, the second phase map can be used to unwrap the other one point by point without spatial phase unwrapping. To obtain the phase of the wider fringe patterns, there are two approaches:

1. use a very wide fringe pattern directly so that the single fringe covers the whole range of measurement; or
2. use two high frequency fringe patterns to generate an equivalent low frequency fringe pattern.

The former is not very commonly used because it is difficult to generate high-quality wide fringe patterns, due to noise and/or hardware limitations. Thus, the latter is more frequently adopted. This subsection will briefly explain the principle of this technique.

This multi-frequency phase-shifting method originated from physical optics, where theoretically, the relationship between the absolute phase Φ, the wavelength λ of light, and the height $h(x, y)$ can be written as:

$$\Phi = [C \cdot h(x, y)/\lambda] \cdot 2\pi \tag{5.9}$$

Here, C is a system constant. Therefore, for $\lambda_1 < \lambda_2$ with absolute phase being Φ_1 and Φ_2, respectively, their difference is:

$$\Delta\Phi_{12} = \Phi_1 - \Phi_2 = \left[C \cdot h(x, y)/\lambda_{12}^{eq}\right] \cdot 2\pi \tag{5.10}$$

Here,

$$\pi\lambda_{12}^{eq} = \lambda_1\lambda_2/|\lambda_2 - \lambda_1| \tag{5.11}$$

is the equivalent wavelength between λ_1 and λ_2. If $\lambda_2 \in (\lambda_1, 2\lambda_1)$, we have $\lambda_{12}^{eq} > \lambda_2$. In reality, we only have the wrapped phase, ϕ_1 and ϕ_2. We know that the relationship between the absolute phase is Φ and the wrapped phase $\phi = \Phi \pmod{2\pi}$ with 2π discontinuities. Here the modulus operator is used to convert the phase to a range of $[0, 2\pi)$. Taking the modulus operation on Equation (5.10) will lead to:

$$\Delta\phi_{12} = [\Phi_1 - \Phi_2] \pmod{2\pi} \tag{5.12}$$

$$= [\phi_1 - \phi_2] \pmod{2\pi} \tag{5.13}$$

$\Delta\phi_{12} = \Delta\Phi_{12} \pmod{2\pi}$. If the wavelengths are properly chosen, so that the resultant equivalent wavelength λ_{12}^{eq} is large enough to cover the whole range of the image (i.e., $|C \cdot h(x, y)/\lambda_{12}^{eq}| < 1$, the modulus operator does not affect anything and, thus, no phase unwrapping is required. However, due to the existence of noise, the two-frequency technique is usually not sufficient [48]. Practically, at least three frequency fringe patterns are required for point by point absolute phase measurement. The multifrequency technique is designed so that the equivalent widest fringe stripe can cover the whole image [51].

Assume another set of fringe patterns with wavelength (λ_3) are used; the equivalent wavelength between λ_1 and λ_3 will be $\lambda_{13}^{eq} = \lambda_1\lambda_3/|\lambda_3 - \lambda_1|$ We will have:

$$\Delta\phi_{13} = [\phi_1 - \phi_3] \pmod{2\pi} = \left\{\left[C \cdot h(x, y)/\lambda_{13}^{eq}\right] \cdot 2\pi\right\} \pmod{2\pi} \tag{5.14}$$

$$\Delta\phi_{123} = (\Delta\phi_{13} - \Delta\phi_{12}) \pmod{2\pi} = \left\{\left[C \cdot h(x, y)/\lambda_{123}^{eq}\right] \cdot 2\pi\right\} \pmod{2\pi} \tag{5.15}$$

Here $\lambda_{123}^{eq} = \lambda_{12}^{eq}\lambda_{13}^{eq}/|\lambda_{13}^{eq} - \lambda_{12}^{eq}|$. Now we only need $|C \cdot h(x, y)/\lambda_{123}^{eq}| < 1$ to ensure that the absolute phase can be obtained without spatial phase unwrapping. Once the absolute phase of the longest equivalent wavelength is obtained, it can unwrap, in reverse, the phase of other wavelengths. The phase of shortest wavelength is usually used to recover 3D information because the measurement accuracy is approximately inversely proportional to the wavelength.

Figure 1.1 Interactive displays and systems of a wide range of form factors and applications are already gaining a large foothold in the market, and some examples are shown above. The displays in many of these systems assume a new role of direct human-interface device, besides the traditional role of displaying visual information to the user.

Figure 1.11 3D segmentation using a depth-sensing imaging device allows easy manipulation of the background. In this illustration, the boy is shown on the left in front of the original background, and on the right on a different background after post-processing. Note that an analysis of the inconsistent shading in the right image would reveal that the background is not original. Depth-sensing imaging devices enable real-time segmentation using the 3D scene information, for use in applications such as video conferencing or video blogging with a custom background.

Interactive Displays: Natural Human-Interface Technologies, First Edition. Edited by Achintya K. Bhowmik.
© 2015 John Wiley & Sons, Ltd. Published 2015 by John Wiley & Sons, Ltd.

Figure 1.12 An example of visual attention indicated by the directions of eye gazes. Left, an image shown to the viewer. Right, regions of interest within the image. cambridgeincolour.com, reproduced with permission from Sean McHugh.

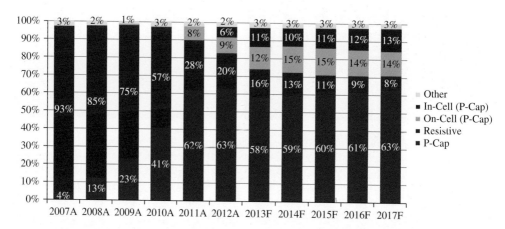

Figure 2.3 Touch technology by share of unit shipments, 2007–2012 actual and 2013–2017 forecast. Analog resistive was dominant with a 93% share in 2007; in 2012 that share had shrunk to 20%, with projected capacitive (in both discrete and embedded forms) taking 78%. DisplaySearch forecasts that in 2017 p-cap's share will increase to 90%, very close to the level of dominance that analog resistive had in 2007. Source: Data from [4].

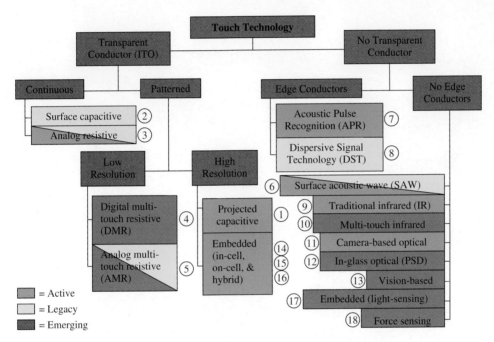

Figure 2.5 In this figure, the 18 touch technologies are classified first by their use of a transparent conductor material (typically ITO). Those technologies that use a transparent conductor are classified by whether the conductor is continuous or patterned; those that are patterned are further classified by the resolution of the patterning. Those technologies that do not use a transparent conductor are classified by their use of printed edge conductors. Note that the touch technologies are numbered to match Table 2.1. The three technologies shown in two colors (i.e., with diagonal lines) have multiple statuses in Table 2.1.

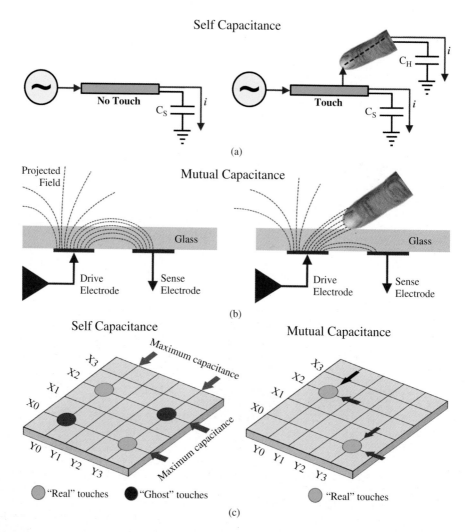

Figure 2.7 These figures illustrate the difference between self-capacitance and mutual-capacitance. **A:** Self-capacitance involves the capacitance of a single electrode to ground (C_S); a touching finger adds capacitance due to the human body capacity to ground (C_H). **B:** Mutual capacitance involves the capacitance between two electrodes; a touching finger reduces the capacitance between the electrodes by "stealing" some of the charge that is stored between the two electrodes. **C:** Self-capacitance measures the capacitance of each electrode on each axis, with the result that it allows "ghost points" because it cannot distinguish between multiple capacitance peaks on an axis (a total of 12 measurements on the 6×6 matrix shown). Mutual capacitance measures the capacitance of every electrode *intersection*, which allows detecting as many touches as there are intersections (36 in the 6×6 matrix shown). Source: Adapted.

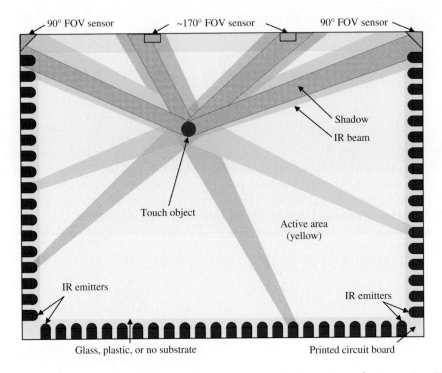

Figure 2.30 An example of a PIN-diode optical touchscreen, which uses two 90° field-of-view (FOV) sensors in the corners, and two 170° FOV sensors along the top edge. The other three edges of the touchscreen contain IR LEDs. The light rays are drawn to illustrate that a sensor may see both the shadow of a touch-object and the peripheral light, depending on the size of the touch-object. Source: Baanto.

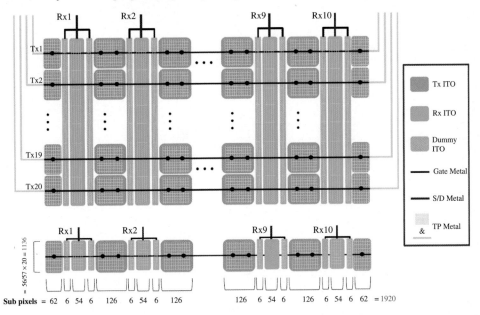

Figure 2.35 A schematic representation of how the Vcom electrodes in the TFT array of the iPhone-5 are grouped to produce both drive and sense electrodes. Source: BOE Technology Group's Central Research Institute.

Table 2.1 In this table, the 18 touch technologies covered in this chapter are classified by device type, size, and status. Within each of four device-type and size categories, each touch technology is shown as in active current use (A, green), in current use but destined to disappear (L, yellow), just beginning to appear in the market (E, blue), or not used (blank, gray). The touch technology numbers (1 through 18) are used throughout this chapter.

Technology number	Touch technology	Mobile (2″–17″)	Stationary commercial (10″–30″)	Stationary consumer (10″–30″)	Large-format (>30″)
1	Projected Capacitive	A	A	A	A
2	Surface Capacitive		L		
3	Analog Resistive	A	A	L	
4	Digital Multi-touch Resistive (DMR)	E			
5	Analog Multi-touch Resistive (AMR)	E		L	
6	Surface Acoustic Wave (SAW)		A	L	A
7	Acoustic Pulse Recognition™ (Elo)		A		
8	Dispersive Signal Technology™ (3M)				L
9	Traditional Infrared		A		A
10	Multi-touch Infrared	E		E	E
11	Camera-based Optical			A	A
12	In-glass Optical Planar Scatter Detection™ (FlatFrog)			E	E
13	Vision-Based				E
14–16	Embedded (on-cell, in-cell, & hybrid)	A			
17	Embedded (light-sensing)				E
18	Force-sensing		E		

A = Active (Green); L = Legacy (Yellow); E = Emerging (Blue)

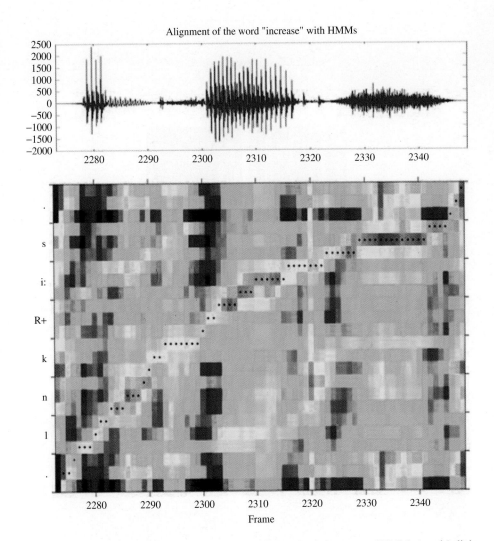

Figure 3.5 Viterbi alignment of a speech signal (x axis); against a sequence of HMMs (y axis); lighter areas indicate higher probabilities as assessed by the HMM for the given frame; dotted line shows alignment.

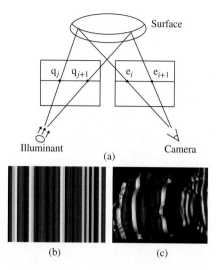

(a)

(b)   (c)

Figure 4.2   Principles of a projected structured light 3D image capture method [7]. (a) An illumination pattern is projected onto the scene and the reflected image is captured by a camera. The depth of a point is determined from its relative displacement in the pattern and the image. (b) An illustrative example of a projected stripe pattern. In practical applications, infrared light is typically used with more complex patterns. (c) The captured image of the stripe pattern reflected from the 3D object. Source: Zhang, Curless and Seitz, 2002. Reproduced with permission from IEEE.

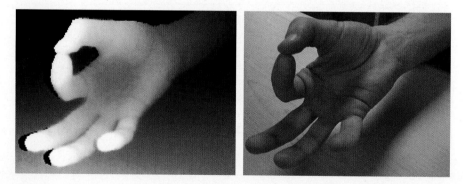

Figure 4.5   Output of a 3D imaging device. Left, range image, also called the depth map, for a hand in front of the 3D sensing device. The intensity values in the depth map decrease with the distance from the sensor, thus the nearer objects are brighter. Right, the corresponding color image. Note that the images are not of the same resolution, and have been scaled and cropped.

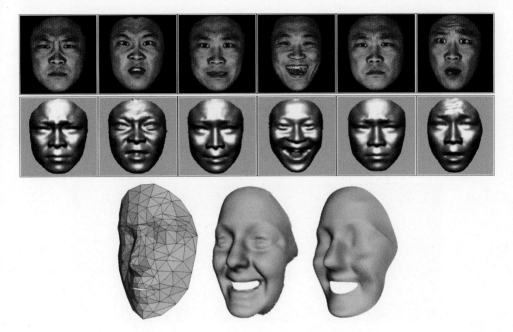

Figure 4.11 Top, an example of 3D facial range models developed by Wang *et al.*, showing six expressions: anger, disgust, fear, happiness, sadness, and surprise (from left to right). Upper row shows the textured models and the row below shows the corresponding shaded models. Source: Wang, Yin, Wei, and Sun 2006. Reproduced with permission from IEEE. Bottom, 3D face model approach used by Mpiperis *et al.* in face and expression recognition. Left, an initial 3D mesh model. Middle, 3D surface of a human face captured using 3D imaging technique. Right, the 3D mesh model fitted to the 3D surface. Source: Mpiperis, Malassiotis, Strintzis 2008. Reproduced with permission from IEEE.

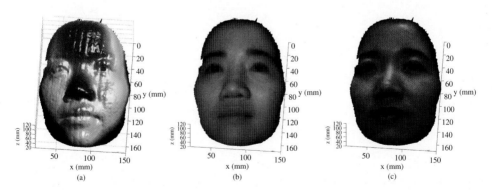

Figure 5.19 Experimental results with capturing 3D geometry and natural color texture in real time. (a) 3D reconstructed face using the infrared fringe patterns; (b) 3D results with b/w texture mapping; (c) 3D results with color texture mapping. Source: Ou P, Li B, Wang Y and Zhang S 2013. Reproduced with permission from the Optical Society of America.

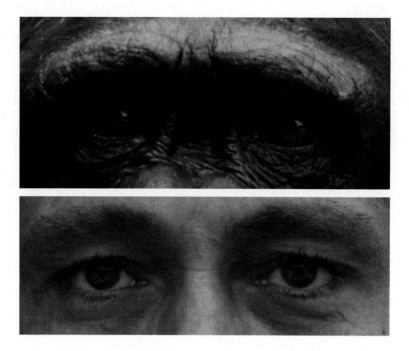

Figure 8.1 The eyes of a chimpanzee and of a human.

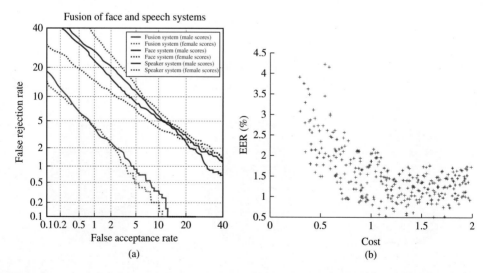

(a) (b)

Figure 10.6 (a) Detection Error Tradeoff curves for unimodal and fused bimodal systems tested on the MoBio database; plots show false acceptance and false rejection rates on a logarithmic scale for a range of decision thresholds, where the lower left point is optimal. The equal error rate (EER) for a given curve occurs at the point where the curve intersects the line $y = x$. (b) EER vs. efficiency for various scaled systems, confirming that better accuracy comes at a cost, defined as the lower of two proportions (memory consumption and time taken) with respect to a baseline system.

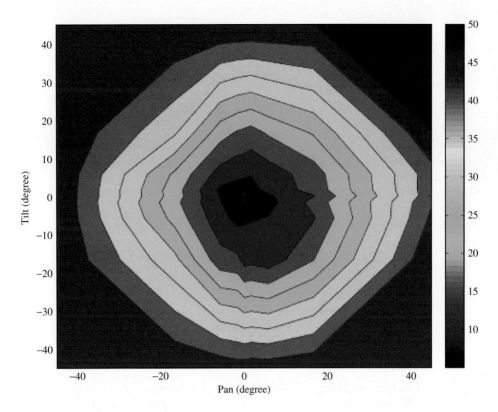

Figure 10.10 EER as a function of pan and tilt directions in degrees.

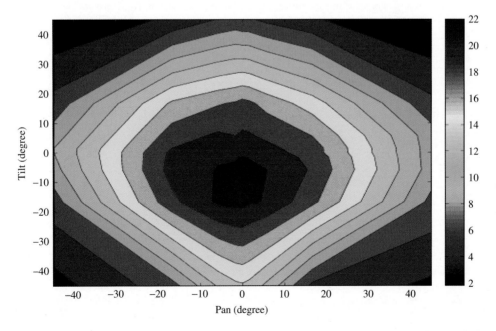

Figure 10.11 A contour plot of the median of face detection output as a function of pan and tilt directions in degrees.

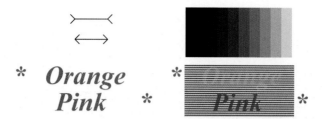

Figure 11.1 Three aspects of the visual system are illustrated in this figure. The Müller-Lyer illusion in the upper left, Mach Bands in the upper right, and a color appearance phenomenon related to spatial frequency and color contrasts in the bottom of the figure, all illustrate how the visual system interacts with the light field signal captured on the retina. There is a detailed discussion of these phenomena in the text.

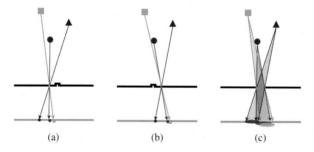

Figure 11.4 A pinhole camera with a variable size and location aperture is illustrated in this diagram. In panel A, a small aperture is located on the left and rays from points on three objects are traced to the projection surface of the camera. The small aperture is moved to the right in panel B illustrating parallax changes, and finally in panel C the aperture is enlarged from location A to B to illustrate blur created by the superposition of all of the pinhole images generated by small pinholes filling the larger aperture.

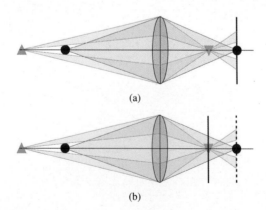

Figure 11.5 In the upper diagram in this figure, a ball with a triangle behind it is imaged onto the projection surface by a lens. In the bottom diagram, the triangle is brought to focus on the solid surface by changing the focal length of the lens. Despite the pyramid being occluded by the ball, it can still be brought to focus, because enough rays are collected through the periphery of the lens aperture in this illustration. The contrast of the triangle would be degraded (i.e., reduced) by the blurred image of the occluding ball. Changing focus would, in this limited way, be "looking through" the ball.

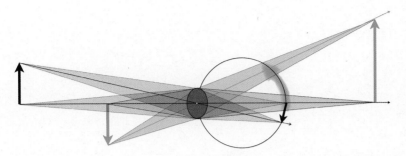

Figure 11.6 This diagram shows a schematic eye focused on the blue arrow. The transparent bundles are the rays from the end points of both arrows that form images of the end points on the retina. The leftmost arrow's image is sharp on the retina and the bundles converge to a point on the retina. The other arrow would come to focus on a projection plane behind the retina. Rays emanating from the this arrow's end points are spread or blurred over a wide area of the retina. When an out-of-focus object occludes an in-focus object, the blur reduces the image contrast of parts of the in-focus object.

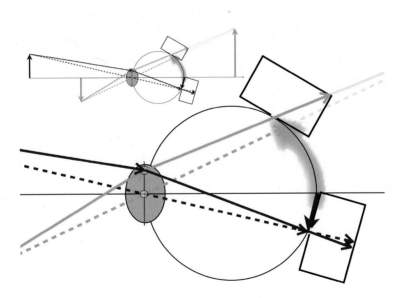

Figure 11.7 Ray tracings for two rays originating at the tip of the left arrow and two rays originating at the tip of the right arrow are followed as they come to focus at different depths in the upper small illustration. The left arrow is in focus on the retina, whereas the right arrow comes to focus behind the arrow. The enlarged illustration traces the rays as they are captured by the apertures, or miss them, in the case of the green dashed ray of two pinhole cameras located on the retina.

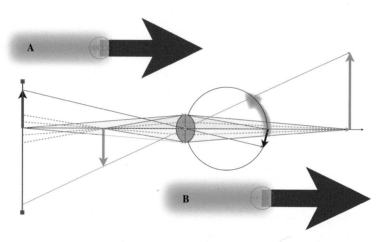

Figure 11.8 The middle of this diagram is a top down view of an eye that has focused on a green arrow pointing towards the observer's right. This arrow is partially occluded by an out of focus green arrow. In this diagram the purple line on the left represents a horizontal cross section of a light field display. The green dashed lines represent rays from the display surface that are discretely reconstructing the blur signal contained in the light field. Suppose the number of rays reconstructing the blur is sufficient to drive the eye's focus, which would mean that the observer viewing this display could focus at arbitrary depths within the viewing volume reconstructed by the display. Nonetheless the reconstructed blur might appear as a series of intense greenish spots as depicted on the retina in the insert labelled B in the lower right corner of the diagram or as overlapping small patches that merge smoothly into a blur as depicted on the left insert labelled A. The point here is that there might be a salient difference in the resolution required to drive the focus control mechanism of human vision and the number or rays required to produce good image quality in a light field display of the future.

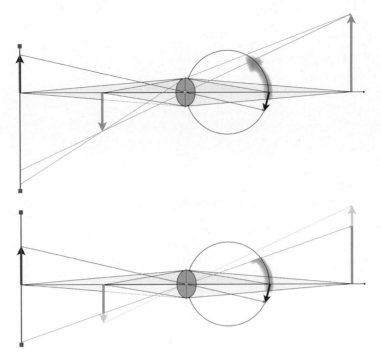

Figure 11.9 The top diagram in this illustration shows side view of a light field display (vertical lines on the extreme left) reconstructing an upward-pointing arrow that is in focus and a downward-pointing arrow partially occluding it and out of focus. Rays from the bottom (in the diagram) of the screen are necessary to render the blurred tip of the out-of-focus downward-pointing arrow. If the image and observer were to shift downward relative to the display screen, then a more general form of an edge violation will occur in this rendering of the light field and, just as in s3D displays, it is likely to appear very strange to an observer as the object that would be occluding the downward-pointing arrow's tip, which would logically be between the observer and the arrow is missing.

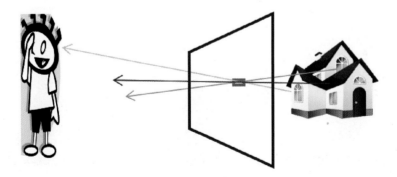

Figure 11.10 Depiction of the light rays going through one "pixel" of a window coming from different points of a toy house behind the window. To the viewer, in the position shown, the pixel will appear to be the color of the walls.

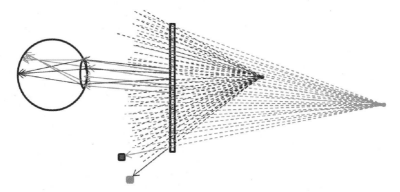

Figure 11.13 Accommodation with high-density of horizontal parallax images. Shown are two objects (dots) to be presented to the viewer by the display screen for a top view of the display screen that here has about 25 pixels. Each dashed line corresponds to the light ray from one of the dots through each pixel of the display. For the uppermost small detector, the bottom pixel on the display would appear the colour of the object further away, while for the lower detector it would appear the colour of the nearer object. For other detector positions, light from the dots would not be detected by those detectors looking at the bottom pixel. Rays from the screen related to the closer dot are more diverging than those from the more distant dot because of their relative distance behind the screen. The eye is shown focusing the rays from the closer dot on the retina. For this case the rays from the more distant dot would not be in focus.

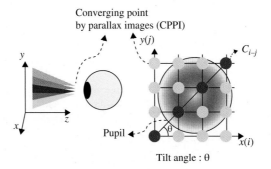

Figure 11.14 Intersection of rays of light from a common point with the pupil. Source: S-K Kim, S-H Kim, D-W Kim 2011. Reproduced with permission from SPIE.

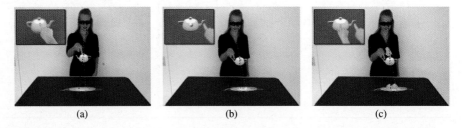

(a) (b) (c)

Figure 11.22 Illustration of visual conflicts during 3D selection of stereoscopically displayed objects: (a) The user is focused on her finger, with the virtual object appearing blurred. (b) Displaying a virtual offset cursor (white marker) at a fixed distance from the real fingertip reduces visual conflicts. (c) A virtual offset hand cursor provides familiar and additional size and distance cues for selection. Source: Bruder, Steinicke, Stuerzlinger 2013. Reproduced with permission from IEEE.

5.3   Structured Light System Calibration

In our review of different structured light techniques, we have talked about being able to triangulate 3D information assuming the system is calibrated. After choosing a codification scheme, points in the camera can be corresponded with a projector point (u^p, v^p) or a line which can then be used with the projector and camera intrinsic and extrinsic matrices to triangulate 3D points using a linear equation solver. The goal of structured light system calibration is to determine these intrinsic and extrinsic matrices for the camera and projector.

Numerous structured light system calibration methods have been developed to achieve high accuracy, and the structured light system calibration involves both the projector and camera calibration. Camera calibration has been well established with the flat checkerboard method developed by Zhang (2000) [52] being extensively used, most likely due to its simplicity and calibration speed. In this method, the camera is treated as the pin-hole camera model. The camera calibration determines its intrinsic parameters (e.g., focal length, principal point) and extrinsic parameters; the transformation between the (x, y, z) coordinates in the real world coordinate system (x^w, y^w, z^w) and that in the camera coordinate system. Figure 5.5 illustrates a simple pinhole system whose intrinsic parameters are described as,

$$A = \begin{bmatrix} \alpha & \gamma & u_0 \\ 0 & \beta & v_0 \\ 0 & 0 & 1 \end{bmatrix} \tag{5.16}$$

Where (u_0, v_0) are the coordinates of the principal point, or the intersection between the optical axis and the image sensor plane, α and β are the focal lengths along the u and v axes of the image plane, and γ is the parameter that represents the skewness of the u, v coordinates. This is typically zero for modern camera sensors, where the u and v directions are perpendicular to each other.

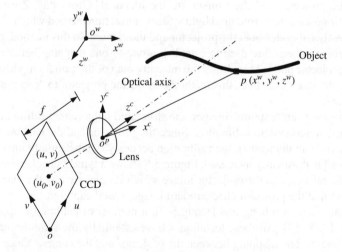

Figure 5.5   Pinhole camera model. The camera model describes that an arbitrary point in 3D space under its own coordinate system is transformed to the camera lens coordinate system, and finally 3D coordinates in lens coordinate system are projected onto a 2D imaging space. Source: Zhang S and Huang PS 2006b. Reproduced with permission from SPIE.

The extrinsic parameters of the pin-hole camera model can be mathematically described as:

$$[R, t] = \begin{bmatrix} r_{11} & r_{12} & r_{13} & t_x \\ r_{21} & r_{22} & r_{23} & t_y \\ r_{31} & r_{32} & r_{33} & t_z \end{bmatrix} \quad (5.17)$$

Here, R is a 3×3 rotation matrix, and t is a 3×1 translation vector.

As illustrated in Figure 5.5, for an arbitrary point p with coordinates (x^w, y^w, z^w) in the world coordinate system $(0^w; x^w, y^w, z^w)$ and (x^c, y^c, z^c) in the camera coordinate system $(0^c; x^c, y^c, z^c)$, its projection onto a uv image plane can be mathematically represented as

$$sI = A[R, t]X^w \quad (5.18)$$

Where $I = [u, v, 1]^T$ is the homogeneous coordinate of the image point in the image plane, $X^w = [x^w, y^w, z^w, 1]^T$ the homogeneous world coordinate for that point in the world coordinate system, and s is a scaling factor. The above equation describes a linear camera model; nonlinear effects can be compensated for by adopting a nonlinear model. For simplicity, this chapter has only described the linear model.

The structured light system differs from a stereo system in that it replaces one of the cameras with a projector. Due to this replacement, the structured light system calibration used to be very difficult, since the projector cannot capture images like a camera. Numerous techniques have been developed to fully calibrate structured light systems [53, 7], yet the majority of them are very time-consuming and it is very difficult to achieve high accuracy. In 2004, Legarda-S'aenz *et al.* (2004) [54] proposed a method that uses absolute phase to find the marker centers of the calibration board for the projector by projecting a sequence of fringe patterns. Through optimization, this method performed well in terms of accuracy. However, it requires the use of the calibrated camera to calibrate the projector and, thus, the calibration errors of the camera will be coupled into the projector calibration, which is not desirable.

Optically, the projector and the camera are the identical. Given this, Zhang and Huang (2006b) [47] proposed a new structured light system calibration method which simultaneously and independently calibrates both the projector and the camera. In this method, the horizontal and vertical fringe patterns are used to establish a one-to-one mapping between the camera pixel and the projector pixel. This allows the intensity map of the camera to virtually generate images for the projector and, thus, this method permits the projector to "capture" images like a regular camera.

Once the projector can "capture" images, the structured light system calibration becomes a well-established stereo system calibration. Since the projector and the camera are calibrated independently and simultaneously, the calibration accuracy is substantially improved, and the calibration speed is drastically increased. Figure 5.6 shows a typical checkerboard image pair captured by the camera, and the projector image which is converted by the mapping method. It clearly shows that the projector checkerboard image is well captured.

Following the work by Zhang and Huang [47], a number of calibration approaches have been developed [55–58]. All these techniques have essentially the same primary objective: to establish a one- to-one mapping between the projector and the camera. Once the system is calibrated, the (x, y, z) coordinates can be calculated using the *absolute* phase as a constraint, which will be discussed next.

The absolute phase map provides a one-to-one mapping between the camera pixel and the projector line. This constraint is sufficient for obtaining unique (x, y, z) coordinates if the

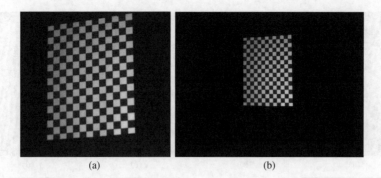

(a)   (b)

Figure 5.6 Checkerboard image pair by using the proposed technique by Zhang and Huang [47]. (a) The checkerboard image captured by the camera; (b) The mapped checkerboard image for the projector, which is regarded as the checkerboard image captured by the projector. Source: Zhang S and Huang PS 2006b. Reproduced with permission from SPIE.

camera and the projector are calibrated in the same world coordinate system. For a structured light system, Equation (5.18) can be re-written to represent the camera pin-hole model.

$$s^c I^c = A^c [R^c, t^c] X^w \tag{5.19}$$

Here, s^c is the scaling factor for the camera, I^c the homogeneous camera image coordinates, A^c the intrinsic parameters for the camera, and $[R^c, t^c]$ is the extrinsic parameter matrix for the camera. This provides a projection from the coordinate system to the camera image plane.

Similarly, the projection from the world coordinate system to the projector image plane can be represented as,

$$s^p I^p = A^p [R^p, t^p] X^w \tag{5.20}$$

Here, s^p is the scaling factor for the projector, I^p the homogeneous projector image coordinates, A^p the intrinsic parameters for the projector, and $[R^p, t^p]$ is the extrinsic parameter matrix for the projector.

Equations (5.19) to (5.21) provide six equations, but have seven unknowns (x^w, y^w, z^w), s^p, s^c, u^p, and v^p. In order to uniquely solve for the world coordinates (x^w, y^w, z^w), one more equation (or constraint) is required, which is provided by the absolute phase information: each point on the camera corresponds to one line with the same absolute phase on the projected image plane [47]. That is, assume the fringe stripe is along the v direction, and we can establish a relationship between the captured fringe image and the projected fringe image,

$$\phi_a(u^c, v^c) = \phi_a(u^p) \tag{5.21}$$

With this constraint equation, the (x^w, y^w, z^w) coordinates can be uniquely solved for pixel by pixel [59].

5.4   Examples of 3D Sensing with DFP Techniques

This section shows some examples of using DFP techniques to perform high-spatial resolution 3D sensing. Figure 5.7 illustrates an example of 3D sensing using a three-step phase-shifting

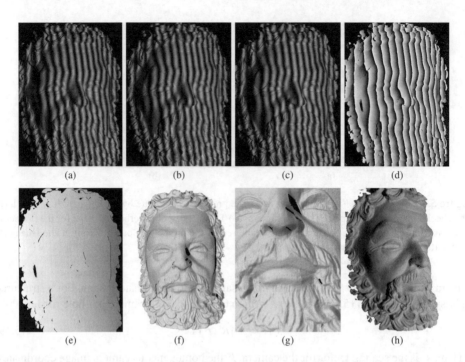

Figure 5.7 Example of 3D sensing using a three-step phase-shifting method. (a) $I_1(-2\pi/3)$; (b) $I_2(0)$; (c) $I_3(2\pi/3)$; (d) Wrapped phase map; (e) Unwrapped phase map; (f) 3D shape rendered in shaded mode; (g) Zoom in view; (h) 3D shape rendered with texture mapping. Source: Zhang S and Huang PS 2006b. Reproduced with permission from SPIE.

method. Figures 5.7(a) to 5.7(c) show three phase-shifted fringe images with $2\pi/3$ phase shift. Figure 5.7(d) shows the phase map after applying Equation (5.7) to these fringe images, and it clearly shows phase discontinuities. Applying the phase unwrapping algorithm discussed in [60], this wrapped phase map can be unwrapped to get a continuous phase map, as shown in Figure 5.7(e). The unwrapped phase map is then converted to a 3D shape by applying the method introduced in Section 5.3. The 3D shape can be rendered by *OpenGL*, as shown in Figures 5.7(f) to 5.7(g). At the same time, by averaging these three fringe images, a texture image can be obtained, which can be mapped onto the 3D shape to for a more realistic visual effect, as seen in Figure 5.7(h).

Figures 5.8 and 5.9 illustrate an example of 3D sensing using a multi-frequency phase-shifting method [61]. We chose three frequency fringe patterns with $\lambda_1 = 60$, $\lambda_2 = 90$, and $\lambda_3 = 102$ pixels. It can be found that the resultant equivalent fringe wavelength is 765 pixels. In other words, if we use the projector to generate 765 pixels-wide fringe patterns, no spatial phase unwrapping is needed to recover absolute phase, and thus 3D shape.

In this experiment, there are two separate objects, as shown in Figure 5.8(a). The diameter of the spherical ball is about 17 mm and the statue is approximately 60 (H) × 45 (W) mm in the viewing volume of the camera. Figures 5.8(b) to 5.8(d) show the distorted fringe patterns with different frequencies. Figures 5.8(e) to 5.8(g) respectively show the wrapped phase maps extracted from those frequencies. The corresponding equivalent wrapped phase for λ_1 and λ_2 is shown in Figure 5.8(h), and that for λ_1 and λ_3 is shown in Figure 5.8(i). Finally, the phase

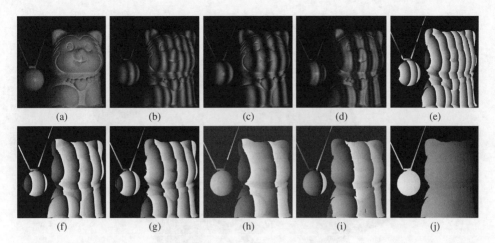

(a)   (b)   (c)   (d)   (e)

(f)   (g)   (h)   (i)   (j)

Figure 5.8 (a) Photograph of the captured scene; (b) One fringe pattern ($\lambda_1 = 60$ pixels); (c) One fringe pattern ($\lambda_2 = 90$ pixels); (d) One fringe pattern ($\lambda_3 = 102$ pixels); (e) Wrapped phase ϕ_1; (f) Wrapped phase ϕ_2; (g) Wrapped phase ϕ_3; (h) Equivalent phase difference $\Delta\phi_{12}$; (i) Equivalent phase difference $\Delta\phi_{13}$; (j) Resultant phase $\Delta\phi_{123}$. Source: Wang Y and Zhang S 2011. Reproduced with permission from the Optical Society of America.

map of the longest equivalent wavelength can be obtained from these two equivalent phase maps, and the result is shown in Figure 5.8(j). It can be seen from this figure that this phase map has no 2π discontinuities and, thus, no spatial phase unwrapping is needed.

From this longest equivalent wavelength phase map, the shortest, with $\lambda_1 = 60$ pixels, can be unwrapped point by point and can then be used to recover 3D information. Figure 5.9 shows the reconstructed 3D result. It can be seen that both 3D profiles are properly recovered, which cannot be achieved using the single frequency phase-shifting method.

5.5 Real-Time 3D Sensing Techniques

To reach real time speeds, a single structured pattern with a random codification or sinusoidal structures could be used. However, these techniques usually have substantial limitations, either on surface property requirements or on achievable spatial resolution. The trade-off usually is to use multiple structured patterns that are switched rapidly, so that the number of patterns required to recover one 3D shape can be captured in a short period of time. Rusinkiewicz *et al.* (2002) [62] developed a real-time 3D model acquisition system using a stripe boundary code [63]. The stripe boundary code is generated by projecting a sequence of binary level structured patterns. As aforementioned, the spatial resolution of such a technology is limited by the projector resolution.

Using color-encoded structured patterns [64–67] could reach higher speed, because a structured pattern can be encoded with three structured patterns. However, the measurement accuracy is affected, to a various degree, by object surface color [68]. For example, with a blue object, information conveyed by red and green colors will be lost because no signal will be captured. Furthermore, the color coupling between red/green regimes and green/blue regimes will affect the measurement quality if no optical filtering is used [67].

Figure 5.9   3D result using the absolute phase map. Source: Wang Y and Zhang S 2011. Reproduced with permission from the Optical Society of America.

To avoid the problems induced by color, monochromatic, or black and white (b/w), structured patterns are typically used. For instance, Zhang and Huang (2006a) [47] have developed a 3D sensing system using b/w sinusoidally varying structured patterns. By adopting a fast three-step phase-shifting algorithm [69], Zhang *et al.* (2006a) developed a system that achieved simultaneous data acquisition, reconstruction, and display in real time. This technique takes advantage of the unique projection mechanism of a single-chip digital-light-processing (DLP) projector; three structured patterns are encoded into the RGB channels of the projector and are switched automatically and naturally by the DLP projector.

By these means, we achieve 3D surface measurement at 60 Hz with more than 300K points per frame [71]. We will discuss the details of this technology in this section.

5.5.1   Fundamentals of Digital-light-processing (DLP) Technology

The digital light processing (*DLP^{TM}*) concept originated from Texas Instruments in the late 1980s. In 1996, Texas Instruments began commercializing its *DLP^{TM}* technology. At the core of every *DLP^{TM}* projection system, there is an optical semiconductor called the digital micro-mirror device, or DMD, which functions as an extremely precise light switch. The DMD chip contains an array of hinged, microscopic mirrors, each of which corresponds to one pixel of light in a projected image.

Figure 5.10 shows the working principle of the micro mirror. Data in the cell controls electrostatic forces that can move the mirror $+\theta_L$ (ON) or $-\theta_L$ (OFF), thereby modulating the light that is incident on the mirror. The rate of a mirror switching ON and OFF determines the brightness of the projected image pixel. An image is created by light reflected from the

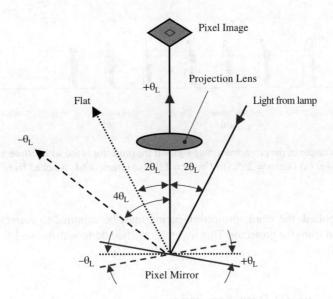

Figure 5.10 Optical switching principle of a digital micromirror device (DMD) (modified from [72] with permission).

ON mirrors passing through a projection lens onto a screen. Grayscale values are created by controlling the proportion of ON and OFF times of the mirror during one frame period – black being 0% ON time and white being 100% ON time.

DLP^{TM} projectors embraced the DMD technology to generate color images. All *DLP^{TM}* projectors include a light source, a color filter system, at least one digital micro-mirror device (DMD), digital light processing electronics, and an optical projection lens. For a single-chip DLP projector, the color image is produced by placing a color wheel into the system. The color wheel, which contains red, green, and blue filters, spins at a very fast speed – thus red, green, and blue channel images will be projected sequentially onto the screen. However, because the refresh rate is so high, human eyes can only perceive a single color image instead of three sequential ones.

A DLP projector produces a grayscale value by time integration [73]. A simple test was performed for a single-chip DLP projector, PLUS U5-632h, that runs under a monochromatic mode with a projection speed of 120 Hz. The output light was sensed by a photodiode (Thorlabs FDS100), and the photocurrent was converted to voltage signal and monitored by an oscilloscope. The photodiode used has a response time of 10 ns, an active area of 3.6 mm × 3.6 mm, and a bandwidth of 35 MHz. The oscilloscope used to monitor the signal was Tektronix TDS2024B and has a bandwidth of 200 MHz.

Figure 5.11 shows some typical results when it was fed with uniform images with different grayscale values. If the pure green, RGB = (0, 255, 0), is supplied, the signal has the duty cycle of almost 100% ON. When the grayscale value is reduced to 128, approximately half of the channel is filled. If the input grayscale value is reduced to 64, a smaller portion of the channel is filled. If the input grayscale value is 0, the signal has the duty cycle of almost 0% ON. These experiments show that if the supplied grayscale value is somewhere between 0 and 255, the output signal becomes irregular. Therefore, if a sinusoidal fringe pattern varying from

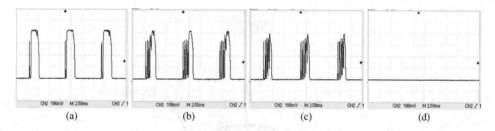

Figure 5.11   Example of the projected timing signal if the projector is fed with different grayscale value of the green image. (a) Green = 255; (b) Green = 128; (c) Green = 64. Adapted from Zhang.S., *et al.* 2013.

0 to 255 is supplied, the whole projection period must be captured to correctly capture the image projected from the projector. This was what we had done with our real-time 3D sensing technology [74].

5.5.2   Real-Time 3D Data Acquisition

As addressed in Section 5.2.4, three structured images can be used to reconstruct one 3D shape if a three-step phase-shifting algorithm is used. This perfectly fits into the DLP technology, where three patterns can be encoded into the three primary color channels of the projector. Since color fringe patterns are not desirable for 3D sensing, we developed a real-time 3D sensing system based on a single-chip DLP projection and white light technique [74]. Figure 5.12 shows the system layout. The computer-generated color-encoded fringe image is sent to a single-chip DLP projector that projects three color channels sequentially and repeatedly in grayscale onto the object. The camera, precisely synchronized with projector, is used to capture three individual channels separately and quickly. By applying the three-step phase-shifting

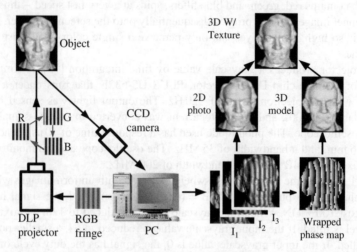

Figure 5.12   Real-time 3D shape measurement system layout. Source: Zhang S 2010. Reproduced with permission from Elsevier.

algorithm to three fringe images, the 3D geometry can be recovered. Averaging three fringe images will result in a texture image that can be further mapped onto the recovered 3D shape to enhance certain visual effect.

The projector projects a monochrome fringe image for each of the RGB channels sequentially; the color is a result of a color wheel placed in front of a projection lens. Each "frame" of the projected image is actually three separate images. By removing the color wheel and placing each fringe image in a separate channel, the projector can produce three fringe images at 120 fps (360 fps individual color channel refreshing rate). Therefore, if three fringe images are sufficient to recover one 3D shape, the 3D measurement speed is up to 120 Hz. However, due to the speed limit of the camera used, it takes two projection cycles to capture three fringe images, and thus the measurement speed is 60 Hz.

Figure 5.13 shows the timing chart for the real-time 3D shape measurement system. Due to the speed limit of the camera we used (maximum speed of 200 fps at full resolution), it takes two projection cycles to capture the three structured images required to recover one 3D shape. Therefore, we have realized a 60 Hz 3D shape measurement speed, which is faster than real time (typically referred to as 24 Hz or better).

5.5.3 Real-Time 3D Data Processing and Visualization

Computing 3D coordinates from the phase is computationally intensive, which is very challenging for a single computer CPU to realize in real time. However, because the coordinate calculations are point by point matrix operations, this can be performed efficiently by a GPU. A GPU is a dedicated graphics rendering device for a personal computer or game console. Modern GPUs are very efficient at manipulating and displaying computer graphics, and their highly parallel structure makes them more effective than typical CPUs for parallel computation algorithms. Since there are no memory hierarchies or data dependencies in the streaming model, the pipeline maximizes throughput without being stalled. Therefore, whenever the GPU is consistently fed by input data, performance is boosted, leading to an extraordinarily scalable architecture [76]. By utilizing this streaming processing model, modern GPUs outperform their CPU counterparts in some general purpose applications, and the difference is expected to increase in the future [77].

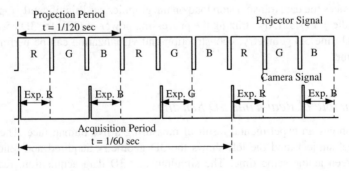

Figure 5.13 Timing chart of the measurement system. Source: Zhang S and Yau ST 2007b. Reproduced with permission from SPIE.

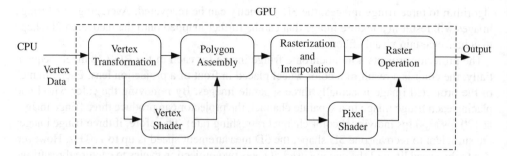

Figure 5.14 GPU pipeline. Vertex data, including vertex coordinates and vertex normal, are sent to the GPU. GPU generates the lighting of each vertex, creates the polygons and rasterizes the pixels, then output the rasterized image to the display screen. Source: Zhang S, Royer D, and Yau ST, 2006. Reproduced with permission from the Optical Society of America.

Figure 5.14 shows the GPU pipeline. The CPU sends the vertex data, including the vertex position coordinates and the vertex normal, to the GPU, which generates the lighting of each vertex, creates the polygons, rasterizes the pixels, and then output the rasterized image to the display screen. Modern GPUs allow user specified code to execute within both the vertex and pixel sections of the pipeline, which are called vertex shaders and pixel shaders, respectively. Vertex shaders are applied for each vertex and run on a programmable vertex processor.

Vertex shaders take vertex coordinates, color, and normal information from the CPU. The vertex data is streamed into the GPU, where the polygon vertices are processed and assembled based on the order of the incoming data. The GPU handles the transfer of streaming data to parallel computation automatically. Although the clock rate of a GPU might be significantly slower than that of a CPU, it has multiple vertex processors acting in parallel and, therefore, the throughput of the GPU can exceed that of the CPU. As GPUs increase in complexity, the number of vertex processors increase, leading to great performance improvements.

By taking advantage of the processing power of the GPU, 3D coordinate calculations can be performed in real time with an ordinary personal computer with an NVidia graphics card [79]. Moreover, because 3D shape data is already on the graphics card, it can be rendered immediately without any lag. Therefore, by these means, real-time 3D geometry visualization can also be realized in real time simultaneously. Also, because only the phase data, instead of the 3D coordinates and the surface normals, are transmitted to graphics card for visualization, this technique reduces the data transmission load on the graphics card significantly (approximately six times smaller). In short, by utilizing the processing power of GPU for 3D coordinates calculations, real-time 3D geometry reconstruction and visualization can be performed rapidly and in real time.

5.5.4   Example of Real-Time 3D Sensing

Figure 5.15 shows an experimental result of measuring a live human face. The right figure shows the real subject, and the left shows the 3D geometry acquired and rendered on the computer screen at the same time. The simultaneous 3D data acquisition, reconstruction, and display speed achieved is 30 frames/sec with more than 300,000 points per frame being acquired.

Figure 5.15 Simultaneous 3D data acquisition, reconstruction, and display at 30 fps. Source: Zhang S 2010. Reproduced with permission from Elsevier.

5.6 Real-Time 3D Sensing for Human Computer Interaction Applications

Real-time 3D sensing brings many advanced technologies together to provide the ability to capture objects in the real world quickly and accurately. This allows software to make precise measurements, recognize predetermined patterns and act on them, control the system via the sensed data, and more. Another way to derive and view the implications this technology can have within the field of human computer interaction is to ask the question, "What is a system like this actually *doing*?"

3D systems, such as the structured light system, *observe*. By observing in 3D, and also in real time, many new dynamic and responsive ways to interact with our personal computers become available. With advancements in equipment and software within the 3D computer vision field, no longer must a user be tied behind a desk, manipulating a two-dimensional plane (their computer workspace) with a mouse, keyboard, or other traditional input device. New methods of input to our machines allow for more creative interactions, more natural interactions and, until recently, interactions of which were previously not possible. This section provides examples of these new interactions and focuses on the details and implications of the interaction.

5.6.1 Real-Time 3D Facial Expression Capture and its HCI Implications

One new interaction tool that 3D sensing provides would be the ability to capture and act on facial expressions and other facial gestures; for example, imagine a computing system where facial expressions had some corresponding action associated with them. Now, while this technology has been researched previously [80], the structured light technology that we have been developing at Iowa State University, as described thus far in this chapter, has the ability to provide much more visual and spatial information. This in turn, makes way for more precise control over the system.

In traditional computer vision, for a computer to recognize facial features and expressions, the system would process a two-dimensional video stream. Then, based on recognizable features in each video frame, an algorithm could determine which facial expression was on the user's face, based on some predetermined and/or learned pattern [80]. Under these same principles, imagine the possible interactions and minute facial features that could be detected via a structured light capturing system as it is being described in this chapter. This sort of system provides another dimensionality of data – depth information – which is difficult to account for in traditional 2D capturing. Another advantage which structured light systems have are their ability to obtain very high quality and high detail spatial coordinates.

No longer must a facial expression be blatantly obvious before it can be deciphered by a capture system in real time. With today's structured light technology, something as discrete as a smirk on one's mouth or a twitch of their eye can be captured and handled, as shown by Figure 5.16. Now, consider for a moment the implications this would have for users with disabilities which make it difficult to use the traditional mouse, keyboard, or microphone as inputs to a machine [81]. Structured light 3D capturing systems provide another means to interact with today's computers via the capture and processing of the 3D object and scene.

5.6.2   Real-Time 3D Body Part Gesture Capture and its HCI Implications

Not only can this technology be applied to capturing and processing very discrete facial features and gestures, but it can also be applied to various other areas of the body. Take, for example, the idea of finger pointing in a capture system. With traditional 2D capture methods, a direct point at the camera is difficult to decipher, due to a lack of an easily discernible silhouette. With high-detail 3D information, however, a pointing finger suddenly becomes more

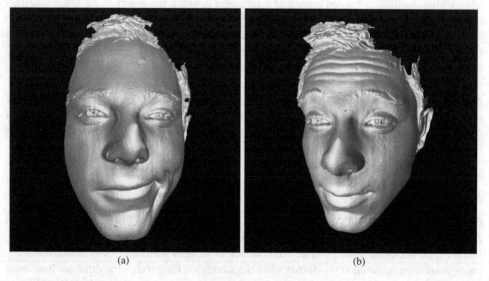

(a)   (b)

Figure 5.16   Different examples of minute facial features and gestures that can be captured by a real-time 3D structured light system. (a) A smirk could be captured and used as a control; (b) Although a very discrete feature, wrinkles can easily be seen on the forehead and can be deciphered.

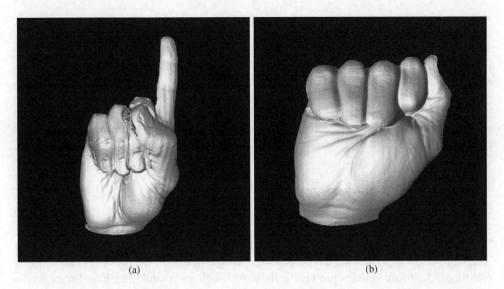

(a)   (b)

Figure 5.17 Different examples of minute hand positions and gestures that can be captured by a real-time 3D structured light system; this discrete detail capturing could be applied to any body part. (a) A pointing finger; (b) A curled fist as a result of bringing the pointer finger inward.

observable, as it has depth and a sense of context in the real world. To afford for more gestures and interactions, these concepts can be expanded to the entire hand, arm, and body. Figure 5.17 displays an example of a finger being captured as it turns from a point into a fist. The level of detail in the structured light capture is high, and therefore it is possible to decipher minor movements and features.

Not too long ago, computer control via motion was a concept that only existed in science fiction and required invasive contact based sensors; it is through the advancements of technologies, such as real-time 3D sensing, that is making this concept an ever-present reality. Body gesture recognition is already at play in some consumer electronics today, but let us consider some of their limitations.

The lack of quality in sensed data and the amount of reliable spatial information provided by today's 2D and 3D capture systems are their limiting factors. This being the case, possible gestures used to control these systems might include a full swipe across one's body with one or both arms. This gesture is drastic enough to be deciphered as a control. Are the majority of users able to make these sorts of gestures [81]? Do these gestures tire users after prolonged use [82]? What about environments where there is not enough spatial room to perform such complex gestures?

When more advanced and precise information provided by a 3D structured light system is introduced, the number of possible gestures increases significantly, as well as their fidelity of control. As these systems provide the ability to capture and process very discrete facial features and gestures, as mentioned previously, they also provide this for body parts. Minute finger and hand controls can be performed in 3D space to control a computer. The ability to perform these minute gestures, as well as today's existing set of large gestures, will make interacting with computer systems a much more natural process. As the actions will be performed with one's body, a potential user will not have much difficulty in learning and using the controls [83].

5.6.3   Concluding Human Computer Interaction Implications

The implications in regard to human computer interaction are significant, as these systems provide yet another way for users to interact with their computing environments and the services provided by them. The new dimensionality of data, along with higher resolution, allows for a greater level of control in gesture based input devices. In addition to the examples provided, there are numerous other areas in HCI that could benefit from this technology. It will be the minds of HCI practitioners who take this superfast 3D sensing technology, as presented in this chapter, and create the next wave of human computer interaction ideals, controls and techniques.

5.7   Some Recent Advancements

5.7.1   Real-Time 3D Sensing and Natural 2D Color Texture Capture

Although real-time 3D sensing has become increasingly available, the majority of the existing real-time 3D sensing systems use visible light. Although successful, they have limitations in applications such as human computer interaction, homeland security and biometrics, where visible light might not be desirable. To address this, Microsoft *Kinect* has used a near infrared (NIR) laser to replace white light; however, using laser light may pose concerns on eye safety. With recent advancements of LED technology, NIR LED light is a viable solution that does not pose the same concerns with eye safety.

Currently, there is little documented study on high-resolution, real-time 3D sensing using an infrared fringe projection technique. Moreover, most of the aforementioned real-time 3D sensing can simultaneously provide a black-and-white (b/w) texture by analyzing the structured patterns that are also used for 3D reconstruction. However, this texture generation approach is not natural, meaning that the texture is not captured without the directional projection light. This usually introduces shadows on the acquired texture, due to the measured geometric shapes. To acquire a natural texture image, the projection light must be turned off, such that only the ambient light is on when the texture image is captured. This can be done by capturing an additional image without the projection light on. This, however, sacrifices measurement speed, often drastically.

Simultaneously acquiring 2D color texture is more difficult, though viable, by using a color camera for both 3D geometry and 2D color texture capture [60, 67, 84–86]. An alternative is to add another color camera to the system which is only used for color texture capture [74, 87, 88], and establish the mapping between the color camera and the b/w camera. The former usually sacrifices 3D measurement quality, due to the inherent color-induced problems (e.g., color coupling). The latter typically requires a complicated hardware setup or a sophisticated calibration routine to create the mapping between the color camera and the b/w camera.

To our knowledge, there is no system that can simultaneously capture natural 2D color texture and high-resolution 3D geometry in real time. Recently, Ou *et al.* (2013) [89] have demonstrated that NIR could, indeed, be used for 3D sensing with the same algorithm we developed. In this study, we used a near infrared camera/projector pair to perform 3D sensing and a secondary color camera to simultaneously capture 2D color images solely illuminated by ambient lights. Since these two light wavelengths do not overlap, simultaneous 3D shape and natural 2D texture capture can be realized without sacrificing speed; and, because the color camera only captures visible light without being interfered with by the infrared light for 3D shape measurement, nature 2D color images can be obtained.

Figure 5.18 shows a photograph of the system we developed. It is composed of an infrared digital-light processing (DLP) projector (LightCommander, Logic PD, Inc.), a high-speed infrared CMOS camera (Phantom V9.1, Vision Research), and a color CCD camera (DFK 21BU04, The Imaging Source). The wavelength of the infrared LED used in this projector is 850 nm. An infrared filter is placed in front of the high-speed infrared CMOS camera to block visible light. Two cameras are triggered by an external circuit board that is synchronized with the projector. The infrared CMOS camera was set to capture images with a resolution of 576×576, and the color CCD camera was set with a resolution of 640×480. Due to the low intensity of the infrared projector, the high-speed projector was set to project binary patterns at 200 Hz, and the infrared camera was precisely synchronized with the projector to capture 2D fringe patterns at 200 Hz. The 3D capturing rate is 20 Hz, since it requires ten phase-shifted fringe patterns to reconstruct one 3D frame. The color camera was, hence, configured to capture 2D color images at 20 Hz to precisely align with the 3D capture rate.

The color camera and the infrared camera were precisely synchronized through the external trigger timing circuit. Figure 5.19(a) shows the 3D shape of a human face. One may notice that the reconstructed 3D geometry is not very smooth (some bumps on the face). This might be because the infrared light penetrates to different depths of skin on different parts of the face. From these fringe images, the b/w texture can be retrieved and mapped on to the 3D geometry, as shown in Figure 5.19(b). The b/w is not natural, due to the direct illumination of the projector (shadows on the side of nose are obvious). In contrast, the color texture was captured under the ambient light and does not have the same shadow problem. Figure 5.19(c) shows the result of mapping natural color texture on the 3D geometry captured by IR system. It is important to notice that the mapping is usually not perfectly aligned with pixels, due to

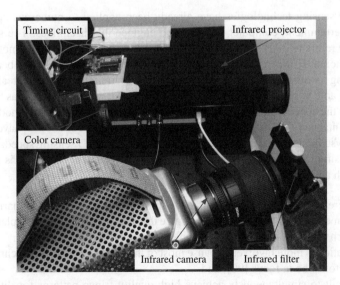

Figure 5.18 System developed for both simultaneous 3D geometry and natural color texture capture. Source: Ou P, Li B, Wang Y and Zhang S 2013. Reproduced with permission from the Optical Society of America.

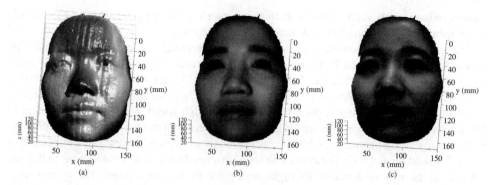

Figure 5.19 Experimental results with capturing 3D geometry and natural color texture in real time. (a) 3D reconstructed face using the infrared fringe patterns; (b) 3D results with b/w texture mapping; (c) 3D results with color texture mapping. Source: Ou P, Li B, Wang Y and Zhang S 2013. Reproduced with permission from the Optical Society of America. (See color figure in color plate section).

the discrete effects of the cameras. Linear interpolation was adopted in this research to map color texture. It is also important to note that the mappings we used in this research are linear, without considering the nonlinear distortion of the camera lens. However, no obvious artifacts were created, even with the linear model.

This technology seems promising, yet the LightCommander has been discontinued, and there are no other commercially available and affordable NIR projectors on the market right now.

5.7.2   Superfast 3D Sensing

Real-time 3D sensing technologies have already been applied to numerous applications in medicine, entertainment, computer science, and engineering disciplines. The speed of real-time typically refers to 30 Hz or higher, which is sufficient to capture slow motions (such as facial expressions). However, for capturing higher speed motions such as a beating heart or even a speaking mouth, the current posed technology has limitations. Speed breakthroughs are required to be able to capture rapidly changing scenes. Our recent research has been focusing on addressing the speed limit issue by developing the binary defocusing method [12].

Taking advantage of the DLP Discovery platform, Zhang *et al*. (2010) [13] have successfully developed a system that could achieve tens of kiloHertz (kHz) 3D sensing speed. However, the binary defocusing technique is not trouble free. The measurement capability is rather limited compared with a conventional DFP method:

(a) the measurement accuracy is lower due the influence of high-frequency harmonics;
(b) the depth measurement range is smaller since the object must be properly placed in a small region such that the binary pattern becomes good-quality sinusoidal [45];
(c) the calibration is more involved because most of the existing calibration techniques require the projector to be in focus [90]; and
(d) it is difficult to simultaneously achieve high-quality fringe patterns for different spatial frequencies [61].

To enhance the binary defocusing method, the pulse width modulation (PWM) technique, developed in the Power Electronics field, has been adopted into the 3D sensing field. Ayubi *et al.* (2010) [91] has introduced the sinusoidal pulse width modulation (SPWM) technique and Wang and Zhang (2010) [92] have developed the optimal pulse width modulation (OPWM) technique to further improve phase quality. However, the improvement of OPWM or SPWM techniques are still limited if the fringe stripes are too narrow or too wide [93].

Due to the discrete nature of the fringe patterns, it is understandable that the PWM techniques have limited improvements. This is because the PWM techniques are, after all, one-dimensional in nature. Therefore, there is large room for further improvement if the optimization can be performed in both dimensions. Lohry and Zhang (2012) [94] recently proposed a technique to locally modulate the pixels to emulate a triangular pattern to reduce the influence of high-frequency harmonics.

Most recently, we note that representing grayscale images with binary images has been extensively studied in the digital image processing field for high-quality printing. This technology is called half-toning or dithering, which has been practiced since the 1960s [95]. Wang and Zhang (2012b) [96] have borrowed concepts from the simple Bayer-dithering technique and later the error-diffusion techniques [97–101] and applied them to the 3D shape measurement field [102]. Our research found that the dithering techniques drastically improve the measurement quality when the fringe stripes are wide, yet the improvement is limited if the fringe stripes are narrow.

The majority of the dithering techniques are simply applying a matrix to the grayscale images, and "binarizes" the image by comparing the original, or the modified intensity, with the matrix. These algorithms were all developed to generate high quality visual effects for generic grayscale images. However, they were not specifically designed to take advantage of some inherent structures of the grayscale images, e.g., sinusoidal structures for fringe patterns. Therefore, great improvements could be achieved if we utilize the inherent unique sinusoidal structures of the fringe patterns. Lately, we have developed a couple of algorithms [103, 104] to optimize the dithering techniques, and substantial improvements have been achieved.

We measured a 3D sculpture with a fringe period of $T = 90$ pixels, with the projector being slightly defocused. Figure 5.20 shows the results. The first row shows the captured structured images, and the second row shows the 3D results. It clearly shows that if the projector is nearly focused, the binary structure is clear for squared binary patterns, but the dithered or optimized dithered patterns are sinusoidal in nature. This experiment also shows that neither the squared binary method (SBM), nor the PWM, can generate high-quality 3D results, despite the small enhancement that PWM has on SBM, while the dithering technique and optimized dithering technique could perform well in reconstructing the phase of the sculpture. Optimized dithering has quite a clear improvement upon the dithering technique.

Wang et al. (2013) [102] developed a system to measure live rabbit hearts that beat at approximately 200 beats per minutes. The system is composed of a digital-light-processing (DLP) projector (DLP LightCrafter, Texas Instruments, TX) and a high-speed CMOS camera (Phantom V9.1, Vision Research, NJ). The camera was triggered by an external electronic circuit that senses the timing signal of the DLP projector. The camera was set to capture images with a resolution of 576×576, and the projector has a resolution of 608×648.

The projector was set to switch binary patterns at 2000 Hz, and the camera was chosen to capture at 2000 Hz. A two-wavelength phase-shifting technique was chosen, with the shorter

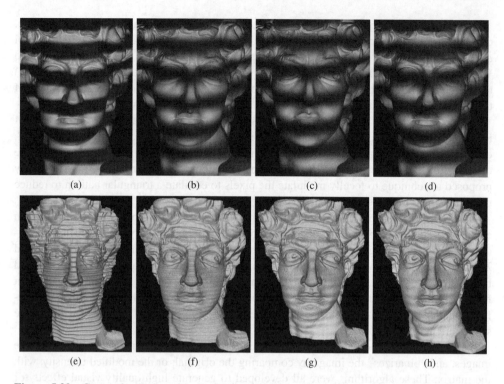

Figure 5.20 Captured patterns with the fringe period $T = 90$ pixels when the projector is slightly defocused. (a) Squared binary pattern; (b) PWM pattern; (c) Dithering pattern; (d) Optimized dithering pattern; (e)–(f) Corresponding reconstructed 3D results of (a)–(d). Source: Li B, Wang Y, Dai J, Lohry W and Zhang S 2013. Reproduced with permission from Elsevier.

wavelength being the OPWM pattern [92] and longer wavelength being Stucki-dithered patterns [101].

Figure 5.21 shows some typical results of a rabbit heart, and clearly shows that the dynamic motion of the heart was well captured. Without the binary defocusing techniques, the live rabbit's heart surface could not been properly measured, as we found that at least 800 Hz was required to properly measure the heart without obvious motion artifacts.

5.8 Summary

This chapter has reviewed generic 3D sensing with structured light techniques, explained the principles behind these methods, detailed the real-time 3D sensing with digital fringe projection (DFP) technique that we have developed over the years, and then presented some of the recent work on superfast 3D sensing using the binary defocusing method we are developing. The advancements within the 3D sensing field are exciting, and have progressed to a point at which they are mature enough to tackle many of the challenges of real environment, every day usage. There are still some challenging problems to face before such a technology can be extensively adopted into our daily life, including the production and purchasing cost of such

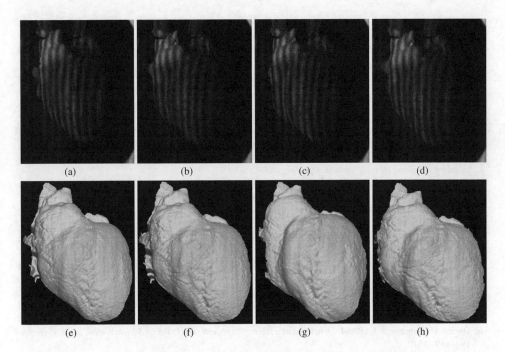

Figure 5.21   Example of capturing live rabbit hearts with binary defocusing techniques. (a)–(d) Captured fringes of a live rabbit heart; (e)–(f) Corresponding reconstructed 3D results of (a)–(d). Source: Li B, Wang Y, Dai J, Lohry W and Zhang S 2013. Reproduced with permission from Elsevier.

equipment. However, with this technology becoming increasingly available to fields such as human computer interaction, the possibilities and applications are endless.

Acknowledgements

The work presented in this chapter was carried out by numerous collaborators: Professors Fu-Pen Chiang from Stony Brook University, Peisen Huang from Stony Brook University, and Shing-Tung Yau from Harvard University; as well as numerous students from our laboratory at Iowa State University, Laura Ekstrand, Shuanyan Lei, Beiwen Li, William Lohry, and Yajun Wang. The authors would like to thank all of them. This study was sponsored by the National Science Foundation (NSF) under grant numbers: CMMI-1150711 and CMMI-1300376. The views expressed in this chapter are those of the authors, and not necessarily those of the NSF.

References

1. Zhang, S. (ed.) (2013). *Handbook of 3D machine vision: Optical metrology and imaging,* 1st edition. CRC Press, New York, NY.
2. Geng, G. (2011). Structured-light 3D surface imaging: a tutorial. *Advances in Opt. and Photonics* **3**(2), 128–160.
3. Gorthi, S., Rastogi, P. (2010). Fringe projection techniques: Whither we are?. *Opt. Laser. Eng.* **48**, 133–140.
4. Zhang, S. (2010). Recent progresses on real-time 3-D shape measurement using digital fringe projection techniques. *Optics and Lasers in Engineering* **48**(2), 149–158.

5. Lei, S., Zhang, S. (2010). Digital sinusoidal fringe generation: defocusing binary patterns vs focusing sinusoidal patterns. *Optics and Lasers in Engineering* **48**(5), 561–569.

6. Guo, H., He, H., Chen, M. (2004). Gamma correction for digital fringe projection profilometry. *Appl. Opt.* **43**, 2906–2914.

7. Hu, Q., Huang, P.S., Fu, Q., Chiang, F.P. (2003). Calibration of a 3-D shape measurement system. *Opt. Eng.* **42**(2), 487–493.

8. Kakunai, S., Sakamoto, T., Iwata, K. (1999). Profile measurement taken with liquid-crystal grating. *Appl. Opt.* **38**(13), 2824–2828.

9. Pan, B., Kemao, Q., Huang, L., Asundi, A. (2009). Phase error analysis and compensation for nonsinusoidal waveforms in phase-shifting digital fringe projection profilometry. *Opt. Lett.* **34**(4), 2906–2914.

10. Zhang, S., Huang, P.S. (2007). Phase error compensation for a three-dimensional shape measurement system based on the phase shifting method. *Opt. Eng.* **46**(6), 063601.

11. Zhang, S., Yau, S.T. (2007a). Generic nonsinusoidal phase error correction for three-dimensional shape measurement using a digital video projector. *Appl. Opt.* **46**(1), 36–43.

12. Lei, S., Zhang, S. (2009). Flexible 3-D shape measurement using projector defocusing. *Opt. Lett.* **34**(20), 3080–3082.

13. Zhang, S., van der Weide, D., Oliver, J. (2010). Superfast phase-shifting method for 3-D shape measurement. *Opt. Express* **18**(9), 9684–9689.

14. Zhang, S., Wang, Y., Laughner, J.I., Efimov, I.R. (2013). Measuring dynamic 3D micro-structures using a superfast digital binary phase-shifting technique *ASME 2013 International Manufacturing Science and Engineering Conference*, Madison, Wisconsin.

15. Li B, Wang Y, Dai J, Lohry W, Zhang S. (2013). Some recent advances on superfast 3D shape measurement with digital binary defocusing techniques. *Optics and Lasers in Engineering* (doi:10.1016/j.optlaseng.2013.07.010).

16. Dhond, U., Aggarwal, J. (1989). Structure from stereo – a review. *IEEE Trans. Systems, Man, and Cybernetics* **19**, 1489–1510.

17. Salvi, J., Pages, J., Batlle, J. (2004). Pattern codification strategies in structured light systems. *Patt. Recogn.* **37**, 827–849.

18. Huang, Y., Shang, Y., Liu, Y., Bao, H. (2013). *Handbook of 3D Machine Vision: Optical Metrology and Imaging* 1st edn; CRC chapter 3D shapes from Speckle, pp. 33–56.

19. Zhang, Z. (2012). Microsoft kinect sensor and its effect. *IEEE Multimedia* **19**(2), 4–10

20. Hartley, R., Zisserman, A. (2000). *Multiple view geometry in computer vision* vol. 2. Cambridge Univ Press.

21. Pan, J., Huang, P., Zhang, S., Chiang, F.P. (2004). Color n-ary gray code for 3-D shape measurement *12th Intl Conf. on Exp. Mech.*

22. Carrihill, B., Hummel, R. (1985). Experiments with the intensity ratio depth sensor. *Computer Vision, Graphics and Image Processing* **32**, 337–358.

23. Chazan, G., Kiryati, N. (1995). Pyramidal intensity-ratio depth sensor. Technical report, Israel Institute of Technology, Technion, Haifa, Israel.

24. Jia, P., Kofman, J., English, C. (2007).Two-step triangular-pattern phase-shifting method for three-dimensional object-shape measurement. *Opt. Eng.* **46**(8), 083201.

25. Huang, P.S., Zhang, S., Chiang, F.P. (2005). Trapezoidal phase-shifting method for three-dimensional shape measurement. *Opt. Eng.* **44**(12), 123601.

26. Takeda, M., Mutoh, K. (1983). Fourier transform profilometry for the automatic measurement of 3-D object shape. *Appl. Opt.* **22**, 3977–3982.

27. Su, X., Zhang, Q. (2010). Dynamic 3-D shape measurement method: A review. *Opt. Laser. Eng.* **48**, 191–204.

28. Kemao, Q. (2004). Windowed Fourier transform for fringe pattern analysis. *Appl. Opt.* **43**, 2695–2702.

29. Guo, H., Huang, P. (2008). 3-D shape measurement by use of a modified fourier transform method. *Proc. SPIE* 7066:70660E.

30. Schreiber, H., Bruning, J.H. (2007). *Optical Shop Testing* 3rd edn. chapter Phase shifting interferometry, pp. 547–655. John Wiley & Sons.

31. Malacara, D. (ed.) (2007). *Optical Shop Testing* 3rd edn. John Wiley and Sons, New York.

32. Ghiglia, D.C., Pritt, M.D. (eds.) (1998). *Two-Dimensional Phase Unwrapping: Theory, Algorithms, and Software*. John Wiley and Sons, New York.

33. Baldi, A. (2003). Phase unwrapping by region growing. *Appl. Opt.* **42**, 2498–2505.

34. Buchland, J.R., Huntley, J.M., Turner, S.R.E. (1995). Unwrapping noisy phase maps by use of a minimum-cost-matching algorithm. *Appl. Opt.* **34**, 5100–5108.

35. Chyou, J.J., Chen, S.J., Chen, Y.K. (2004). Two-dimensional phase unwrapping with a multichannel least-mean-square algorithm. *Appl. Opt.* **43**, 5655–5661.
36. Cusack, R., Huntley, J.M., Goldrein, H.T. (1995). Improved noise-immune phase unwrapping algorithm. *Appl. Opt.* **34**, 781–789.
37. Flynn, T.J. (1997). Two-dimensional phase unwrapping with minimum weighted discontinuity. *J. Opt. Soc. Am. A.* **14**, 2692–2701.
38. Ghiglia, D.C., Romero, L.A. (1996). Minimum *lp*-norm two-dimensional phase unwrapping. *J. Opt. Soc. Am. A.* **13**, 1–15.
39. Goldstein, R.M., Zebker, H.A., Werner, C.L. (1988). Two-dimensional phase unwrapping. *Radio Sci.* **23**, 713–720.
40. Hung, K.M., Yamada, T. (1998). Phase unwrapping by regions using least-squares approach. *Opt. Eng.* **37**, 2965–2970.
41. Huntley, J.M. (1989). Noise-immune phase unwrapping algorithm. *Appl. Opt.* **28**, 3268–3270.
42. Merraez, M.A., Boticario, J.G., Labor, M.J., Burton, D.R. (2005). Agglomerative clustering-based approach for two dimensional phase unwrapping. *Appl. Opt.* **44**, 1129–1140.
43. Salfity, M.F., Ruiz, P.D., Huntley, J.M., Graves, M.J., Cusack, R., Beauregard, D.A. (2006). Branch cut surface placement for unwrapping of undersampled three-dimensional phase data: Application to magnetic resonance imaging arterial flow mapping. *Appl. Opt.* **45**, 2711–2722.
44. Zhang, S., Li, X., Yau, S.T. (2007a). Multilevel quality-guided phase unwrapping algorithm for real-time three-dimensional shape reconstruction. *Appl. Opt.* **46**(1), 50–57.
45. Xu, Y., Ekstrand, L., Dai, J., Zhang, S. (2011). Phase error compensation for three-dimensional shape measurement with projector defocusing. *Appl. Opt.* **50**(17), 2572–2581.
46. Zhang, S., Yau, S.T. (2006). High-resolution, real-time 3-D absolute coordinate measurement based on a phase-shifting method. *Opt. Express* **14**(7), 2644–2649.
47. Zhang, S., Huang, P.S. (2006b). Novel method for structured light system calibration. *Opt. Eng.* **45**(8), 083601.
48. Creath, K. (1987). Step height measurement using two-wavelength phase-shifting interferometry. *Appl. Opt.* **26**(14), 2810–2816.
49. Cheng, Y.Y., Wyant, J.C. (1985). Multiple-wavelength phase shifting interferometry. *Appl. Opt.* **24**, 804–807.
50. Sansoni, G., Carocci, M., Rodella, R. (1999). Three-dimensional vision based on a combination of gray-code and phase-shift light projection: Analysis and compensation of the systematic errors. *Appl. Opt.* **38**, 6565–6573.
51. Towers, C.E., Towers, D.P., Jones, J.D. (2003). Optimum frequency selection in multifrequency interferometry. *Opt. Lett.* **28**(11), 887–889.
52. Zhang, Z. (2000). A flexible new technique for camera calibration. *IEEE Trans. Pattern Anal. Mach. Intell.* **22**(11), 1330–1334.
53. Cuevas, F.J., Servin, M., Rodriguez-Vera, R. (1999). Depth object recovery using radial basis functions. *Opt. Commun.* **163**(4), 270–277.
54. Legarda-Sáenz, R., Bothe, T., Jüptner, W.P. (2004). Accurate procedure for the calibration of a structured light system. *Opt. Eng.* **43**(2), 464–471.
55. Gao, W., Wang, L., Hu, Z. (2008). Flexible method for structured light system calibration. *Opt. Eng.* **47**(8), 083602.
56. Huang, P., Han, X. (2006). On improving the accuracy of structured light systems *Proc. SPIE.* **6382**, 63820H.
57. Li, Z., Shi, Y., Wang, C., Wang, Y. (2008). Accurate calibration method for a structured light system. *Opt. Eng.* **47**(5), 053604.
58. Yang, R., Cheng, S., Chen, Y. (2008). Flexible and accurate implementation of a binocular structured light system. *Opt. Lasers Eng.* **46**(5), 373–379.
59. Zhang, S., Yau, S.T. (2006). High-resolution, real-time 3-D absolute coordinate measurement based on a phase-shifting method. *Opt. Express* **14**(7), 2644–2649.
60. Zhang, Z., Towers, C.E., Towers, D.P. (2007b). Phase and colour calculation in color fringe projection. *J. Opt. A: Pure Appl. Opt.* **9**, S81–S86.
61. Wang, Y., Zhang, S. (2011). Superfast multifrequency phase-shifting technique with optimal pulse width modulation. *Opt. Express* **19**(6), 5143–5148.
62. Rusinkiewicz, S., Hall-Holt, O., Levoy, M. (2002). Real-time 3D model acquisition. *ACM Trans. Graph.* **21**(3), 438–446.
63. Hall-Holt, O., Rusinkiewicz, S. (2001). Stripe boundary codes for real-time structured-light range scanning of moving objects *The 8th IEEE International Conference on Computer Vision* **II**, 359–366.
64. Geng, Z.J. (1996). Rainbow 3-D camera: New concept of high-speed three vision system. *Opt. Eng.* **35**, 376–383.

65. Harding, K.G. (1988). Color encoded morié contouring *Proc. SPIE* **1005**, 169–178.
66. Huang, P.S., Hu, Q., Jin, F., Chiang, F.P. (1999). Color-encoded digital fringe projection technique for high-speed three-dimensional surface contouring. *Opt. Eng.* **38**, 1065–1071.
67. Pan, J., Huang, P.S., Chiang, F.P. (2006). Color phase-shifting technique for three-dimensional shape measurement. *Opt. Eng.* **45**(12), 013602.
68. Zhang, S., Huang, P. (2004). High-resolution, real-time 3-D shape acquisition *IEEE Comp. Vis. and Patt Recogn Workshop* **3**, 28–37, Washington DC, MD.
69. Huang, P.S., Zhang, S. (2006). Fast three-step phase shifting algorithm. *Appl. Opt.* **45**(21), 5086–5091.
70. Zhang, S., Royer, D., Yau, S.T. (2006a). GPU-assisted high-resolution, real-time 3-d shape measurement. *Opt. Express* **14**(20), 9120–9129.
71. Zhang, S., Yau, S.T. (2007b). High-speed three-dimensional shape measurement using a modified two-plus-one phase-shifting algorithm. *Opt. Eng.* **46**(11), 113603.
72. Gong, Y., Zhang, S. (2010). *Opt. Express* **18**, 19743.
73. Hornbeck, L.J. (1997). Digital light processing for high-brightness, high-resolution applications *Proc. SPIE* **3013**, 27–40.
74. Zhang, S., Huang, P.S. (2006a). High-resolution, real-time three-dimensional shape measurement. *Opt. Eng.* **45**(12), 123601.
75. Ekstrand, L. *et al.* (2013). *Handbook of 3-D machine vision: Optical metrology and imaging* **9**, 233.
76. Ujaldon, M., Saltz, J. (2005). Exploiting parallelism on irregular applications using the gpu. *Intl. Conf. on Paral. Comp.*, **13–16**.
77. Khailany, B., Dally, W., Rixner, S., Kapasi, U., Owens, J., Towles, B. (2003). Exploring the vlsi scalability of stream processors. *Proc. 9th Symp. on High Perf. Comp. Arch.*, 153–164.
78. Zhang, S., Royer, D., Yau, S.T. (2006a). GPU-assisted high-resolution, real-time 3-d shape measurement. *Opt. Express* **14**(20), 9120–9129.
79. Zhang, S., Royer, D., Yau, S.T. (2006b). Gpu-assisted high-resolution, real-time 3-D shape measurement. *Opt. Express* **14**(20), 9120–9129.
80. Bartlett, M.S., Littlewort, G., Fasel, I., Movellan, J.R. (2003). Real time face detection and facial expression recognition: Development and applications to human computer interaction. *Conference on Computer Vision and Pattern Recognition Workshop, 2003, CVPRW '03*, 53–53.
81. Mauri, C., Granollers, T., Lorés, J., García, M. (2006). Computer vision interaction for people with severe movement restrictions. *Human Technology* **2**(1), 38–54.
82. Wachs, J.P., Kölsch, M., Stern, H., Edan, Y. (2011). Vision-based hand-gesture applications. *Commun. ACM* **54**(2), 60–71.
83. Nielsen, M., StÃrring, M., Moeslund, T., Granum, E. (2004). A procedure for developing intuitive and ergonomic gesture interfaces for hci. In Camurri, A., Volpe, G. (eds) *Gesture-Based Communication in Human-Computer Interaction*, vol. 2915 of *Lecture Notes in Computer Science,* pp. 409–420. Springer Berlin Heidelberg.
84. Zhang, L., Curless, B., Seitz, S.M. (2002). Rapid shape acquisition using color structured light and multi-pass dynamic programming. *The 1st IEEE International Symposium on 3D Data Processing, Visualization, and Transmission*, 24–36.
85. Zhang, S., Yau, S.T. (2008). Simultaneous three-dimensional geometry and color texture acquisition using single color camera. *Opt. Eng.* **47**(12), 123604.
86. Zhang, Z., Towers, C.E., Towers, D.P. (2006c). Time efficient color fringe projection system for 3D shape and color using optimum 3-frequency selection. *Opt. Exp.* **14**, 6444–6455.
87. Liu, X., Peng, X., Chen, H., He, D., Gao, B.Z. (2012). Strategy for automatic and complete three-dimensional optical digitization. *Opt. Lett.* **37**, 3126–3128.
88. Notni, G.H., Kühmstedt, P., Heinze, M., Notni, G. (2002). Simultaneous measurement of 3-D shape and color of objects *Proc. SPIE* **4778**, 74–82.
89. Ou, P., Li, B., Wang, Y., Zhang, S. (2013). Flexible real-time natural 2d color and 3D shape measurement. *Optics Express* **21**(14), 16736–16741.
90. Merner, L., Wang, Y., Zhang, S. (2013). Accurate calibration for 3D shape measurement system using a binary defocusing technique. *Opt. Laser Eng.* **51**(5), 514–519.
91. Ayubi, G.A., Ayubi, J.A., Martino, J.M.D., Ferrari, J.A. (2010). Pulse-width modulation in defocused 3-D fringe projection. *Opt. Lett.* **35**, 3682–3684.
92. Wang, Y., Zhang, S. (2010). Optimum pulse width modulation for sinusoidal fringe generation with projector defocusing. *Opt. Lett.* **35**(24), 4121–4123.

93. Wang, Y., Zhang, S. (2012a). Comparison among square binary, sinusoidal pulse width modulation, and optimal pulse width modulation methods for three-dimensional shape measurement. *Appl. Opt.* **51**(7), 861–872.
94. Lohry, W., Zhang, S. (2012). 3D shape measurement with 2D area modulated binary patterns. *Opt. Laser Eng.* **50**(7), 917–921.
95. Schuchman, T.L. (1964). Dither signals and their effect on quantization noise. *IEEE Trans. Communication Technology* **12**(4), 162–165.
96. Wang, Y., Zhang, S. (2012b). Three-dimensional shape measurement with binary dithered patterns. *Appl. Opt.* **51**(27), 6631–6636.
97. Bayer, B. (1973). An optimum method for two-level rendition of continuous-tone pictures. *IEEE International Conference on Communications* **1**, 11–15.
98. Floyd, R., Steinberg, L. (1976). An adaptive algorithm for spatial gray scale. *Proc. Society for Information Display* **17**, 75–77.
99. Kite, T.D., Evans, B.L., Bovik, A.C. (2000). Modeling and quality assessment of halftoning by error diffusion. *IEEE International Conference on Image Processing* **9**(5), 909–922.
100. Purgathofer, W., Tobler, R., Geiler, M. (1994). Forced random dithering: improved threshold matrices for ordered dithering. *IEEE International Conference on Image Processing* **2**, 1032–1035.
101. Stucki, P. (1981). Meccaa multiple-error correcting computation algorithm for bilevel hardcopy reproduction. Technical report, IBM Res. Lab., Zurich, Switzerland.
102. Wang, Y., Laughner, J.I., Efimov, I.R., Zhang, S. (2013). 3D absolute shape measurement of live rabbit hearts with a superfast two-frequency phase-shifting technique. *Opt. Express* **21**(5), 5822–5632.
103. Dai, J., Zhang, S. (2013). Phase-optimized dithering technique for high-quality 3D shape measurement. *Opt. Laser Eng.* **51**(6), 790–795.
104. Lohry, W., Zhang, S. (2013). Genetic method to optimize binary dithering technique for high-quality fringe generation. *Opt. Lett.* **38**(4), 540–542.
105. Huang, P.S., Zhang, C., Chiang, F.P. (2002). High-speed 3-D shape measurement based on digital fringe projection. *Opt. Eng.* **42**(1), 163–168.

6

Real-Time Stereo 3D Imaging Techniques

Lazaros Nalpantidis

Robotics, Vision and Machine Intelligence lab., Department of Mechanical and Manufacturing Engineering, Aalborg University Copenhagen, Denmark

6.1 Introduction

Stereo vision involves the simultaneous use of two vision sensors and leads to the recovery of depth. It is based on the principal, first utilized by nature, that two spatially differentiated views of the same scene provide enough information to enable perception of the depth of the portrayed objects. This fact was first realized by Sir Charles Wheatstone, about two centuries ago, who stated: " ... the mind perceives an object of three dimensions by means of the two dissimilar pictures projected by it on the two retinae ... " [1].

Calculating the depth of various points, or any other primitive, in a scene relative to the position of a camera, is one of the important tasks of computer and robot vision systems. The most common method for extracting depth information from intensity images is by means of a pair of synchronized camera images acquired by a stereo rig, as the ones shown in Figure 6.1. Matching pixels of one image to pixels of the second image (also known as the stereo correspondence problem) results in the so called disparity map [2]. Disparity is defined as the difference of the coordinates of a matched pixel in the two images, which is proportional to its depth value. For rectified stereo pairs, the vertical disparity is always zero and, as a result, a disparity map consists of the horizontal disparity values for the matched image pixels. Nevertheless, the accurate and efficient estimation of disparity maps is a long-standing issue for the computer vision community [3].

The importance of stereo vision is apparent in the fields of machine vision [4], computer vision [5], virtual reality, robot navigation [6], simultaneous localization and mapping [7, 8], depth measurements [9] and 3D environment reconstruction [4]. The objective of this chapter is to provide a comprehensive survey of real-time stereo imaging algorithms and systems. The main characteristics of stereo vision algorithms are highlighted, with references to more detailed analyses where needed, and the state of the art concerning real-time implementations is given. The categorization followed in the rest of this Chapter follows the one shown in Figure 6.2.

Interactive Displays: Natural Human-Interface Technologies, First Edition. Edited by Achintya K. Bhowmik.
© 2015 John Wiley & Sons, Ltd. Published 2015 by John Wiley & Sons, Ltd.

A rigid stereo camera   A human-like active stereo head

(a)   (b)

Figure 6.1   Stereo vision sensors.

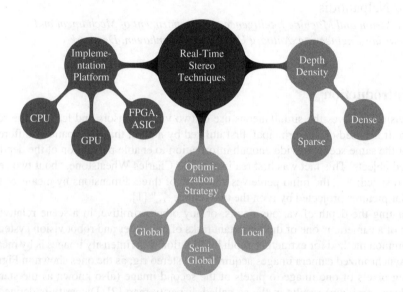

Figure 6.2   Categorization of real-time stereo 3D imaging techniques.

6.2   Background

Even if stereo vision has been the focus of attention for many researchers during the past decades, it is still a very active and popular topic. The richness of the related literature and the number of newly published works every year reveal that there is room for improvement in the current state of the art. A consequence of this is that every attempt to capture the state of the art is destined to be outdated soon. However, what one can still deduce from the study of such surveys is the shift of focus and the new trends that arise over time.

Historically, the first attempts to gather, survey and compare stereo vision algorithms resulted in the review papers of Barnard and Fischler [10], Dhond and Aggarwal [11], and Brown [12].

However, the work that has had the biggest impact on the topic and defined clear directions is probably the seminal taxonomy and review paper of Scharstein and Szeliski [13]. In that work, apart from a detailed snapshot of the contemporary algorithms, a taxonomy framework, and an openly available objective test bench were proposed. The test bench consists a standard dataset of stereo images, metrics for evaluating the accuracy of the results, and also a website that hosts all the aforementioned, as well as an ever-updating list with the results of evaluated algorithms [14].

While the focus of Scharstein and Szeliski was the definition of a framework that would allow the computer vision community to pursue further accuracy of stereo algorithms, the rapid evolution of robotic applications and real-time vision systems in general, has indicated the execution speed as a factor of equal (if not greater) importance as the accuracy of the depth estimations. This trend is reflected and captured in subsequent reviews, such as in [15], in [16], where the branch of real-time hardware implementations is additionally covered, and in [17], where the focus is on real-time algorithms targeting resource-limited systems.

Apart from reviews and comparative presentations of complete stereo algorithms, there are also very useful works that compare the various alternatives for implementing each basic block of a stereo algorithm. As will be discussed in Section 3 of this chapter, the structure of stereo algorithms can be very well defined, and the various constituent parts can be clearly identified. Hirschmuller and Scharstein, in [18, 19] compare different (dis-)similarity measures (also referred to as matching cost functions) for local, semi-global, and global stereo algorithms.

Additionally, Gong and his colleagues, in [20], consider various alternatives for matching cost aggregation in real-time GPU-accelerated systems. Considering global methods for disparity assignment, Szeliski *et al.*, in [21] propose some energy minimization benchmarks and use them to compare the quality of the results and the speed of several common energy minimization algorithms.

The stereo correspondence problem can be efficiently addressed by considering the geometry of the two stereo images and by applying image rectification to them. In the general case, the image planes of the two capturing cameras do not belong to the same plane. While stereo algorithms can deal with such cases, the demanded calculations are considerably simplified if the stereo image pair has been rectified. The process of rectification, as shown in Figure 6.3, involves the replacement of the initial image pair I_l, I_r by another projectively equivalent pair $I_{l,rect}, I_{r,rect}$ [5, 2]. The initial images are reprojected on a common plane P that is parallel to the baseline B joining the optical centers of the initial images.

Epipolar geometry provides tools in order to solve the stereo correspondence problem, i.e., to recognize the same feature in both images. If no rectification is performed, the matching procedure involves searching within two-dimensional regions of the target image. However, this matching can be done as a one-dimensional search, if accurately rectified stereo pairs are assumed in which horizontal scan lines reside on the same epipolar line, as shown in Figure 6.4. A point P_1 in one image plane may have arisen from any of points in the line C_1P_1, and may appear in the alternate image plane at any point on the epipolar line E_2 [4]. Thus, the search is theoretically reduced within a scan line, since corresponding pair points reside on the same epipolar line. The difference of the horizontal coordinates of these points is the disparity value. The disparity map consists of all the disparity values of the image.

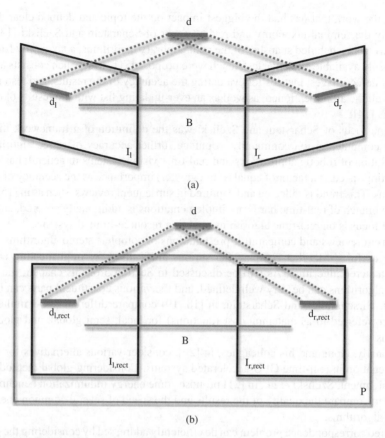

Figure 6.3 Rectification of a stereo pair. The two images I_l, I_r of an object d (a) are replaced by two pictures $I_{l,\,rect}$, $I_{r,\,rect}$, that lie on a common plane P (b).

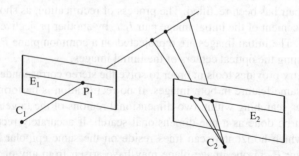

Figure 6.4 Geometry of epipolar lines, where C_1 and C_2 are the left and right camera lens centers, respectively. (a) Point P_1 in one image plane may have arisen from any of the points in the line C_1P_1, and may appear in the alternate image plane at any point on the epipolar line E_2.

6.3   Structure of Stereo Correspondence Algorithms

The majority of the reported stereo correspondence algorithms can be described using more or less the same structural set [13]. The basic building blocks are the following:

1. Computation of a matching cost function for every pixel in both the input images.
2. Aggregation of the computed matching cost inside a support region for every pixel and every potential disparity value.
3. Finding the optimum disparity value for every pixel of one picture.
4. Refinement of the resulted disparity map.

Every stereo correspondence algorithm makes use of a matching cost function, so as to establish correspondence between two pixels, as discussed in Section 6.3.1. The results of the matching cost computation comprise the Disparity Space Image (DSI). DSI can be visualized as a 3D matrix containing the computed matching costs for every pixel and for all its potential disparity values [22]. The structure of a DSI is illustrated in Figure 6.5.

Usually, the matching costs are aggregated over support regions. These regions could be 2D or even 3D [23, 24] ones within the DSI cube. The selection of the appropriate disparity value for each pixel is performed afterwards. It can be a simple Winner-Takes-All (WTA) process or a more sophisticated one. In many cases, this is an iterative process, as depicted in Figure 6.6. An additional disparity refinement step is frequently adopted. It is usually intended to filter the calculated disparity values, to give sub-pixel accuracy, or to assign values to not calculated pixels. The general structure of the majority of stereo correspondence algorithms is shown in Figure 6.6. Each block is discussed in more detailed in the rest of this section.

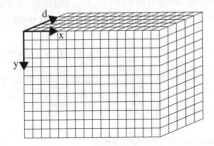

Figure 6.5   DSI containing matching costs for every pixel of the image and for all its potential disparity values.

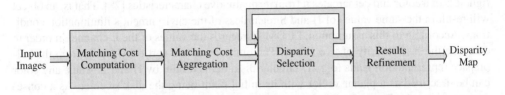

Figure 6.6   General structure of a stereo correspondence algorithm.

6.3.1   Matching Cost Computation

Detecting conjugate pairs in stereo images is a challenging research problem known as the correspondence problem, i.e., to find for each pixel in the left image the corresponding pixel in the right one [25]. The pixel to be matched without any ambiguity should be distinctly different from its surrounding pixels. To determine that two pixels form a conjugate pair, it is necessary to measure the similarity of these pixels, using some kind of a matching cost function. Among the most common ones are the absolute intensity differences (AD), the squared intensity differences (SD), and the normalized cross correlation (NCC). Evaluation of various matching costs can be found in [13, 18, 26].

AD is the simplest measure of all. It involves simple subtractions and calculations of absolute values. As a result, it is the most commonly used measure found in literature. The mathematical formulation of AD is:

$$AD(x, y, d) = |I_{left}(x, y) - I_{right}((x - d), y)| \tag{6.1}$$

where I_{left} and I_{right} denote the intensity values for the left and right image respectively, d is the value of the disparity under examination ranging from 0 to $D-1$ and x, y are the coordinates of the pixel on the image plane.

SD is somewhat more accurate in expressing the dissimilarity of two pixels. However, the higher computational cost of calculating the square of the intensities difference is not usually justified by the accuracy gain. It can be calculated as:

$$SD(x, y, d) = (I_{left}(x, y) - I_{right}((x - d), y))^2 \tag{6.2}$$

The normalized cross correlation calculates the dissimilarity of image regions instead of single pixels. It produces very robust results, on the cost of computational load. Its mathematical expression is:

$$NCC(x, y, d) = \frac{\sum_{x,y \in W} I_{left}(x, y) \cdot I_{right}(x, y - d)}{\sqrt{\sum_{x,y \in W} I_{left}^2(x, y) \cdot \sum_{x,y \in W} I_{right}^2(x, y - d)}} \tag{6.3}$$

where W is the image region under consideration.

The luminosity-compensated dissimilarity measure (LCDM), as introduced in [27], provides stereo algorithms with tolerance to difficult lighting conditions. The images are initially transformed from the RGB to the HSL colorspace. The HSL colorspace inherently expresses the lightness of a color and demarcates it from its qualitative characteristics [28]. That is, an object will result in the same values of H and S regardless of the environment's illumination conditions. According to this assumption, LCDM disregards the values of the L channel in order to calculate the dissimilarity of two colors. The omission of the vertical (L) axis from the colorspace representation leads to a 2D circular disk, defined only by H and S. Thus, any color can be described as a planar vector with its initial point being the disk's center. As a consequence, each color P_k can be described as a polar vector or, equivalently, as a complex number with modulus equal to S_k and argument equal to H_k. As a result, the difference, or equivalently the luminosity-compensated dissimilarity measure (LCDM), of two colors P_1 and P_2 can be calculated as the difference of the two complex numbers:

$$LCDM_{P_1,P_2} = |P_1 - P_2|$$

$$= |S_1 e^{iH_1} - S_2 e^{iH_2}|$$

$$= \sqrt{S_1^2 + S_2^2 - 2S_1 S_2 \cos(H_1 - H_2)} \qquad (6.4)$$

On the other hand, the rank transform replaces the intensity of a pixel with its rank among all pixels within a certain neighborhood, and then calculates the absolute difference between possible matches [29]. Another matching cost used for pixel correlation is Mutual Information. This can express how similar two parts of an image are by calculating and evaluating the entropies of the joint and the individual probability distributions. The probability distributions are derived from the histograms of the corresponding image parts. Finally, there are also phase-based methods that treat images as signals and perform matching of some phase-related function.

A useful review and evaluation of various matching costs for establishing pixel correlation can be found in [19].

6.3.2 Matching Cost Aggregation

Usually, the matching costs are aggregated over support regions. Those support regions, often referred to as support or aggregating windows, could be square or rectangular, of fixed size or adaptive ones. The aggregation of the aforementioned cost functions leads to the core of most of the stereo vision methods, which can be mathematically expressed as follows, for the case of the sum of absolute differences (SAD):

$$SAD(x, y, d) = \sum_{x,y \in W} |(I_{left}(x, y) - I_{right}((x - d), y)| \qquad (6.5)$$

And for the case of the sum of squared differences (SSD):

$$SSD(x, y, d) = \sum_{x,y \in W} (I_{left}(x, y) - I_{right}(x - d, y)^2 \qquad (6.6)$$

where I_{left}, I_{right} are the intensity values in the left and right image, x, y are the pixel's coordinates, d is the disparity value under consideration and W is the aggregated support region.

More sophisticated aggregation is traditionally considered as time-consuming. Methods based on extended local methodologies sacrifice some of their computational simplicity in order to obtain more accurate results [30]. Methods based on adaptive support weights (ASW) [31, 32] achieved this by using fixed-size support windows whose pixels' contribution in the aggregation stage varies, depending on their degree of correlation to the windows' central pixel. Despite the acceptance that these methods have enjoyed, the determination of a correlation function is still an open topic.

An ASW-based method for correspondence search is presented in [31]. The support-weights of the pixels in a given support window are adjusted based on color similarity and geometric proximity to reduce the image ambiguity. A similar approach is also adopted in [27], where stereo vision in non-ideal lighting conditions is examined, and in [33], where a psychophysically inspired stereo algorithm is presented.

On the other hand, some matching cost functions do not need an additional aggregation step, since this is an integral part of the cost calculation. As an example, the NCC and the rank transform require the *a priori* definition of a region where they are to be calculated.

An in-depth review and evaluation of different matching cost aggregation methods, suitable for implementation in programmable GPUs and for use in real-time systems can be found in [20].

6.4   Categorization of Characteristics

As already mentioned, there are numerous papers about stereo vision algorithms and systems published every year. This ever-growing list needs to be categorized according to some meaningful criteria to become tractable. Traditionally, the most common characteristics to divide the stereo algorithms literature has been the density of the computed disparity maps and the strategy used for the final assignment of disparity values. The rest of this Section examines the classes of such a categorization and discusses some indicative algorithms.

6.4.1   Depth Estimation Density

Stereo correspondence algorithms can be grouped into two broad groups; those producing dense and those producing sparse results. A dense stereo algorithm does not necessarily have to obtain disparity estimation for every single pixel. In practice, it is often impossible to achieve 100% density, due to occlusions in the stereo pair (even if certain techniques allow information propagation to such areas). As a result, the borderline between dense and sparse algorithms can be drawn depending on whether they actually attempt to match all the pixels, or only focus on a certain subset of them.

6.4.1.1   Dense

As more computational power becomes available in vision systems, dense stereo algorithms become more and more popular. Nowadays, published works about dense algorithms are dominating the relevant literature. Apart from the availability of the required computational resources, another factor that has boosted this class of algorithms is the existence of a standard test bench for the evaluation and objective comparison of their results accuracy. Thus, contemporary research in stereo matching algorithms, in accordance with the ideas of Maimone and Shafer [34], is reigned by the online tool available at the site maintained by Scharstein and Szeliski [13, 14]. The website provides a common dataset, and the produced disparity maps can be uploaded, automatically evaluated and listed together with the evaluation results, for ease of comparison.

As discussed already, not every pixel of one image has a corresponding one in its stereo pair image, mainly due to occlusions. However, algorithms that apply some kind of global optimization or "hole filling" mechanism can provide 100% coverage. In [35], an algorithm based on a hierarchical calculation of a mutual information-based matching cost is proposed. Its goal is to minimize a proper global energy function, not by iterative refinements, but by aggregating matching costs for each pixel from all directions. The final disparity map is sub-pixel accurate and occlusions are detected. The processing speed for the Teddy image set is 0.77 fps. The error in non-noccluded regions is found to be less than 3% for all the standard image sets. Calculations are made on an Intel Xeon processor running at 2.8 GHz.

An enhanced version of the previous method is proposed by the same author in [36]. Mutual information is, once again, used as cost function. The extensions applied in it result in intensity-consistent disparity selection for untextured areas and discontinuity-preserving interpolation for filling holes in the disparity maps. It treats successfully complex shapes and uses planar models for untextured areas. Bidirectional consistency check and sub-pixel estimation, as well as invalid disparities interpolation, are also performed. The experimental results indicate that the percentages of bad matching pixels in non-noccluded regions are 2.61, 0.25, 5.14 and 2.77 for the Tsukuba, Venus, Teddy and Cones image sets, respectively, with 64 disparity levels searched each time. However, the reported running speed on a 2.8 GHz PC is less than 1 fps.

6.4.1.2 Sparse

Sparse stereo algorithms provide depth estimations for only a limited subset of all the image pixels. The sparse disparity map is produced either by extracting and matching distinctive image features, or by filtering out the results of a dense method according to some reliability measure. In many cases, they stem from human vision studies and are based on matching for example segments or edges between two images.

Algorithms resulting in sparse, or semi-dense, disparity maps tend to be less attractive, as most of the contemporary applications require dense disparity information. However, they are very useful when:

- very rapid execution times are demanded;
- detail in the whole image is not required;
- reliability of the obtained depth estimations is more important than density.

Commonly, these types of algorithms tends to focus on striking characteristics of the images, leaving occluded and poorly textured areas unmatched. Veksler, in [37], presents an algorithm that detects and matches dense features between the left and right images of a stereo pair, producing a semi-dense disparity map. A dense feature is a connected set of pixels in the left image and a corresponding set of pixels in the right image, such that the intensity edges on the boundary of these sets are stronger than their matching error. All of these are computed during the stereo matching process. The algorithm computes disparity maps, with 14 disparity levels for the Tsukuba pair producing 66% density and 0.06% average error in the non-occluded regions.

Another method developed by Veksler [38] is based on the same basic concepts as the former one. The main difference is that this one uses the graph cuts algorithm for the dense feature extraction. As a consequence, this algorithm produces semi-dense results, with significant accuracy in areas where features are detected. The results are significantly better, considering density and error percentage, but they require longer running times. For the Tsukuba pair, it achieves a density up to 75%, the total error in the non-occluded regions is 0.36% and the running speed is 0.17 fps. For the Sawtooth pair, the corresponding results are 87%, 0.54% and 0.08 fps. All the results are obtained from a Pentium III PC running at 600 MHz; thus, significant speedup can be expected using a more powerful computer system.

In [39], a sparse stereo algorithm is initialized by extracting interesting points using the Harris corner detector. These points are used to construct a graph, and then the labeling problem is solved minimizing a global energy by means of graph cuts. Furthermore, a robust illumination changes structure tensor descriptor is used as similarity measurement to obtain even more accurate results.

More recently, Schauwecker *et al.*, in [40], have incorporated a modified version of the very efficient FAST feature detector within a sparse stereo-matching algorithm. An additional consistency check filters out possible mismatches, and the final implementation can achieve processing rates above 200 fps on a simple dual core CPU computer platform.

6.4.2   Optimization Strategy

Stereo matching algorithms can be divided in three general classes, according to the way they assign disparities to pixels. First, there are algorithms that decide the disparity of each pixel according to the information provided by its local, neighboring pixels – thus called local or area-based methods. They are also referred to as window-based methods, because disparity computation at a given point depends only on intensity values within a finite support window. Second, there are algorithms that assign disparity values to each pixel, depending on information derived from the whole image – thus called global algorithms. These are also sometimes referred to as energy-based algorithms, because their goal is to minimize a global energy function, which combines data and smoothness terms, taking into account the whole image. Finally, the family of algorithms called semi-global select disparity values minimizing an energy function along scan-lines [41]. Of course, there are many other methods [42] that are not strictly included in either of these broad classes. The issue of stereo matching has recruited a variation of computation tools. Advanced computational intelligence techniques are not uncommon and constitute interesting and promiscuous directions [43, 44].

6.4.2.1   Local

After the computation and aggregation (if necessary) of matching costs, the actual selection of the best candidate matching pixel (and, thus, disparity value) has to be performed. Most local algorithms just choose the candidate match that exhibits the minimum matching cost. This simple winner-takes-all approach is very efficient in computational terms, but is also often inaccurate, resulting in faulty and incoherent disparity values.

Ogale and Aloimonos, in [45–47], propose a compositional approach to unify many early visual modules, such as segmentation, shape and depth estimation, occlusion detection, and local signal processing. The dissimilarity measure used is the phase differences from various frequency channels. As a result, this method can process images with contrast, among other mismatches.

6.4.2.2   Global

Contrary to local methods, global ones produce very accurate results at the expense of computation time. They formalize the disparity assignment step as a labeling problem, and their goal is to find the optimum disparity function $d = d(x, y)$ that minimizes a global cost function E, which combines data and smoothness terms.

$$E(d) = E_{data}(d) + \lambda E_{smooth}(d) \tag{6.7}$$

where E_{data} takes into consideration the x, y pixel's value throughout the image, E_{smooth} provides the algorithm's smoothening assumptions and λ is a weight factor.

The minimization of the energy function is performed using a suitable iterative algorithm. Commonly used such algorithms include, among others, graph cuts [48, 49] and loopy belief propagation [50]. However, the main disadvantage of the global methods is that they are more time-consuming and computationally demanding.

An informative review and comparison of various energy minimization methods for global stereo matching can be found in [21].

6.4.2.3 Semi-global

The most popular semi-global stereo algorithms are based on Dynamic Programming (DP). DP has been used widely as an optimization method for the estimation of disparity d along image scanlines [41, 51], and this is the reason for being called semi-global. The general idea behind a DP stereo algorithm is to treat the correspondence problem again as an energy minimization problem, but limited on scanlines. Thus, an energy function is formulated, again as in Equation 6.7, where the smoothness term can be formulated to handle depth discontinuities and occlusions along each scanline.

DP stands somewhere between local and global algorithms, providing good accuracy of results in acceptable frame rates. Moreover, recent advances in DP-based stereo algorithms seem to be able to significantly improve both of these two aspects. Hardware implementations of DP have been reported [52, 53] that provide high execution speeds. Additionally, the incorporation of adaptive support weight aggregation schemes has been shown to further improve the accuracy and detail of the produced depth maps [54].

6.5 Categorization of Implementation Platform

Many stereo vision algorithms can achieve real-time or near-real-time operation. Such execution speeds can be obtained either by optimizing local stereo algorithms, or by employing custom hardware for speeding up the computations. In this section, implementations based on central processing unit (CPU), graphics processing unit (GPU), as well as field-programmable gate array (FPGA), or application-specific integrated circuits (ASIC), are covered that can be used in real-time systems.

6.5.1 CPU-only Methods

Gong and Yang, in [55], propose a reliability-based dynamic programming (RDP) algorithm that uses a different measure to evaluate the reliabilities of matches. According to this, the reliability of a proposed match is the cost difference between the globally best disparity assignment that includes the match and the one that does not include it. The inter-scanline consistency problem, common to the DP algorithms, is reduced through a reliability thresholding process. The result is a semi-dense unambiguous disparity map, with 76% density, 0.32% error rate and 16 fps for the Tsukuba, and 72% density, 0.23% error rate and 7 fps for the Sawtooth image pair. Accordingly, the results for Venus and Map pairs are 73%, 0.18%, 6.4 fps and 86%, 0.7%, 12.8 fps. As a result, the reported execution speeds, tested on a 2 GHz Pentium 4 PC, are encouraging for real-time operation if a semi-dense disparity map is acceptable.

Tombari et al. [56] present a local stereo algorithm that maximizes the speed-accuracy trade-off by means of an efficient segmentation-based AD cost aggregation strategy. The reported processing speed for the Tsukuba image pair is 5 fps and 1.7 fps for the Teddy and Art stereo image pairs.

Finally, in [57], a local stereo matching algorithm that computes dense disparity maps in real-time is presented. Two sizes for the used support windows improve the accuracy of the underlying SAD core, while keeping the computational cost low.

6.5.2 GPU-accelerated Methods

The active use of a GPU on a computer system can significantly improve execution speeds by exploiting the capabilities of GPUs to parallelize computations.

A hierarchical disparity estimation algorithm implemented on programmable 3D graphics processing unit (GPU) is reported in [58]. This method can process either rectified or uncalibrated image pairs. Bidirectional matching is utilized in conjunction with a locally aggregated sum of absolute intensity differences. This implementation on an ATI Radeon 9700 Pro can achieve up to 50 fps for 256×256 pixel input images.

In [59], a GPU implementation of a stereo algorithm that is able to achieve real-time performance with 4839 Millions of Disparity Estimation per Second (MDE/s) is presented. Highly accurate results are obtained, using a modified version of the SSD and filtering the results with a reliability criterion in the matching decision rule.

Kowalczuk and his colleagues, in [60], present a real-time stereo matching method that uses a two-pass approximation of adaptive support-weight aggregation and a low-complexity iterative disparity refinement technique. The implementation on a programmable GPU using CUDA is able to achieve 152.5 MDE/s, resulting in 62 fps for 320×240 images with 32 disparity levels.

The work of Richardt et al. [61] proposes a variation of Yoon and Kweon's [31] algorithm, implemented on a GPU. The implementation can achieve more than 14 fps, retaining the high quality of the original algorithm.

An algorithm that generates high quality results in real time is reported in [62]. It is based on the minimization of a global energy function. The hierarchical belief propagation algorithm iteratively optimizes the smoothness term, but achieves fast convergence by removing redundant computations involved. In order to accomplish real-time operation, the authors take advantage of the parallelism of a GPU. Experimental results indicate 16 fps processing speed for 320×240 pixel self-recorded images with 16 disparity levels. The percentages of bad matching pixels in non-occluded regions for the Tsukuba, Venus, Teddy and Cones image sets are found to be 1.49, 0.77, 8.72 and 4.61. The computer used is a 3 GHz PC and the GPU is an NVIDIA GeForce 7900 GTX graphics card with 512M video memory.

Additionally, another GPU implementation of a global stereo algorithm based on hierarchical belief propagation is presented in [63]. An approximate disparity map calculation or motion estimation, in higher levels, is used to limit the search space without significantly affecting the accuracy of the obtained results.

Wang *et al.*, in [54], present a stereo algorithm that combines high quality results with real-time performance. DP is used in conjunction with an adaptive aggregation step. The per-pixel matching costs are aggregated in the vertical direction only, resulting in improved inter-scanline consistency and sharp object boundaries. This work exploits the color and distance proximity-based weight assignment for the pixels inside a fixed support window, as reported in [31]. The real-time performance is achieved due to the parallel use of the CPU and the GPU of a computer. This implementation can process 320×240 pixel images with 16 disparity levels at 43.5 fps, and 640×480 pixel images with 16 disparity levels at 9.9 fps. The test system is a 3.0 GHz PC with an ATI Radeon XL1800 GPU.

Finally, a near-real-time stereo matching technique is presented in [64], which is based on the RDP algorithm. This algorithm can generate semi-dense disparity maps. Two orthogonal RDP passes are used to search for reliable disparities along both horizontal and vertical scanlines. Hence, the inter-scanline consistency is explicitly enforced. It takes advantage of the computational power of programmable graphics hardware, which further improves speed. The algorithm is tested on an Intel Pentium 4 computer running at 3 GHz, with a programmable ATI Radeon 9800 XT GPU equipped with 256MB video memory. It results in 85% dense disparity map with 0.3% error rate at 23.8 fps for the Tsukuba pair, 93% density, 0.24% error rate at 12.3 fps for the Sawtooth pair, 86% density, 0.21% error rate at 9.2 fps for the Venus pair and 88% density, 0.05% error rate at 20.8 fps for the Map image pair. If needed, the method can also be used to generate more dense disparity maps, deteriorating the execution speed.

6.5.3   Hardware Implementations (FPGAs, ASICs)

The use of dedicated and specifically designed hardware for the implementation of stereo algorithms can really boost their performance. However, not all stereo algorithms are easy to implement efficiently in hardware [16]. Furthermore, the effort and time required for such implementations are significant, while further changes are difficult to incorporate.

The FPGA-based system developed by [65] is able to compute dense disparity maps in real time, using the SAD method over fixed windows. The whole algorithm, including radial distortion correction, Laplacian of Gaussian (LoG) filtering, correspondence finding, and disparity map computation, is implemented in a single FPGA. The system can process 640×480 pixel images with 64 disparity levels and 8-bit depth precision at 30 fps speed, and 320×240 pixel images at 50 fps.

On the other hand, a slightly more complex implementation than the previous one is proposed in [66]. It is based on the SAD using adaptive sized windows. The proposed method iteratively refines the matching results by hierarchically reducing the window size. The results obtained by the proposed method are 10% better than that of a fixed-window method. The architecture is fully parallel and, as a result, all the pixels and all the windows are processed simultaneously. The speed for 64×64 pixel images with 8-bit grayscale precision and 64 disparity levels is 30 fps. The resources consumption is 42.5K logic elements, i.e. 82% of the considered FPGA device.

SAD aggregated using adaptive windows is the core of the work presented in [67]. A hardware-based CA parallel-pipelined design is realized on a single FPGA device. The achieved speed is nearly 275 fps, for 640×480 pixel image pairs with a disparity range of

80 pixels. The presented hardware-based algorithm provides very good processing speed at the expense of accuracy. The device utilization is 149K gates, that is 83% of the available resources.

The work of [52] implement an SAD algorithm on a FPGA board featuring external memory and a Nios II embedded processor clocked at 100 MHz. The implementation produces dense 8-bit disparity maps of 320×240 pixels with 32 disparity levels at a speed of 14 fps. Essential resources are about 16K logic elements whereas, by migrating to more complex devices, the design can be scaled up to provide better results.

The same authors, in [53], present an improved SAD-based algorithm with a fixed 3×3 aggregation window and a hardware median enhancement filter. The presented system can process 640×480 images with 64 disparity levels at 162 fps speed. The implementation requires 32K logic elements, equivalent to about 63K gates.

The work of Ambrosch and Kubinger in [68] presents an FPGA implementation of a local stereo algorithm. The algorithm combines the use of gradient-based census transform with adaptive support window SAD. The implementation is able to process 750×400 pixel images at 60 fps.

Zicari et al., in [69], implemented a SAD algorithm with an additional consistency check on an FPGA. The algorithm can process 1280×780 grayscale images with 30 disparity levels at 97 fps.

Furthermore, the work of Kostavelis et al. [70] includes a FPGA implementation of a dense SAD-based stereo algorithm, as part of a vision system for planetary autonomous robots. The implementation on a Xilinx Virtex 6 FPGA device can process 1120×1120 images, with 200 disparity level and quarter-pixel accuracy in 0.59 fps, which is found to be more than sufficient for the needs of the considered space exploratory rover.

The use of DP has been explored as well. The implementation presented in [71] uses the DP search method on a trellis solution space. It can cope with the case of vergent cameras, i.e., cameras with optical axes that intersect. The images received from a pair of cameras are rectified using linear interpolation, and then the disparity is calculated. The architecture has the form of a linear systolic array, using simple processing elements. The design is canonical and simple to be implemented in parallel. The implementation requires 208 processing elements and the resulting system can process 1280×1000 pixel images with up to 208 disparity levels at 15 fps.

An extension of the previous method is presented in [72]. The main difference is that data from the previous line are incorporated, so as to enforce better inter-scanline inconsistency. The running speed is 30 fps for 320×240 pixel images with 128 disparity levels. The number of utilized processing elements is 128. The percentage of pixels with disparity error larger than 1 in the non-noccluded areas is 2.63, 0.91, 3.44 and 1.88 for the Tsukuba, Map, Venus and Sawtooth image sets, respectively.

Finally, the work of [53] also presents a custom parallelized DP algorithm. Once again, a fixed 3×3 aggregation window and a hardware median enhancement filter is used. Moreover, inter-scanline support is utilized. The presented system can process 640×480 images with 65 disparity levels at 81 fps speed. The implementation require 270K logic elements, equivalent to about 1.6M gates.

The implementation of stereo algorithms on ASICs can result in very fast systems – even faster than when using FPGAs. However, this alternative is much more expensive, apart from the case of massive production. The prototyping times are considerable longer, and the result is highly process-dependent. Any further changes are difficult and, additionally, time- and money-consuming. As a result, the performance supremacy of ASIC implementations does not, in most cases, justify choosing them over the other possibilities. These are the main reasons that make recent ASIC implementation publications rare in contrast to the FPGA-based ones.

Published works concerning ASIC implementations of stereo matching algorithms [73, 74] are restricted to the use of SAD. The reported architectures make extensive use of parallelism and seem promising.

6.6 Conclusion

Stereo vision remains an attractive solution to the 3D imaging issue. This chapter has covered the basics of the theory behind stereo vision algorithms and has tried to categorize them according to their main characteristics and implementation platforms. A snapshot of the state of the art has been given.

One of the conclusions that can be deduced from this work is that the center of attention regarding stereo vision seems to have shifted; it is not only accuracy that is pursued, as it used to be, but also real-time performance that is now demanded. The solution to this issue is coming, either by simple algorithms implemented on powerful contemporary CPU systems, or by exploiting the power of programmable GPU co-processors, or by developing hardware-friendly implementations for FPGAs. The two latter options seem to be gaining popularity, as they can combine real-time execution speeds with very accurate depth estimations.

Another interesting observation is that local algorithms are not any more the only ones able to be implemented and executed in real time. Semi-global, or even purely global stereo algorithms, are reported to achieve acceptable frame rates. This trend seems to be growing, due to the fact that more powerful platforms are becoming constantly available, and more efficient and optimized versions of those algorithms are regularly being proposed.

The maturity of stereo vision techniques and their capability to operate in both indoor and outdoor environments give stereo vision a solid foundation in the harsh competition against other 3D sensing technologies. As a result, real-time implementations of stereo vision algorithms find their way in modern human-machine interaction systems, home entertainment systems, and autonomous robots, just to name a few.

References

1. Wheatstone, C. (1838). Contributions to the physiology of vision – part the first. on some remarkable, and hitherto unobserved, phenomena of binocular vision. *Philosophical Transactions of the Royal Society of London*, 371–394.
2. Faugeras, O. (1993). *Three-dimensional computer vision: a geometric viewpoint*. MIT Press, Cambridge.
3. Marr, D., Poggio, T. (1976). Cooperative computation of stereo disparity. *Science* **194**(4262), 283–287.
4. Jain, R., Kasturi, R., Schunck, B.G. (1995). *Machine vision*. McGraw-Hill, New York, USA.
5. Forsyth, D.A., Ponce, J. (2002). *Computer Vision: A modern Approach*. Prentice Hall, Upper Saddle River, NJ, USA.
6. Metta. G, Gasteratos, A., Sandini, G. (2004). Learning to track colored objects with log-polar vision. *Mechatronics* **14**(9), 989–1006.
7. Murray, D., Little, J.J. (2000). Using real-time stereo vision for mobile robot navigation. *Autonomous Robots* **8**(2), 161–171.
8. Murray, D., Jennings, C. (1997). *Stereo vision based mapping and navigation for mobile robots*. IEEE International Conference on Robotics and Automation, **2**, 1694–1699.

9. Manzotti, R., Gasteratos, A., Metta, G., Sandini, G. (2001). Disparity estimation on log-polar images and vergence control. *Computer Vision and Image Understanding* **83**(2), 97–117.

10. Barnard, S.T., Fischler, M.A. (1982). Computational stereo. *ACM Computing Surveys* **14**(4), 553–572.

11. Dhond, U.R., Aggarwal, J.K. (1989). Structure from stereo – a review. *IEEE Transactions on Systems, Man, and Cybernetics* **19**(6), 1489–1510.

12. Brown, L.G. (1992). A survey of image registration techniques. *Computing Surveys* **24**(4), 325–376.

13. Scharstein, D., Szeliski, R. (2002). A taxonomy and evaluation of dense two-frame stereo cor- respondence algorithms. *International Journal of Computer Vision* **47**(1–3), 7–42.

14. http://vision.middlebury.edu/stereo/ (2010).

15. Sunyoto, H., van der Mark, W., Gavrila, D.M. (2004). *A comparative study of fast dense stereo vision algorithms.* IEEE Intelligent Vehicles Symposium, 319–324.

16. Nalpantidis, L., Sirakoulis, G.C., Gasteratos, A. (2008). Review of stereo vision algorithms: from software to hardware. *International Journal of Optomechatronics* **2**(4), 435–462.

17. Tippetts, B., Lee, D.J., Lillywhite, K., Archibald, J. (2013). Review of stereo vision algorithms and their suitability for resource-limited systems. *Journal of Real-Time Image Processing.*

18. Hirschmuller, H., Scharstein, D. (2007). *Evaluation of cost functions for stereo matching.* IEEE Computer Society Conference on Computer Vision and Pattern Recognition, Minneapolis, Minnesota, USA.

19. Hirschmuller, H., Scharstein, D. (2009). Evaluation of stereo matching costs on images with radiometric differences. *IEEE Transactions on Pattern Analysis and Machine Intelligence* **31**(9), 1582–1599.

20. Gong, M., Yang, R., Wang, L., Gong, M. (2007). A performance study on different cost aggregation approaches used in real-time stereo matching. *International Journal of Computer Vision* **75**(2), 283–296.

21. Szeliski, R., Zabih, R., Scharstein, D., Veksler, O., Kolmogorov, V., Agarwala, A., Tappen, M., Rother, C. (2008). A comparative study of energy minimization methods for markov random fields with smoothness-based priors. IEEE transact. pattern anal. mach. intell. *IEEE Transactions on Pattern Analysis and Machine Intelligence* **30**(6), 1068–1080.

22. Muhlmann, K., Maier, D., Hesser, J., Manner, R. (2002). Calculating dense disparity maps from color stereo images, an efficient implementation. *International Journal of Computer Vision* **47**(1–3), 79–88.

23. Zitnick, C.L., Kanade, T. (2000). A cooperative algorithm for stereo matching and occlusion detection. *IEEE Transactions on Pattern Analysis and Machine Intelligence* **22**(7), 675–684.

24. Brockers, R., Hund, M., Mertsching, B. (2005). Stereo vision using cost-relaxation with 3D support regions. *Image and Vision Computing New Zealand*, 96–101.

25. Barnard, S.T., Thompson, W.B. (1980). Disparity analysis of images. *IEEE Transactions on Pattern Analysis and Machine Intelligence* **2**(4), 333–340.

26. Mayoral, R., Lera, G., Perez-Ilzarbe, M.J. (2006). Evaluation of correspondence errors for stereo. *Image and Vision Computing* **24**(12), 1288–1300.

27. Nalpantidis, L., Gasteratos, A. (2010). Stereo vision for robotic applications in the presence of non-ideal lighting conditions. *Image and Vision Computing* **28**, 940–951.

28. Gonzalez, R.C., Woods, R.E. (1992). *Digital Image Processing.* Addison-Wesley Longman Publishing Co, Inc, Boston, MA, USA.

29. Zabih, R., Woofill, J. (1994). *Non-parametric local transforms for computing visual correspondence.* European Conference of Computer Vision, 151–158.

30. Mordohai, P., Medioni, G.G. (2006). Stereo using monocular cues within the tensor voting framework. *IEEE Transactions on Pattern Analysis and Machine Intelligence* **28**(6), 968–982.

31. Yoon, K.-J., Kweon, I.S. (2006). Adaptive support-weight approach for correspondence search. *IEEE Transactions on Pattern Analysis and Machine Intelligence* **28**(4), 650–656.

32. Gu, Z., Su, X., Liu, Y., Zhang, Q. (2008). Local stereo matching with adaptive support-weight, rank transform and disparity calibration. *Pattern Recognition Letters* **29**(9), 1230–1235.

33. Nalpantidis, L., Gasteratos, A. (2010). Biologically and psychophysically inspired adaptive support weights algorithm for stereo correspondence. *Robotics and Autonomous Systems* **58**, 457–464.

34. Maimone, M.W., Shafer, S.A. (1996). *A taxonomy for stereo computer vision experiments.* ECCV Workshop on Performance Characteristics of Vision Algorithms, 59–79.

35. Hirschmuller, H. (2005). *Accurate and efficient stereo processing by semi-global matching and mutual information.* IEEE Computer Society Conference on Computer Vision and Pattern Recognition, **2**, 807–814.

36. Hirschmuller, H. (2006). *Stereo vision in structured environments by consistent semi-global matching.* IEEE Computer Society Conference on Computer Vision and Pattern Recognition, **2**, 2386–2393.

37. Veksler, O. (2002). Dense features for semi-dense stereo correspondence. *International Journal of Computer Vision* **47**(1–3), 247–260.

38. Veksler, O. (2003). Extracting dense features for visual correspondence with graph cuts. *IEEE Computer Vision and Pattern Recognition* **1**, 689–694.

39. Mu, Y., Zhang, H., Li, J. (2009). *A global sparse stereo matching method under structure tensor constraint*. International Conference on Information Technology and Computer Science, **1**, 609–612.

40. Schauwecker, K., Klette, R., Zell, A. (2012). *A new feature detector and stereo matching method for accurate high-performance sparse stereo matching*. IEEE/RSJ International Conference on Intelligent Robots and Systems, 5171–5176.

41. Cox, I.J., Hingorani, S.L., Rao, S.B., Maggs, B.M. (1996). A maximum likelihood stereo algorithm. *Computer Vision and Image Understanding* **63**, 542–567.

42. Liu, C., Pei, W., Niyokindi, S., Song, J., Wang, L. (2006). Micro stereo matching based on wavelet transform and projective invariance. *Measurement Science and Technology* **17**(3), 565–571.

43. Binaghi, E., Gallo, I., Marino, G., Raspanti, M. (2004). Neural adaptive stereo matching. *Pattern Recognition Letters* **25**(15), 1743–1758.

44. Kotoulas, L., Gasteratos, A., Sirakoulis, G.C., Georgoulas, C., Andreadis, I. (2005). *enhancement of fast acquired disparity maps using a 1-D cellular automation filter*. IASTED International Conference on Visualization, Imaging and Image Processing, Benidorm, Spain, 355–359.

45. Ogale, A.S., Aloimonos, Y. (2005). *Robust contrast invariant stereo correspondence*. IEEE International Conference on Robotics and Automation, 819–824.

46. Ogale, A.S., Aloimonos, Y. (2005). Shape and the stereo correspondence problem. *International Journal of Computer Vision* **65**(3), 147–162.

47. Ogale, A.S., Aloimonos, Y. (2007). A roadmap to the integration of early visual modules. *International Journal of Computer Vision* **72**(1), 9–25.

48. Boykov, Y., Veksler, O., Zabih, R. (2001). Fast approximate energy minimization via graph cuts. *IEEE Transactions on Pattern Analysis and Machine Intelligence* **23**(11), 1222–1239.

49. Boykov, Y., Kolmogorov, V. (2001). An experimental comparison of min-cut/max-ow algo- rithms for energy minimization in vision. *IEEE Transactions on Pattern Analysis and Machine Intelligence* **26**, 359–374.

50. Yedidia, J.S., Freeman, W., Weiss, Y. (2000). *Generalized belief propagation*. Conference on Neural Information Processing Systems, 689–695.

51. Bobick, A.F., Intille, S.S. (1999). Large occlusion stereo. *International Journal of Computer Vision* **33**, 181–200.

52. Kalomiros, J.A., Lygouras, J. (2008). Hardware implementation of a stereo co-processor in a medium-scale field programmable gate array. *IET Computers and Digital Techniques* **2**(5), 336–346.

53. Kalomiros, J., Lygouras, J. (2009). Comparative study of local SAD and dynamic programming for stereo processing using dedicated hardware. *EURASIP Journal on Advances in Signal Processing* 1–189.

54. Wang, L., Liao, M., Gong, M., Yang, R., Nister, D. (2006). *High-quality real-time stereo using adaptive cost aggregation and dynamic programming*. Third International Symposium on 3D Data Processing, Visualization, and Transmission, 798–805.

55. Gong, M., Yang, Y.–H. (2005). Fast unambiguous stereo matching using reliability-based dynamic programming. *IEEE Transactions on Pattern Analysis and Machine Intelligence* **27**(6), 998–1003.

56. Tombari, F., Mattoccia, S., Di Stefano, L., Addimanda, E. (2008). *Near real-time stereo based on effective cost aggregation*. International Conference on Pattern Recognition, 1–4.

57. Gupta, R.K., Cho, S.-Y. (2010). *A correlation-based approach for real-time stereo matching*. International Symposium on Visual Computing, **6454**, 129–138.

58. Zach, C., Karner, K., Bischof, H. (2004). *Hierarchical disparity estimation with programmable 3D hardware*. International Conference in Central Europe on Computer Graphics, Visualization and Computer Vision, 275–282.

59. Drazic, V., Sabater, N. (2012). *A precise real-time stereo algorithm*. 27th Conference on Image and Vision Computing New Zealand, 138–142.

60. Kowalczuk, J., Psota, E.T., Perez, L.C. (2013). Real-time stereo matching on cuda using an iterative refinement method for adaptive support-weight correspondences. *IEEE Transactions on Circuits and Systems for Video Technology* **23**(1), 94–104.

61. Richardt, C., Orr, D., Davies, I., Criminisi, A., Dodgson, N. (2010). A. *Real-time spatiotemporal stereo matching using the dual-cross-bilateral grid*. European Conference on Computer Vision (ECCV), ser. Lecture Notes in Computer Science, **6313**, 510–523.

62. Yang, Q., Wang, L., Yang, R. (2006). *Real-time global stereo matching using hierarchical belief propagation*. British Machine Vision Conference, **3**, 989–998.

63. Grauer-Gray, S., Kambhamettu, C. (2009). *Hierarchical belief propagation to reduce search space using cuda for stereo and motion estimation.* Workshop on Applications of Computer Vision, 1–8.

64. Gong, M., Yang, Y.–H. (2005). *Near real-time reliable stereo matching using programmable graphics hardware.* IEEE Computer Society Conference on Computer Vision and Pattern Recognition, **1**, 924–931.

65. Jia, Y., Xu, Y., Liu, W., Yang, C., Zhu, Y., Zhang, X., An, L. (2003). A miniature stereo vision machine for real-time dense depth mapping. International Conference on Computer Vision Systems, ser. Lecture Notes in Computer Science, **2626**, 268–277.

66. Hariyama, M., Kobayashi, Y., Sasaki, H., Kameyama, M. (2005). FPGA implementation of a stereo matching processor based on window-parallel-and-pixel-parallel architecture. *IEICE Transactions on Fundamentals of Electronics, Communications and Computer Science* **88**(12), 3516–3522.

67. Georgoulas, C., Kotoulas, L., Sirakoulis, G.C., Andreadis, I., Gasteratos, A. (2008). Real-time disparity map computation module *Journal of Microprocessors and Microsystems* **32**(3), 159–170.

68. Ambrosch, K., Kubinger, W. (2010). Accurate hardware-based stereo vision. *Computer Vision and Image Understanding* **114**(11), 1303–1316.

69. Zicari, P., Perri, S., Corsonello, P., Cocorullo, G. (2012). Low-cost FPGA stereo vision system for real time disparity maps calculation. *Microprocessors and Microsystems* **36**(4), 281–288.

70. Kostavelis, I., Nalpantidis, L., Boukas, E., Rodrigalvarez, M., Stamoulias, I., Lentaris, G., Diamantopoulos, D., Siozios, K., Soudris, D., Gasteratos, A. (in press). SPARTAN: Developing a vision system for future autonomous space exploration robots. *Journal of Field Robotics*.

71. Jeong, H., Park, S. (2004). Generalized trellis stereo matching with systolic array. International Symposium on Parallel and Distributed Processing and Applications, **3358**, 263–267. Springer Verlag.

72. Park, S., Jeong, H. (2007). *Real-time stereo vision FPGA chip with low error rate.* International Conference on Multimedia and Ubiquitous Engineering, 751–756.

73. Hariyama, M., Takeuchi, T., Kameyama, M. (2000). *Reliable stereo matching for highly-safe intelligent vehicles and its VLSI implementation.* IEEE Intelligent Vehicles Symposium, 128–133.

74. Hariyama, M., Sasaki, H., Kameyama, M. (2005). Architecture of a stereo matching VLSI processor based on hierarchically parallel memory access. *IEICE Transactions on Information and Systems* **E88-D**(7), 1486–1491.

7

Time-of-Flight 3D-Imaging Techniques

Daniël Van Nieuwenhove
SoftKinetic Sensors, Belgium

7.1 Introduction

In the last decade, awareness of the unique advantages of 3D imaging, resulting in new and compelling applications enabled by real-time 3D user interactions, has grown significantly. Most known techniques for 3D sensing include stereo vision, time-of-flight, and structured light methods. In this chapter, we will discuss the time-of-flight (TOF) techniques.

In the first part of the chapter, we will explain the techniques in detail, distinguishing between different time-of-flight 3D imaging approaches, such as pulsed time-of-flight and continuous time-of-flight. We will then present the operating principles and the main equations, and discuss the accuracies of these methods. Finally, we will detail the challenges and required improvements, as well as some typical values of a camera system, and the current state-of-the-art with respect to the resolutions across research centers worldwide.

7.2 Time-of-Flight 3D Sensing

Only recently, time-of-flight (TOF) 3D imaging techniques have proven their reliability for more widespread 3D applications [1–3]. Generally, in all the TOF 3D imaging schemes, a modulated light wave is sent out to the scene and its reflections are detected and used to determine the light's round trip time and distance [4]. By focusing the reflected light on a matrix of pixels, complete range images are obtained at once. Challenges in this approach are the sensitivity and dynamic range necessary to be able to measure the weak reflected signals in the presence of ambient light of up to several orders of magnitude larger than the modulated light. Several efforts in this direction have been made [2, 5].

Interactive Displays: Natural Human-Interface Technologies, First Edition. Edited by Achintya K. Bhowmik.
© 2015 John Wiley & Sons, Ltd. Published 2015 by John Wiley & Sons, Ltd.

In time-of-flight measurements, the distance is obtained by measuring the time that light needs to travel back and forth between the source and the target. This round trip time is translated to distance by multiplication with the speed of light:

$$c = 3 \times 10^8 \text{ m/s} = 2 \times 150 \text{ m/}\mu\text{s} = 2 \times 0.15 \text{ m/ns} = 2 \times 0.15 \text{ mm/ps}.$$

This optical radar technique was first implemented in LIght Detection And Ranging (LIDAR) devices. Such devices use a laser with a single point detector to obtain the time-of-flight or the distance. Scanning the scene results in a complete 3D image. This was, and still is, used in various domains, and has been perfected over the years, but it is slow due to the scanning process. It is also expensive and fragile, as it requires mechanically moving parts for the scanning of the scene.

In recent years, due to improvements made on integrated circuit technology, building matrices of miniaturized time-of-flight sensors has become feasible. This can be used to make complete 3D camera systems based on time-of-flight principles [6, 7]. In these systems, the whole scene is illuminated at once, usually using light emitting diodes (LEDs), and the reflected light is focused on an array of time-of-flight detectors (see Figure 7.1). Each of these detectors simultaneously measures the distance to a point, so that the range information is obtained for a whole image at once. In order to minimize errors, the active light source and the receiver are located very close to each other. This enables a compact setup and avoids shadowing effects.

This technique, called indirect or continuous time-of-flight, has speed and robustness as advantages. It has become a popular and promising technique as, in contrast to the other time-of-flight techniques, it requires no moving parts. According to [8], indirect time-of-flight 3D cameras are on the verge of becoming the natural substitutes for their 2D vision counterparts. The basic operating principle is illustrated in Figure 7.2. For this technique, we can distinguish between pulsed and continuous operation.

In the rest of this chapter, after a brief review of the pulsed time-of-flight technique, we will discuss mostly the continuous time-of-flight technique. Therefore, we have taken the liberty of referring to this using the general term "time-of-flight" ("TOF" for short). Whenever pulsed time-of-flight is discussed, it is explicitly mentioned.

Figure 7.1 Time-of-flight range imaging technique. The active illumination reflects a wave front of the whole scene containing the needed time-of-flight information to deduce distances for all points in the scene (see text for details).

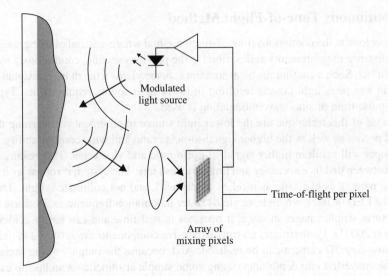

Figure 7.2 Illustration of the time-of-flight measurement principle. In every pixel, the distance is recovered by measuring the time-of-flight for each pixel, dividing it by 2 and multiplying it by the speed of light c.

7.3 Pulsed Time-of-Flight Method

In pulsed time-of-flight method, a light pulse is projected onto the scene while, at the same moment, a high-accuracy stopwatch is started in every pixel to measure the time of flight. The light pulse travels to the target and back. At the detection of the returning light pulse, a mechanism in the pixel ends the stopwatch, now showing the time of flight of the light pulse.

Because the light pulse travels the path twice (forth and back) a measured time of 6.67 ns corresponds to a distance of one meter, while a time accuracy of better than seven picoseconds is required for a distance resolution of 1 mm. The accuracy can be improved by repeating the measurement a number of times and averaging the obtained results.

A major disadvantage of this time-of-flight approach is that, at the receiver side, both high dynamic range and large bandwidth need to be available. In this technique, it is difficult for the receiving path to detect the arrival time of the back-scattered light pulse precisely, because:

1. The optical threshold is not a fixed value but changes with background and distance of the object and with the object reflectivity.
2. Atmospheric attenuation leads to the dispersion of the light pulse and flattens the slope of the received pulse. Therefore a high power pulsed light source will be required.

On the other hand, besides emitting high power, the light source also has to be able to produce very short light pulses with fast rise time. This is necessary to assure an accurate detection of the incoming light pulse. Current commercially viable lasers or laser diodes, the only optical elements offering the required short pulse widths at sufficiently high optical power, can typically repeat the pulses at some 10 kHz. Such low repetition rates drastically restrict the frame rate for pulsed TOF systems [9].

7.4   Continuous Time-of-Flight Method

We will now look at the continuous time-of-flight method where, instead of using single pulses to obtain distance measurements as described in the previous section, continuously modulated light is sent out. Such a continuous measurement scheme yields a much higher signal-to-noise ratio, using less peak light power, resulting in lower light source requirements. Typically, a repeating pulse train or sine-wave modulation is used.

Advantages of this technique are the lower light source requirements concerning the band-width and power, as well as the higher signal-to-noise ratio and wide configurability. Measuring for longer will result in higher signal-to-noise ratio and vice versa. This results in a nice trade-off between distance accuracy and image refresh rate. The system is robust, as it contains no moving parts. It is eye-safe, as it relies on diffused, and not collimated, light. The source itself can be LED or laser, where laser yields faster modulation frequencies. Because the tech-nique captures whole images at once, it operates in real time and can easily achieve frame rates of over 200 Hz. Furthermore, no truly expensive components are involved in the system, so that a low-cost 3D camera can be realized. And, because the outputs of the image sensor chip can be converted into depth maps using some simple arithmetic formulas, no exhaustive post-processing is required in achieving this.

A disadvantage is that the distances calculated can be ambiguous, as the measurement of an object further than the unambiguous distance range from the camera will be projected back, thus making the object appear closer than it actually is. The unambiguous distance is, in most cases, defined by the modulation frequency or pulse rate. For example, a typical 20 MHz mod-ulation wave yields an unambiguous distance range of 7.5 m. This ambiguity in measured distance is often referred to as aliasing.

7.5   Calculations

In this chapter we will deduce some general TOF equations. The chapter assumes sinusoidal modulation. Similar equations can be obtained for other modulation waves. The time-of-flight t_d translates, for the continuous time-of-flight, into a phase difference α between the sent and the received modulation signal:

$$\alpha = t_d \omega. \tag{7.1}$$

where ω is the angular frequency of modulation. Finding this phase difference is the goal in the time-of-flight technique. The time delay and distance can then be obtained using:

$$\text{distance} = t_d c = \frac{\alpha}{\omega} c = \frac{\alpha}{2\pi f ML} c \tag{7.2}$$

where f_{ML} is the modulation frequency (e.g. 20 MHz) and c is the speed of light, 3×10^8 m/s. We assumed, for simplification reasons, the index of refraction for air to be 1. We can find the only unknown in this formula, the phase difference α, by measuring the in-phase (I) and quadrature (Q) parameters, as shown in Figure 7.3. The required phase difference is then found by:

$$\alpha = \text{atan}\left(\frac{Q}{I}\right) \tag{7.3}$$

In Figure 7.4, a typical signal path for a continuous time-of-flight measurement is shown. First, the modulated light is emitted onto the scene, then its reflections are focused on the

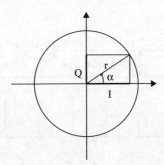

Figure 7.3 Goniometric circle, showing I, Q and α, used for the TOF distance calculation based on sinusoidal modulation signals (see text).

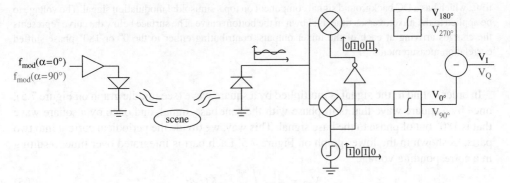

Figure 7.4 Schematic representation of the different components involved in obtaining distance measurements using continuous time-of-flight technique. An LED emits light to the scene at the left side, which is reflected and detected at the right side. Subsequently, it is mixed with 0° and 180° phase-shifted versions of the original modulation signal, integrated over time and subtracted to result in a voltage corresponding to either V_I or V_Q (= $V_{0°} - V_{180°}$). Then, a 90° phase-shifted modulated illumination is used to obtain a voltage signal for Q (= $V_{180°} - V_{270°}$).

detector node and converted to the shown current signal. This current signal is mixed, typically with 0° and 180° phase shifted versions of the original modulation signal. After that, the resulting signals are integrated over time and subtracted from each other. For reasons explained in detail later in this section, the whole procedure is done twice, once using 0°, then 90° phase-shifted modulated illumination to obtain a voltage signal for *I*.

We will now look into the mathematics of this approach in more detail. In Figure 7.5, the top graph shows the current amplitude versus time for the detector signal. This signal has a background light component which, for simplicity, is assumed to be constant over time, and a modulated light component:

$$I_{det} = I_{BL} + I_{ML}\sin(\omega_{ML}t + \alpha) \tag{7.4}$$

where ω_{ML} is $2\pi f_{ML}$, I_{BL} corresponds to the current induced by the background light, and I_{ML} corresponds to the current induced by the modulated light.

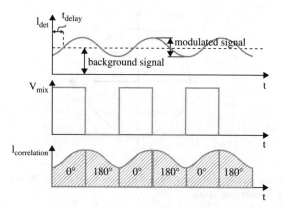

Figure 7.5 Principle of continuous time-of-flight method. Top curve shows the detector current versus time, which has a DC background signal component on top a sinusoidal modulation signal. The voltage to be applied to the mixer, versus time, is shown in the bottom curve. The surface below the curve represents the charges arriving at each of the mixer outputs, contributing either to the 0° or 180° phase-shifted correlation measurement.

In order to find α, the signal is multiplied by a square wave (see middle graph on Figure 7.5), once by a square wave that is in-phase with the sent base signal, and once by a square wave that is 180° out of phase to the base signal. This way, we divide the periodical current into two parts, as shown in the lower graph on Figure 7.5. Each part is integrated over time, resulting in a corresponding voltage:

$$V_{0°} = \frac{1}{C_{int}} \int_0^{t_{int}} I_{det} K(t)dt \tag{7.5}$$

where $K(t)$ describes the square wave multiplication performed by the mixer:

$$K(t) = 1 \text{ for } (n-1)T < t < n\frac{T}{2}$$

$$= 0 \text{ for } n\frac{T}{2} < t < nT \quad \text{where } n \in N \tag{7.6}$$

assuming $t_{int} = zT$ and $z \in N$:

$$\Rightarrow V_{0°} = \frac{1}{2C_{int}} \left(t_{int}I_{BL} + \frac{t_{int}}{T}I_{ML} \int_0^{\frac{T}{2}} \sin\left(\omega_{ML}t + \alpha\right)dt \right)$$

$$= \frac{t_{int}}{2C_{int}} \left(I_{BL} + \frac{I_{ML}}{\omega_{ML}T}(-\cos(\pi + \alpha) + \cos(0 + \alpha)) \right)$$

$$= \frac{t_{int}}{2C_{int}} \left(I_{BL} + I_{ML}\frac{\cos\alpha}{\pi} \right)$$

$$= \frac{V_{BL}}{2} + \frac{V_{ML}}{2\pi}\cos\alpha \tag{7.7}$$

In the same way, an expression for $V_{180°}$ is obtained:

$$V_{180°} = \frac{V_{BL}}{2} - \frac{V_{ML}}{2\pi} \cos \alpha \tag{7.8}$$

Subtracting Equation 7.8 from Equation 7.7, we get a background-level independent measure for I :

$$V_{0°} - V_{180°} = \frac{V_{ML}}{\pi} \cos \alpha \propto I \tag{7.9}$$

Continuing the measurement cycle, mixing with a 90° and 270° phase shifted signal, we obtain a value proportional to Q:

$$V_{90°} - V_{270°} = \frac{V_{ML}}{\pi} \sin \alpha \propto Q \tag{7.10}$$

So that the desired phase delay can be found using:

$$\alpha = \text{atan} \left(\frac{V_{90°} - V_{270°}}{V_{0°} - V_{180°}} \right) \tag{7.11}$$

Together with Equation 7.2, we can then find a general expression for obtaining distances using the continuous time-of-flight method:

$$\text{distance} = \frac{c}{2\pi f_{ML}} atan \left(\frac{V_{90°} - V_{270°}}{V_{0°} - V_{180°}} \right) \tag{7.12}$$

We can conclude that to acquire a distance estimate, four measurements have to be obtained, namely $V_{0°}$, $V_{90°}$, $V_{180°}$ and $V_{270°}$. It is to be noted that, when V_0 and V_{180} are nearly equal, Equation 7.12 will produce poor results because, in that case, the denominator is very small. The solution to this difficulty is obtained using the inverse cotangent instead of the inverse tangent, so a small nominator on a large denominator is obtained.

7.6 Accuracy

It is important to have an overview of the different noise contributions that have an effect on the final depth accuracy. This is done by studying the influence of the stochastic noise on measurements of $V_{0°}, V_{90°}, V_{180°}$ and $V_{270°}$ on the phase error $\delta\alpha$. From Equation 7.12 and using the rules of error propagation we get a general expression for $\delta\alpha$:

$$\sqrt{ \left(\frac{\delta\alpha}{\delta V_{0°}} \right)^2 \Delta^2 V_{0°} + \left(\frac{\delta\alpha}{\delta V_{90°}} \right)^2 \Delta^2 V_{90°} + \left(\frac{\delta\alpha}{\delta V_{180°}} \right)^2 \Delta^2 V_{180°} + \left(\frac{\delta\alpha}{\delta V_{270°}} \right)^2 \Delta^2 V_{270°} } \tag{7.13}$$

Solving this equation for special phase values, such as 0°, 45°, 90°, 135°, 180°, we get [6]:

$$\delta D_f = \frac{D_u}{2\pi} \delta\alpha = \frac{D_u}{\sqrt{8}} \frac{\sqrt{B}}{2A} = \frac{D_u}{\sqrt{8}} \frac{1}{2\text{SNR}'} \tag{7.14}$$

where D_u is the unambiguous distance, determined by the modulation frequency, A is the amplitude of the modulated signal in number of photoelectrons and $\sqrt{B}$ corresponds to the number of photo-electrons induced by the background light shot-noise. It is to be noted that the value for the signal-to-noise ratio (SNR) in this case is not expressed in dB. Filling in $D_u = \frac{c}{2f_{ML}}$, we get:

$$\delta D_f = \frac{c}{2f_{ML}} \frac{1}{\sqrt{8}} \frac{1}{2\text{SNR}} \tag{7.15}$$

We see that two parameters influence the accuracy of the camera system – the signal-to-noise ratio and the modulation frequency. Maximizing these will result in optimal camera accuracy. The noise performance is limited by the unavoidable shot noise in most cases, so we conclude from this that a key method to improve the system performance is by optimizing the signal amplitude and using high modulation frequencies. Both the light source and the pixel need to be able to handle these higher frequencies. LEDs can typically support up to several tens of MHz (e.g. 20 MHz), where laser tends to support up to several 100s of MHz.

7.7 Limitations and Improvements

7.7.1 TOF Challenges

The TOF technique offers many advantages and addresses a lot of markets, but the system requires a couple of challenges to be overcome. A short overview of the parameters for practical considerations is given in Figure 7.6, which shows the TOF "spider web".

In order to target high accuracy, very small time shifts need to be detectable in the TOF system. Due to the light wave propagation speed of 3×10^8 m/s, a distance of 15 cm will correspond to a round trip time of one nanosecond. Therefore, in order to achieve mm resolution in depth measurements, we will need to be able to distinguish picoseconds of time-of-flight for the modulated light wave.

As the power budget of the camera is limited, the sensitivity of the camera sensor is very important as well, in order to be able to see as far as possible with the available light. Furthermore, a strong background light with several orders of magnitude higher intensity is present in

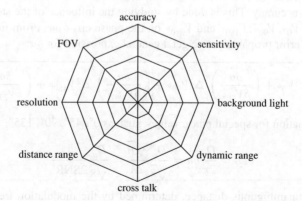

Figure 7.6 The Time-of-Flight parameter "spider web".

the scene besides the modulated light originating from the light source in the camera. In that case, loss of information due to extra noise and/or saturation has to be avoided. Next to these challenges, the dynamic range of the camera system needs to be optimized so that objects, both close and far, can be measured.

Also of significant importance is the avoidance of cross-talk, where both inter-pixel and inter-camera cross-talk have to be considered. The first is caused by the detected IR photons migrating to the adjacent pixels, and the second by interference of the light signals in situations where more than one camera is observing, and thus illuminating, the same scene.

Another challenge is optimization of the distance range limit caused by the aliasing of the emitted modulation signal, as briefly discussed earlier. One must take care of the recurring phase values as multiples of 2π.

Last, but not the least, all challenges mentioned above have to be tackled in such a way that we can still achieve small pixel sizes, so that building high resolution 3D pixel arrays to obtain detailed 3D images is feasible.

The optical design also impacts the TOF system, where the important parameters are Field-of-View (FOV), as well as the lens characteristics (F#, distortion, etc.).

These challenges need to be tackled on one or multiple levels in the system, such as the pixel design, the imager design, the control logic, or the application (see also Figure 7.7).

7.7.2   Theoretical Limits

The limits of the TOF system can be attributed to various components (e.g., the limit on the speed and the intensity available in the LEDs). In this section, however, we want to identify the theoretical maximal accuracy attainable by the system. Of the noise sources in silicon imaging, the photon shot-noise is unavoidable. This kind of noise is due to the statistical fluctuations of the discrete charge carriers, and it is accurately modelled as a Poisson process. The noise amplitude is therefore defined as the square root of the number of photo-generated carriers Q:

$$\delta Q_{xhot} = \sqrt{Q} \tag{7.16}$$

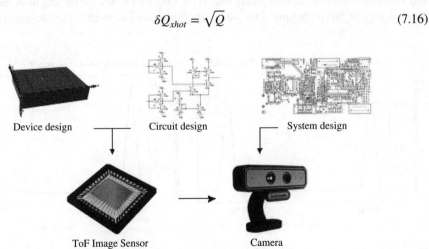

Figure 7.7   Overview of a TOF system.

This noise source is key when designing TOF systems. The goal will always be to bring the image circuit and camera induced noise lower or equal to the shot noise, so that the overall noise performance of the system is at least $\approx 70\%$ of the theoretical maximum.

7.7.3 Distance Aliasing

In continuous time-of-flight ranging, the time-of-flight is translated into a phase difference. Therefore an object at a distance corresponding to slightly more than one period phase delay will be measured to be located nearer. This range is defined by the half of the wavelength and called the "unambiguous range" or aliasing limit. This continuous time-of-flight related ambiguity problem can be solved or improved in several ways, as will be discussed in this section.

A first way to improve the unambiguous range is by taking two measurements using slightly different modulation frequencies. Combining both measurements, the unambiguous distance becomes the least common multiple of the maximal distance intervals defined by each frequency. This result can be obtained mathematically:

$$D_1 = \left(\frac{\alpha_1}{360} + n_1\right)\frac{c}{f_1 2}$$

$$D_2 = \left(\frac{\alpha_2}{360} + n_2\right)\frac{c}{f_2 2} \quad \Rightarrow \quad D = \left(\frac{\alpha_1}{360} + n_1\right)\frac{c}{f_1 2}$$

$$D_1 = D_2 \qquad\qquad \left(\frac{\alpha_1}{360} + n_1\right)f_2 = \left(\frac{\alpha_2}{360} + n_2\right)f_1 \qquad (7.17)$$

Another way to circumvent the aliasing limit is found in the continuous pseudo-noise modulation approach. In this technique, a limited length word is used as a modulation base. If a word is chosen having an autocorrelation, as shown in Figure 7.8, non-zero in the first bit shift and zero elsewhere, the unambiguous interval is extended times the bit length of the word. A disadvantage of this technique is the necessity for higher bandwidth. Pseudo-noise codes can

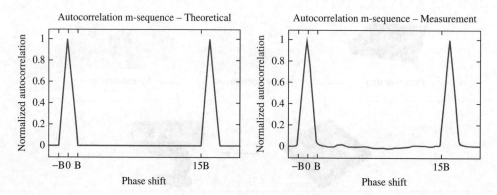

Figure 7.8 Autocorrelation graphs for a chosen pseudo noise sequence. Simulation (left), measurement (right).

be stored in a ROM or even generated using linear shift feedback registers. These are built up from standard digital cells, like inverters and flip-flops, which makes them ideally suited to be generated on-chip or on a PLD.

7.7.4   Multi-path and Scattering

Another side-effect in the whole-scene TOF imaging approaches are the multi-path and scattering effects.

Multi-path effect is caused by the multiple, direct and indirect, return paths that a light beam can follow before entering the optical system of the TOF camera. The output of the sensor will become a weighted average of the distances of the different pathways, together with their intensity.

Scattering is a similar effect, but on a smaller scale inside the lens. As the sensor surface is not 100% absorbent, rays will bounce of the surface, partially reflect on the lens and possibly re-enter the sensor array on a different pixel location.

Both these effects result in errors on the distance measure, especially for the pixels receiving a weaker return signal from the scene. As this work is written, various research centers are working on solutions to compensate or eliminate these effects.

7.7.5   Power Budget and Optimization

Major segments of 3D imaging based applications are in embedded or battery-powered devices, which often require power budgets to be optimized to the minimum possible levels. In a TOF system, power is primarily consumed by the illumination unit that is responsible for the depth measure quality under certain conditions. The main parameters contributing to the required illumination power are the modulation frequency and modulation contrast. Therefore, we define a specific metric, called the modulation efficiency (ME). It is defined, as shown in Equation 7.18, as the multiplication of the modulation frequency with the demodulation contrast at this frequency.

$$\text{Modulation Efficiency (ME)} = f_{mod} \times C_{\text{demod}} \qquad (7.18)$$

This metric is influenced both by system design aspects, such as rise and fall of the illumination source, and quality of the modulation signals, as the native properties of the TOF sensor. Figure 7.9 shows a graph of power consumption of a long range TOF system for different ME values. Pixel pitch is assumed to be 10 μm, and all other parameters are kept constant, as shown in the figure caption.

The curves in Figure 7.9 represent 2 cm, 1 cm and 5 mm noise targets (1σ). The lower the noise, the more precise are the movements that can be detected. As can be seen, the power consumption for TOF systems strongly depends on the modulation efficiency. In today's systems, values of around 70–100 are achievable, depending on the vendor, resulting in a 4–20 W power requirement, linked to the precision needed for detecting hand- or finger-level movements. Continuous improvement of this parameter results in total power consumption levels below 1 W. Hence, in the future, TOF systems will become sufficiently low-power to be part of everyday life inside battery-powered devices such as laptop, tablet or mobile devices.

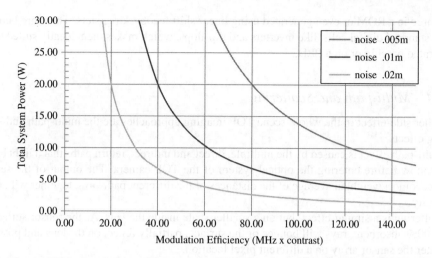

Figure 7.9 Power consumption versus modulation efficiency for three different 1σ noise targets: 2 cm, 1 cm and 5 mm. System configuration parameters: distance = 4 m, TOF sensor resolution = VGA, pixel pitch = 10 μm, FOV = 70° × 50° (H × V), F# lens = 1, reflectivity = 50%).

7.8 Time-of-Flight Camera Components

In a time-of-flight camera system, many different components, such as digital, analog and optical, need to operate in harmony.

The optical elements determine the same properties as for normal imaging, such as captured light budget and field of view. Because time-of-flight imagers typically have a much lower pixel count than traditional imaging sensors (e.g. 10 k pixels), the optical requirements are somewhat lower. Nevertheless, to reduce optical power on the illumination side, "fast" low F# lenses are desired. The illumination has to be optimized to emit high frequency waves, up to several hundreds of MHz. Furthermore, the angle of illumination needs to be matched to the field of view for the imager.

The heart of the camera is the TOF image sensor chip, where the reflections of the modulated light from the scene are focused and converted to depth information. The output of the imager is digitized and, using a communication protocol (e.g., USB), brought to the outside world.

Furthermore, some digital logic is needed to send the required mixer and modulation signals to the time-of-flight imager chip and the illumination board.

7.9 Typical Values

In this section, we will present some typical values for the key parameters, such as the light power, detector current, and background light level, for a continuous time-of-flight 3D imaging system.

7.9.1 Light Power Range

We will now calculate some typical values for the light power budget, impinging on each pixel. If we assume Lambertian reflection for the objects in the scene, as shown in Figure 7.10, the

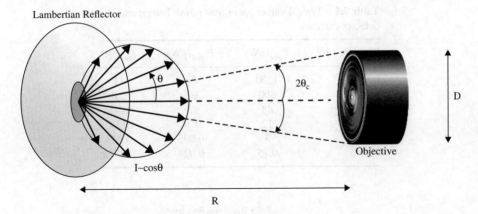

Figure 7.10 Principle of Lambertian reflection, having a cosine relation between the intensity of the reflection and the angle. Most everyday life objects have a surface dominated by Lambertian reflection [9].

reflection intensity is proportional to $\cos \theta$. By taking the ratio of the cosine-weighted integral over the spherical cap of diameter D (i.e., the share of optical power entering the lens), to the cosine-weighted integral over the complete hemisphere, we get the optical power impinging on the lens [9]:

$$P_{\text{lens}} = P_{\text{sent}} \frac{\int_0^{2\pi} \int_0^{\theta_c} \sin \theta \cos \theta d\theta d\alpha}{\int_0^{2\pi} \int_0^{\tau} \sin \theta \cos \theta d\theta d\alpha} = P_{\text{sent}} (\sin \theta_c)^2 = P_{\text{sent}} \left(\frac{D}{2R} \right)^2 \qquad (7.19)$$

We can then calculate the light power impinging on each pixel:

$$P_{\text{pixel}} = \frac{P_{\text{lens}}}{\#\text{pixels}} S_R \cdot v_{total} \qquad (7.20)$$

where S_R is the surface reflectivity and v_{total} is the multiplication of the fill factor (typically 0.7) with the lens and filter efficiency (typically 0.9). The light power impinging per pixel thus depends on the distance to the object R and the aperture of the lens D.

As a typical example, let us consider the emitted light power to be 300 mW and the lens aperture 2 mm. If we further consider a typical sensor responsivity of 0.3 A/W, a fill factor of 70%, a typical reflectivity of 50%, and pixel count of 10k, we can calculate values for the detector-induced currents for different distances, as shown in Table 7.1. The maximum possible current for 100% reflection at 0.5 m, when using this configuration, is 23 pA. This value defines the upper end of the dynamic range required for the system.

7.9.2 Background Light

The background light present in the scene is distinguished from the modulated light through the lock-in principle. However, because all light impinges on the detector, the induced shot noise level is caused by the modulated light as well as the background light. Therefore, the camera will always either require more power, or have poorer performance, in the presence of strong ambient light.

Table 7.1 Typical values for optical power budget and detector currents

Distance (m)	P_{lens}(nW)	P_{pixel}(pW)	I_{det}(pA)
0.5	1200	38	11
1	300	9.5	2.8
2.5	48	1.5	0.45
5	12	0.38	0.11
10	3	0.095	0.028
20	0.75	0.024	0.0071

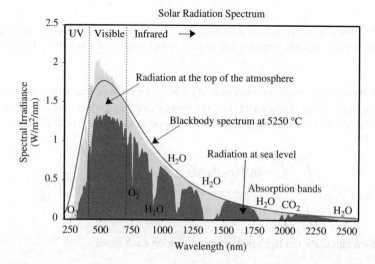

Figure 7.11 Solar energy spectrum, in and out of the atmosphere [10]. Source: By Nick84 [CC-BY-SA-3.0 (http://creativecommons.org/licenses/by-sa/3.0)], via Wikimedia Commons.

An easy way to reduce the light power that impinges on the detector is by using an optical filter, so that the used wavelength (within a certain spectral bandwidth) is selected and light of other wavelengths is attenuated. Using such a filter, the background light can be attenuated typically by a factor of 20.

In Figure 7.11, the spectrum of the sun is shown. It can be seen that around 930 nm, the spectrum shows a local minimum, due to atmospheric absorption. Therefore, this would be a good wavelength to use for the light source in a time-of-flight 3D camera.

The maximal optical power density of the sun for the 250–1100 nm region is 1006.9 W/m², according to the American Society for Testing and Materials (ASTM E 490, AM 0) [10]. In Table 7.2, the optical power densities for various conditions are shown. Indoor and cloudy outdoor values were obtained using a Newport OPM840 optical power measurement unit. We also noted the detector currents in a typical silicon detector having 0.3 A/W responsivity, 70% fill factor, 30 μm × 30 μm dimensions, and used in conjunction with a lens having a typical f-number of 2. Further, we assume the scene reflection to be maximal (100%). To obtain these

Table 7.2 Optical power density values for different background light situations, with and without optical filter, and corresponding detector currents for a typical silicon detector and lens setup

	PD_{BL} (W/m^2)	PD_{BL} filtered (W/m^2)	I_{det} filtered (pA)
Indoor	3	0.15	0.8
Outdoor (cloudy)	30	1.5	8
Outdoor (sunlight)	1006.9	50.3	270

values, we reshape Equation 7.19 and Equation 7.20 into:

$$P_{\text{pixel}} = \frac{PD_{BL}}{2M_f} A_{\text{pixel}} \left(\frac{1}{2 \cdot f/\#} \right)^2 S_R \cdot V_{total} \tag{7.21}$$

where PD_{BL} is the power density of the background light, divided by two because the DC light is spread over the differential detector nodes in a typical TOF pixel, $f/\#$ is the f-number of the lens used, A_{pixel} is the pixel area and M_f is the attenuation coefficient of the optical filter, for which a typical value of 20 is used. Note that this formula is distance-independent. Table 7.2 shows the typical current values induced by the background light. Comparing them with the values obtained from reflections of the modulated light, shown in Table 7.1, we see that the induced background light current values are up to almost five orders of magnitude higher than those of the modulated light.

7.10 Current State of the Art

Time-of-flight technique is widely researched in academia and industrial labs and, as a result, a substantial body of scientific work and a number of products are available in the field. In this section, we discuss the pixel count of the currently available solutions, and the outlook going forward.

From a research and development point of view, VGA resolution is not uncommon today, giving up to 300k TOF pixels on one image sensor. As shown in Figure 7.12, the TOF sensor resolution reported in research papers has been increasing approximately by a factor of four every four years. Following this general trend, sensor resolution of 720p should be up and running in research labs before 2016. The product availability in the market follows a similar trend, delayed by one four-year period from the research results. Currently available products are delivering QVGA resolutions, and are expected to improve it in the future.

7.11 Conclusion

In this chapter, we have reviewed the basic principles of the time-of-flight 3D imaging techniques. We have derived the salient equations and typical system parameters for the continuous time-of-flight setup. We have shown that the most important parameter in the TOF 3D imaging

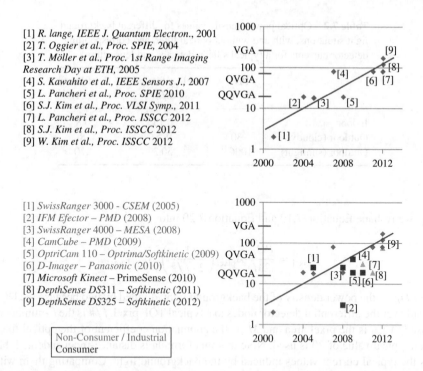

Figure 7.12 Overview and trend of TOF image sensor resolution (Source: [11]). Top figure shows the results from scientific literature and bottom figure shows resolutions of products that are available. Kinect resolution extrapolated from measuring smallest detectable feature size, note that this is much lower than the specified VGA resolution.

system is the modulation efficiency that enables the system to save power and achieve high repeatability and accuracy.

Following this, we reviewed the key challenges and issues with the technique and discussed their resolutions. Finally, we presented the state-of-the-art results and trends from the scientific developments and products that are available in the industry.

In summary, the TOF 3D imaging technique has many advantages, among which are real-time operation and resolution. We believe that TOF systems will penetrate many segments requiring 3D imaging capabilities, to add real-time user interactions within a wide range of applications.

References

1. Gulden, P., Vossiek, M., Heide, P., Schwarte, R. (2002). Novel opportunities for optical level gauging, 3-D-imaging with the photoelectronic mixing device. *IEEE Trans Instrum Meas* **51**, 679–684.
2. Viarani, L., Stoppa, D., Gonzo, L., Gottardi, M., Simoni, A. (2004). A CMOS smart pixel for active 3-D vision applications. *IEEE Sensors J* **4**, 145–152.
3. Van der Tempel, W., Van Nieuwenhove, D., Grootjans, R., Kuijk, M. (2007). Lock-in pixel using a currentas-sisted photonic demodulator implemented in 0.6m standard CMOS. *Japanese Journal of Applied Physics* **46**, 2377–2380.

4. De Nisi, F., Stoppa, D., Scandiuzzo, M., Gonzo, L., Pancheri, L., Betta, G. (2005). Design of electro-optical demodulating pixel in CMOS technology. *Proc. IEEE International Symposium on Circuits, Systems* **1**, 572–575.
5. Oggier, T., Kaufmann, R., Lehmann, M., Buttgen, B., Neukom, S., Richter, M., Schweizer, M., Metzler, P., Lustenberger, F., Blanc, N. (2005). Novel pixel architecture with inherent background suppression for 3D time-of-flight imaging. *Proc of SPIE Electronic Imaging* 1–8.
6. Lange, R., Seitz, P. (2001). Solid-state time-of-flight range camera. *IEEE J Quantum Electron* **37**(3), 390–397.
7. Oggier, T., Lehmann, M., Kaufmann, R., Schweizer, M., Richter, M., Metzler, P., Lang, G., Lustenberger, F., Blanc, N. (2004). An all-solidstate optical range camera for 3D real-time imaging with subcentimeter depth resolution (SwissRanger). *SPIE Optical Design, Engineering* **5249**, 534–545.
8. Hosticka, B., Seitz, P., Simoni, A. (2006). Optical Time-of-Flight Sensors for Solid-State 3D-Vision. *Encyclopedia of Sensors* **7**, 259–289.
9. Lange, R. (2000). *3D Time-of-Flight Distance Measurement with Custom Solid-State Image Sensors in CMOS/CCD-Technology*. PhD thesis, University of Siegen; September 2000.
10. Newport Corporation (2008). *Introduction to Solar Radiation*. http://www.newport.com/Introduction-to-Solar-Radiation/411919/1033/catalog.aspx#.
11. Stoppa, D., Pancheri, L., Perenzoni, M. (2012). *Sensors Architectures for 3D Time-of-Flight Imaging*. 6th Annual Global Conference on Image Sensors.

8

Eye Gaze Tracking

Heiko Drewes
LFE Medieninformatik, Ludwig-Maximilians-Universität, München, Germany

8.1 Introduction and Motivation

The amount of our interaction with computer devices increases steadily. Therefore, research in the field of HCI (Human Computer Interaction) looks for interaction methods which are more efficient, more intuitive, and easier. "Efficiency" means that we can do our interaction as quick as possible. We also try to avoid lectures or training for being able to operate computer devices, which means we like to have intuitive interfaces. Finally, we dislike efforts, either physical or mental. We expect that our interaction with a device should be easy.

Traditional interaction devices, such as mouse and keyboard, can cause physical injuries like carpal tunnel syndrome if used extensively. Keyboard and mouse are also not very practical in a mobile context – and finally, we always look for alternatives or even something better. Using our gaze for computer interaction seems to be a promising idea. The eyes are quick and we move them intuitively and with ease, and therefore eye gaze interaction would fulfill all criteria given above.

In addition, eye trackers could be manufactured at low cost if produced in big quantities. A minimal eye tracker consists of a video camera, an LED, a processor and software. In a smart device, such as smart phones, tablets, laptops and even some new TV sets, all these components are already present but, even if produced separately, the costs will be comparable to the costs of manufacturing an optical mouse. Advanced eye trackers use two cameras for a stereoscopic view and sometimes several LEDs, but still, especially if produced for a mass market, the cost is affordable.

A further reason to consider gaze as a future interaction technology is the importance of gaze in human-human interaction.

When comparing human eyes with animal eyes (see Figure 8.1), the white in the human eyes is very distinctive. In animal eyes – eyes of mammals in particular, which work in a similar way to human eyes – no white eyeball is visible. For this reason, it is more difficult to determine the direction of an animal's gaze than of a human's. What role the awareness of the gaze's direction played in the evolution of our species is a field for speculation. What is certain

Interactive Displays: Natural Human-Interface Technologies, First Edition. Edited by Achintya K. Bhowmik.
© 2015 John Wiley & Sons, Ltd. Published 2015 by John Wiley & Sons, Ltd.

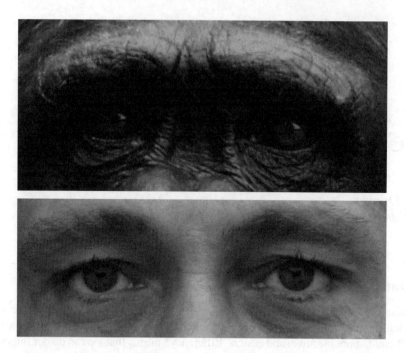

Figure 8.1   The eyes of a chimpanzee and of a human. (See color figure in color plate section).

is that we use our eyes for communication. Typically, we address persons or objects with the gaze. If somebody asks somebody else: "Can I take this?" the person knows at which object the asking person is looking and therefore knows what "this" is. If we want human-computer interaction to be similar to human-human interaction, the computer needs gaze awareness.

Eye gaze interaction brings along further advantages. It works without physical contact and therefore is a very hygienic way of interaction. Without touch, the device also does not need cleaning. An eye tracker does not have moving parts, meaning it does not need maintenance. If combined with a zoom lens, an eye tracker works over distances longer than our arms and can serve as a remote control.

Additionally, eye trackers can make our computer interaction safer, because they explicitly require our attention. A mobile phone which requires eye contact for an outgoing call will not initiate a call due to accidentally pressed buttons while in a pocket. Finally, eye trackers have the potential for more convenient interaction, as they can detect our activities. For example, when we are reading, the system could postpone non-urgent notifications like the availability of a software update.

This chapter continues with basic knowledge of human eyes and an overview on eye-tracking technologies. The next section explains objections and obstacles for eye gaze interaction, followed by a short summary of eye-tracking research during the last three decades. The next three sections present different approaches to gaze interaction.

The first and most obvious approach is eye pointing, which is similar to pointing with a mouse, albeit with less accuracy. The eye pointing section includes a comparison of mouse and gaze pointing, and discusses the coordination of hand and eyes.

The second approach is the use of gaze gestures. Gaze gestures are not very intuitive, but gestures belong to the standard repertoire of interaction methods. Beside gesture detection and gesture alphabets, this section deals also with the separation of gestures from natural eye movements.

The third approach is eye gaze as context information. Here, the eye movements do not trigger intentional commands, but the system observes and analyzes eye movements with the goal of assisting and supporting the user in a smart way. The section deals with activity recognition in general and, in particular, with reading and attention detection. The chapter closes with an outlook on further development of eye gaze interaction.

8.2   The Eyes

There is an immense knowledge on the eyes in fields such as medicine, biology, neurology and psychology. The knowledge on eyes presented here is simplified and only presents facts necessary for understanding this chapter.

From a technical point of view, we can see the eyes as a pair of motion-stabilized cameras moving synchronously. Each eyeball has three pairs of antagonistic muscles (see Figure 8.2), which can compensate the head's three degrees of freedom – horizontal and vertical movements and rotations around the line of sight.

Figure 8.3 depicts the simplified schematics of the eye. The eye is similar to a camera – the iris is the aperture, the retina is a photo-sensitive surface, and both camera and eye have a lens. In contrast to a camera, the eye focuses on an object by changing the form of the lens, not by changing its position. The photo-sensitive surfaces of camera and eye differ a lot. The camera has a planar photosensitive surface with equally distributed light receptors, typically for red, green, and blue light. The photo-sensitive surface of the eye is round, and the light receptors are not equally distributed. Beside receptors for the three light colors (cones) the eye has additional receptors which are unspecific for the color of light but more sensitive (rods). The rods give us some ability of night vision. The cones on the retina have a low density, except for a small spot opposite to the pupil, called the fovea. As a consequence, we do not see clearly except in a narrow field of about 1–2 degrees, which is about a thumbnail size in arm length distance. The high-resolution picture of our mental eye is an illusion generated by the brain.

The small field of vision in which we can see with high resolution has consequences. It is the reason why we move our eyes. We always move our eyes to a position that projects whatever we are looking at directly on our fovea. There are two types of movement to achieve this:

- One type of movement is a compensation movement. It occurs when we move our head but keep our gaze on an object. This motion stabilization is necessary, because we need a stable projection on the fovea to see. Such motion stabilization takes place also when we watch a moving object. Movements for keeping an image stable are smooth.
- The other type of eye movements is an abrupt quick movement, which is called a saccade. Typically, the eye moves as quickly as possible to an interesting spot and rests there for a while, which is called a fixation. After this, the eye does another saccade, and so on. Most of the time, our eyes do saccadic movements.

When the eye moves, it does not change its position but it rotates around its center. Therefore the length of a saccade is defined by the angle of the pupil's normal at the beginning and end

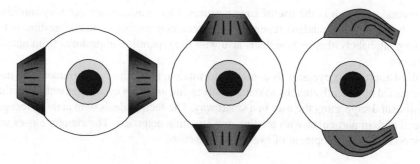

Figure 8.2   Three pairs of muscles can compensate all movements of the head.

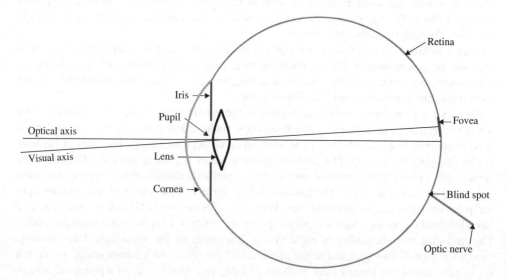

Figure 8.3   Schematics of the eye.

of the saccade. Figure 8.4 shows saccade times versus angle. It is clear that there is a minimum time for a saccade, depending on the angle. However, for large angles, the time only increases marginally. A saccade for angles above 5° lasts about 100–150 milliseconds.

The saccadic movement is so quick – up to 700°/s – that the receptors on the retina do not have time enough to detect a picture and, in consequence, we are blind during the saccade. Therefore, there is no control-feedback loop to steer the eye into the target. Psychology calls the saccadic eye movements ballistic. This means that saccadic eye movements do not obey Fitt's law [1], even if some publications from the HCI community state the opposite [2–5]. In contrast to movements obeying Fitt's law, the time of a ballistic movement does not depend on the target size.

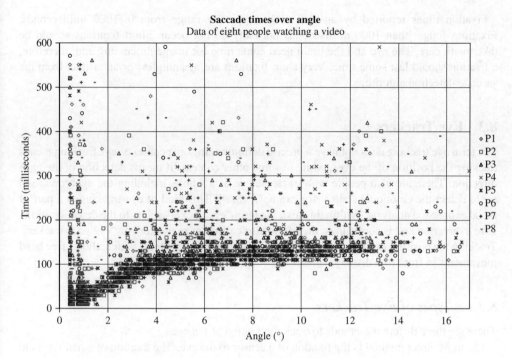

Figure 8.4   Saccade times over their length.

Carpenter measured the rotation amplitude and duration for saccades in 1977 [6]. He used a linear approximation to express the relation between the time T of a saccade and its amplitude A:

$$T = 2.2 \text{ ms/}° \cdot A + 21 \text{ ms}$$

In 1989, Abrams, Meyer, and Kornblum [7] suggested a model where the muscle force is constantly increasing over time. As the mass and shape of the eye does not change, the acceleration $a(t)$ is proportional to the force and also increasing constantly over time:

$$a(t) = k \cdot t, \text{ with constant } k.$$

Integrating twice over time and solving the equation for the amplitude shows that there is a cubic root relation between time and amplitude:

$$T = c \cdot A^{1/3}$$

The constant c depends on constant k and the eye's polar moment of inertia.

Looking at Figure 8.4 shows that a linear approximation, assuming the data are on a straight line, as done by Carpenter, is legitimate in a certain range. The model from [7], however, fits better with experimental data.

Fixation times reported by an eye tracker typically range from 0–1000 milliseconds. Fixations longer than 1000 milliseconds normally do not occur. Short fixations should be taken with care. The eye and the brain need some time for image processing and, therefore, a fixation should last some time. Very short fixations are meaningless or an artifact from the saccade detection algorithm.

8.3   Eye Trackers

The term eye tracking does not have a precise definition. In some contexts, eye tracking means tracking the position of the eyes, while in some other contexts it means detection of the gaze direction. There are also people who track the eye as a whole, including the eyebrows, and try to detect the emotional state – for example, [8, 9]. This kind of eye tracking is a part of facial expression analysis. Within this text, the term "eye tracker" refers to the detection of the gaze's direction and is sometimes, therefore, also called "gaze tracker" or "eye gaze tracker". Tracking the position of the eyes is a subtask of gaze-tracking systems which allows free head movements in front of a display.

8.3.1   Types of Eye Trackers

There are three different methods to track the motion of the eyes.

The most direct method is the fixation of a sensor to the eye. The fixation of small levers to the eyeball belongs to this category, but is not recommended because of high risk of injuries. A safer way of applying sensors to the eyes is using contact lenses. An integrated mirror in the contact lens allows measuring reflected light [10]. Alternatively, an integrated coil in the contact lens allows detection of the coil's orientation in a magnetic field [11]. The thin wire connecting the coil with the measuring device is not comfortable for the subject. The big advantage of such a method is high accuracy and the nearly unlimited resolution in time. For this reason, medical and psychological research uses this method.

Another method is electrooculography (EOG), where sensors attached to the skin around the eyes measure an electric field. Initially, it was believed that the sensors measure the electric potential of the eye muscles. It turned out that it is the electric field of the eye which is an electric dipole. The method is sensitive to electro-magnetic interferences, but works well as the technology is advanced and has already existed for a long time; therefore there is a lot of knowledge and standards on the topic [12]. The big advantage of the method is its ability to detect eye movements even when the eye is closed, e.g., while sleeping. Modern hardware allows integrating the sensors into glasses to build wearable EOG eye trackers [13].

Both methods explained so far are obtrusive and are not suited well for interaction by gaze. The third and preferred method for eye-gaze interaction is video-based. The central part of this method is a video camera connected to a computer for real-time image processing. The image processing takes the pictures delivered from the camera and detects the pupil to calculate the direction of the gaze. The big advantage of video-based eye tracking is its unobtrusiveness. Consequently, it is the method of choice for building eye-gaze interfaces for human-computer interaction. A detailed description of the video-based corneal reflection method follows in the next section.

There are two general types of video-based eye trackers: stationary systems and mobile systems.

Figure 8.5 Stationary eye trackers from Tobii, one stand-alone system and one integrated into a display. Source: Reproduced with permission from Tobii.

Stationary eye trackers, such as the one shown in Figure 8.5, deliver the gaze's direction relative to the space around the user, typically as screen coordinates. Simple stationary eye trackers need a stable position for the eye and, therefore, require head fixation. Disabled people who cannot move anything except the eyes use such systems. People without disabilities prefer systems which allow free movement in front of the display. Such systems do not only track the gaze direction, but also the position and orientation of the head, typically by a pair of cameras providing a stereoscopic view. Stationary eye trackers can be a stand-alone device which allows tracking the gaze on physical objects. However, as many eye-tracking applications track the gaze in front of a display, there are eye trackers integrated into displays which may not even be noticed by the user.

Mobile eye trackers are attached to the head of the user. This type of eye tracker delivers the gaze's direction relatively to the head's orientation. Typically, mobile eye trackers come with a head-mounted camera which captures what the user sees. With the data from the eye tracker, it is possible to calculate the point the user is looking at and mark this point in the images delivered by the head-mounted camera. With the recent progress in building small cameras, it is possible to build eye trackers integrated into glasses. As an example, Figure 8.6 shows a pair of eye tracking glasses.

There is a further class of video-based eye trackers introduced by Vertegaal *et al.*, which are called ECS for eye contact sensors [14]. An ECS does not deliver coordinates for the gaze direction but only signals eye contact. This works up to a distance of ten meters. The eye contact sensor Eyebox2 from xuuk inc. is shown in Figure 8.7.

8.3.2 Corneal Reflection Method

The general task of a video-based eye tracker is to estimate the direction of gaze from the picture delivered by a video camera.

Figure 8.6 Eye tracking glasses from SMI. Source: SMI Eye Tracking Glasses. Reproduced by permission of SensoMotoric Instruments.

Figure 8.7 The Eye Contact Sensor Eyebox2 from xuuk inc. Source: xuuk.inc. Reproduced with permission.

A possible way is to detect the iris using the high contrast of the white of the eye and the dark iris. This method results in good horizontal, but bad vertical, accuracy as the upper and lower part of the iris is covered by the eyelid. For this reason, most video-based eye trackers detect the pupil instead. Detecting the pupil in the camera image is a task for image recognition, typically edge detection, to estimate the elliptical contour of the pupil [15]. Another algorithm for pupil-detection is the starburst algorithm explained in [16].

There are two methods to detect the pupil – the dark and the bright pupil method. With the dark pupil method, the image processing locates the position of a black pupil in the camera

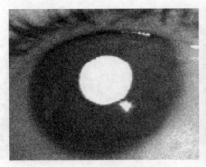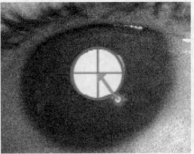

Figure 8.8   The vector from the glint to the pupil's center is the basis to calculate the gaze direction.

image. This can be problematic for dark brown eyes, where the contrast between the brown iris and the black pupil is very low. The bright pupil method uses additional illumination, with infrared light coming from the same direction as the view of the camera. Therefore, an infrared LED has to be mounted inside or close to the camera, which requires mechanical effort. The retina reflects the infrared light and this makes the pupil appear white in the camera image. The effect is well known as "red eye" when photographing faces with a flash. For differences in the infrared bright pupil response among individuals, see [17].

Most eye trackers use a reflection on the cornea, also called first Purkinje image, to estimate the gaze direction. As the cornea has a perfect spherical shape, the glint stays in the same position, independent of the gaze direction (Figure 8.9).

The eye tracker's image processing software detects the position of the glint and the center of the pupil. The vector from the glint to the center of the pupil is the basis for the calculation of the gaze's direction (Figure 8.8), and finally the position of the gaze on the screen. A direct calculation would not only need the spatial geometry of the eye tracker, the infrared LED and the display and the eye, but also the radius of the eyeball, which is specific to the subject using the eye tracker. For this reason, a calibration procedure estimates the parameters for the mapping of the glint-pupil vector to positions on the screen. A calibration procedure asks the user to look at several calibration points displayed during the calibration. A calibration with four points uses calibration points close to the four corners of the display.

The corneal reflection method does not work for people with eyeball deformations. It can also be problematic when people wear contact lenses. Glasses are less problematic; although the glasses may change the position of the glint, the reflection stays in the same position. The calibration compensates for the optical distortion of the glasses.

The corneal reflection method explained so far requires that the eye stays in a stable position, which means it requires a head fixation. For human-computer interaction, it is desirable to allow free movement in front of the device. Such eye trackers use stereoscopic views, with two cameras. Alternatively, it is possible to use only one camera and multiple light sources for multiple glints. For a description of how such system works see [18–20]. The way how commercial eye-tracking systems work, however, is normally a business secret.

The eye contact sensor presented above uses the corneal reflection method, too. For the illumination, a set of infrared LEDs is mounted on-axis around an infrared camera. When the camera delivers a picture with a glint inside the pupil, it means that the onlooker is looking directly towards the camera. A big advantage of this method is that it does not need calibration.

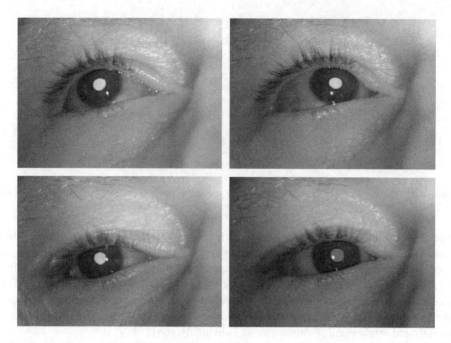

Figure 8.9   Looking to the four corners of the screen – the reflection stays in the same position.

8.4   Objections and Obstacles

In the introduction, gaze interaction seemed promising. However, not all promises are easy to fulfill, and there are some obstacles on the way to gaze interaction.

8.4.1   Human Aspects

There is some general concern that gaze interfaces will conflict with the primary task of the eye, which is vision. The eye could get an input-output conflict, where vision and interaction require different eye movements. Zhai *et al*. wrote in 1999: "*Second, and perhaps more importantly, the eye, as one of our primary perceptual devices, has not evolved to be a control organ. Sometimes its movements are voluntarily controlled while at other times it is driven by external events.*" [21]

Changes in the field of vision can trigger an eye movement. If a gaze-aware interface displays a blinking object, the eyes will most probably move to the blinking object. If this eye movement triggers a new command, the user will invoke this command unintentionally. In general, it is possible to create such input-output conflict in gaze interfaces. However, a developer of gaze interfaces has to construct such conflicts. Typically, such conflicts do not occur, and no scientist reported serious problems of this type.

There are also objections that we are not able to control our eye movements. However, although our eye movements are driven by the task of vision, we are aware of our gaze and we are able to controlling it. Otherwise, we would violate our social protocols.

A further argument against gaze interaction is that the eye could fatigue and problems such as repetitive stress injuries could occur. However, we move our eyes constantly, even while we sleep. If a person does not move the eyes for a minute, we start to worry whether they have lost consciousness. Fatigue seems not to be a problem for the eyes.

Finally, another concern against eye-tracking is acceptance. Video cameras and internet connections in all electronic devices would provide the infrastructure for Orwell's vision of "big brother is watching you". The fact that an eye tracker has a video camera is something we may get used to, with all the other cameras around us already. However, the users of an eye-tracking system could feel observed, and perhaps will not accept it in private spaces like the bathroom. It seems that an analysis of eye movements can judge our reading skills and therefore tell a lot about our intelligence, or at least our formation. Such conclusions, which are possible from gaze data, could also frighten people, especially employees.

8.4.2   Outdoor Use

Stationary eye trackers normally work well and reliably, as they are typically located in an indoor environment where the light conditions are relatively stable. In outdoor environments, the light varies in a wide range, which means there can be extreme contrasts. Additionally, the light can change also very quickly, for example by moving shadows, especially if sitting in a car. Such a situation is still a challenge for cameras. Infrared light-based systems, which most commercial systems are, can have problems, as the sun is a bright infrared radiator. This makes it difficult to reliably detect the pupil and the glint within a camera picture. Approaches with differential pictures and infrared illumination synchronized with the frame rate of the video camera or the use of polarized light are promising [22]. Therewith, eye tracking for most light conditions, except the extremes, should be achievable.

8.4.3   Calibration

Eye trackers need calibration to achieve a good accuracy. Methods using the glint of an infrared LED depend on the radius of the eyeball and, therefore, need calibration for the user. Even eye trackers which are able to determine the orientation of the pupil in space without a glint, and therefore do not depend on the eyeball's radius, need a calibration. The reason is that the position of the fovea differs from individual to individual, and the optical axis (the normal on the pupil's center) is not exactly the visual axis (line from the fovea to the pupil's center).

The positive aspect of the calibration procedure is that it is only necessary to do it once. A personal system equipped with an eye tracker requires a calibration only once. For public systems which require reasonable accuracy for the gaze detection, for example ATMs, the calibration procedure is a real obstacle.

One way to avoid the calibration issue is not to use absolute gaze positions, but only relative gaze movements. Detecting only relative movements means using gestures. This is an option, but it is also a severe restriction.

8.4.4   Accuracy

The accuracy issue has two aspects. One is the accuracy of eye tracker; the other is the accuracy of the eye movements.

The accuracy of available eye trackers is still far from physical limits. Eye trackers have a spatial and a temporal resolution. The temporal resolution depends on camera and processor speed and algorithms. For stationary eye trackers the spatial resolution is mainly a question of camera resolution. As camera resolutions and processor speed will continue to increase, we can expect that we can build eye trackers with higher accuracy in near future. For mobile systems, the spatial accuracy also depends on mechanical stability of the head-mounted system.

Eye trackers nowadays typically state an accuracy of $\pm 0.5°$. In arm length distance, which is the typical distance of a display from the eyes, the eye tracker's accuracy is about the size of a thumbnail. This accuracy is not sufficient to use eye gaze as a substitute for a mouse. Typical graphical user interfaces use interaction elements which are smaller than a thumbnail.

The accuracy of our eyes is a subtle problem. The question is not only how accurately we can position our gaze, but also how accurately we do. Even if we can fixate quite accurately, the question is how much concentration we need for this. For recognizing something, it is sufficient that the object has a projection inside the fovea and, therefore, we should expect that the eye positions itself with this accuracy. The situation seems to be comparable with spotting an insect at night with a torch – it is enough if the insect is in the circle of light and it is not worth the effort to center it.

Ware and Mikaelian expressed it this way:

> "The research literature on normal eye movements tells us that it is possible to fixate as accurately as 10 minutes of visual angle (0.16 deg) but that uncontrolled spontaneous movements cause the eye to periodically jerk off target. It is likely, however, that when an observer is attempting to fixate numerous targets successively, accuracy is considerably reduced and misplacements as large as a degree may become common." [2]

8.4.5   Midas Touch Problem

Although gaze pointing is very similar to pointing with a finger, there is an important difference: we cannot lift our gaze like a finger. On a touch-sensitive display, we can point on an interaction element and trigger the command by touching the surface. With the gaze, we can also point to an interaction element, but we do not have a touch event. If we trigger the command just by looking at it, we have the problem that even if we only want to see what is there, we will trigger a command. Jacob called this the Midas Touch problem, which he explained with these words:

> "At first, it is empowering simply to look at what you want and have it happen. Before long, though, it becomes like the Midas Touch. Everywhere you look, another command is activated; you cannot look anywhere without issuing a command" [23].

When Jacob identified the Midas Touch problem, he had eye pointing in his mind. On a more general level, the Midas Touch problem is the problem of deciding whether an eye activity was made to issue a command, or happened within the task of vision. The problem remains, even if changing the interaction method. Gaze gestures also bear the danger that they can occur within natural eye movements and invoke unintended commands. As gestures do not imply touches, the term Midas Touch problem in this context would be misleading, and it is better to speak about the problem of separation of gestures from natural movements.

8.5   Eye Gaze Interaction Research

The idea of using eye gaze for interaction is already more than 30 years old, and a lot of research has been done since then. Therefore, the following overview only goes a short way through the historic development of this field of research.

Gaze interaction became possible with the invention of the electronic video camera and computers which are powerful enough to do real-time image processing. The first systems were built in the 1970s to assist disabled people. The typical application for these people is eye-typing [24]. For eye-typing, the display shows a standard keyboard. If the gaze hits a key on this virtual keyboard, the corresponding key is highlighted. If the gaze stays on this key longer than a predefined dwell time – typically around 500 milliseconds – it means the key was pressed.

In 1981 Bolt [25] provided a vision of multi-modal gaze interaction for people without disabilities. He described a Media Room with a big display where 15–50 windows display dynamic content simultaneously, which he named "World of Windows". His idea was to zoom in upon some window the user is looking at. He described an interface method based on dwell time and he also discussed multi-modal interface techniques. One year later, Bolt published the paper "Eyes at the Interface" [26], summing up the importance of gaze for communication, and concludes there is a need of gaze-awareness for the interface. His vision has not yet been realized completely.

In 1987, Ware and Mikaelian did a systematic research on eye pointing [2]. In their paper, "An evaluation of an eye tracker as a device for computer input", they introduced three different selection methods, which they called "dwell time button", "screen button", and "hardware button", and measured the times the eyes need for selection. For the dwell-time button method, the gaze has to stay for a certain time (the dwell time) on the button to trigger the action associated with the button. The screen button method is a two-target task. The gaze moves to the chosen button and, afterwards, to the screen key to trigger the action. The hardware button method uses a key to press with the finger in the moment when the gaze is on the chosen button. The first two methods are gaze-only, while the hardware button uses an extra modality. Ware and Mikaelian fitted their data against a "modified" Fitts' law to compare their results with the experiments done by Card *et al.* [27] for mouse devices. However, Fitts' law does not hold for the eye (see also the chapter on the eyes on page 252).

In 1990, Jacob [23] systematically researched the interactions needed to operate a GUI (graphical user interface) with the eyes – object selection, moving objects, scrolling text, invoking menu commands and setting the keyboard focus to a window. One big contribution of this paper is the identification of the Midas Touch problem (explained in section 8.4). The generality of Jacob's paper caused all further research to focus on single or more specialized problems of eye-gaze interaction.

In 1999, Zhai *et al.* made a suggestion to handle the low accuracy intrinsic to eye pointing (explained in section 8.4) and named it MAGIC (Mouse And Gaze Input Cascaded) pointing [21]. MAGIC pointing uses the gaze for raw positioning and a traditional mouse device for the fine positioning.

In 2005, Vertegaal *et al.* introduced Media EyePliances, where a single remote control can interact with several media devices [14]. The selection of the device to control happens by looking at it. For this, they augmented the devices with a simple form of an eye tracker, called eye contact sensor ECS. In another paper from the same year [28], Vertegaal *et al.* used the ECS in combination with mobile devices. It is typical for mobile devices that they do not get full attention all the time, as the user of the mobile device has to pay some attention to her or his surroundings. With two applications, *seeTXT* and *seeTV*, they demonstrated how to use the eye gaze to detect attention and how to use this context information to control the device. The *seeTV* application is a video player, which automatically pauses when the user is not looking at it. *seeTXT* is a reading application, which advances text only when the user is looking.

In recent years, eye-tracking interaction research has become popular and the number of publications has increased enormously. There has been a special conference for the topic since the year 2000 called ETRA (Eye-Tracking Research and Application). Since 2005, the COGAIN (Communication by Gaze Interaction) initiative, supported by the European Commission, has also organized conferences and offers a bibliography of research papers on the internet.

8.6   Gaze Pointing

The most obvious way to utilize gaze as computer input is gaze pointing. Looking at something is intuitive, and the eyes perform this task quickly and with ease. Pointing is also the basic operation to interact with graphical user interfaces, and performing pointing actions with the eyes would speed up our interaction. In consequence, most of the research done on gaze interaction deals with eye gaze pointing. However, as mentioned already in the chapter on objections and obstacles (section 8.4), there are intrinsic problems in eye pointing, such as the Midas Touch problem and the low accuracy.

8.6.1   Solving the Midas Touch Problem

A gaze-aware interface which issues commands when looking at an interaction object would trigger actions even when we only want to look what is there. The problem is known as Midas Touch problem, and there are several ways solving it. Gaze-only systems, as typically used by disabled people, introduce a dwell time. This means that the user has to look for a certain time, the dwell time, at an interaction object for triggering a command. The dwell time is normally in the range of 500–1000 milliseconds and eats up the time saved by the quick eye movement.

Another method for solving the Midas Touch problem is using a further modality like a gaze key. Pressing the gaze key invokes the command at which the eye is looking at that moment. Using a gaze key allows quick interaction, but also destroys some of the benefits of gaze interfaces. An additional key also means that the interface is not hygienic any more, as there is again something to touch. Additionally, it does not work over distance, as there has to be a key in reach – and finally, it does not work for disabled people as they use a gaze interface, because they are not able to press a key. Thinking more deeply about a gaze key reveals that it only makes sense if the intention is to enter two-dimensional coordinates. Entering a command (e.g., a save operation) by looking at the "save interaction object" and pressing the gaze key provokes the question, "Why not simply press ctrl-s without looking at a special interaction object?"

The frequently mentioned suggestion to blink with the eye for triggering a command does not seem to be an option. The eyes blink from time to time to keep the eyes moist and, therefore, a blink command has to take longer than a natural blink and any speed benefit is lost. It is also a little bit paradoxical to trigger a command at the gaze position in the moment the eye is closed. The main reason, however, for not using blinking is that it does not feel comfortable. The blinking would be a substitute for a mouse click. The number of mouse clicks we perform when operating a graphical user interface is enormous, and easily exceeds thousand clicks per hour. Blinking a thousand times with the eye typically creates a nervous eye.

8.6.2   Solving the Accuracy Issue

The accuracy of current eye trackers, and also the accuracy of our eye movements, does not allow addressing small objects, or even a single pixel. Much research has been done on solving the accuracy problem. The easiest solution is enlarging the interaction objects. Assuming an accuracy of half inch on a 72 dpi display means that interaction objects should not be smaller than 36×36 pixels. Existing graphical user interfaces use buttons of 16×16 pixels in size, and the height of a menu item or line of text is around $8-12$ pixels. This means that the graphical user interface has to be enlarged by a factor of about three in each dimension, or that we need displays roughly ten times bigger. For most situations, such a waste of display area is not acceptable.

There are several suggestions from research how to solve the accuracy problem – raw and fine positioning, adding intelligence, expanding targets, and the use of further input modalities – which are discussed below.

Zhai *et al.* made a suggestion for handling low accuracy which the called MAGIC (Mouse And Gaze Input Cascaded) pointing [21]. MAGIC pointing uses the gaze for raw positioning and a traditional mouse device for the fine positioning. The basic idea of MAGIC pointing is positioning the mouse cursor at the gaze position on the first mouse move after some time of inactivity. With MAGIC pointing, the gaze positions the mouse pointer close to the target and the mouse is used for fine positioning. In their research, Zhai *et al.* found out that there is a problem of overshooting the target, because hand and mouse are already in motion at the moment of positioning. They suggested a compensation method calculated from the distance and the initial motion vector.

Drewes *et al.* suggested an enhancement of this principle and called it MAGIC Touch [29]. They built a mouse with a touch-sensitive sensor and positioned the mouse pointer at the gaze position when touching the mouse key. The improvement lies in the absence of compensation methods, because the mouse does not move when putting the finger on the mouse key. A further advantage is that the user can choose the moment of gaze positioning and does not need a period of mouse inactivity.

When watching people working with big screens or dual monitor setups, it is easy to observe that they sometimes have problems with finding the mouse pointer. In addition, the mouse pointer is very often far away from the target intended to click. In consequence, it is necessary to draw the mouse a long way across the screen. The principle of raw positioning by gaze and fine positioning with the mouse not only solves the accuracy problem, but also avoids the problem with finding the mouse pointer. Additionally, it saves a lot of the distance covered by the mouse and therefore helps to prevent repetitive stress injuries.

In 2000, Salvucci and Anderson presented their intelligent gaze-added interfaces [30]. This starts with a system where the gaze tracker delivers x-y positions to a standard GUI that highlights the interaction object the user is looking at. A gaze key, analogous to a mouse key, offers the user the possibility of triggering the action. To solve the accuracy issue, the system interprets the gaze input in an intelligent way – it maps the gaze points to the items which the user is likely attending. To find these items, the system uses a probabilistic algorithm which determines the items by the location of the gaze (i.e., the items close to the reported gaze point) and the context of the task (e.g., the likelihood of a command after a previous command).

Another approach to solve the accuracy problem is the use of expanding targets. Balakrishnan [31] and Zhai [32] researched the use of expanding targets for manual pointing and showed that this technique facilitates the pointing task. Miniotas, Špakov and MacKenzie applied this technique to eye gaze pointing [33]. In their experiments, the target expansion was not visually presented to the users, but the interface responded to an expanded target area. They called this technique static expansion. In a second publication, Miniotas and Špakov studied dynamically expanding targets [34], i.e., means targets where the expansion is visible to the users. The research was done for menu targets, and the results showed that the error rate for selecting a menu item reduces drastically at the cost of an increased selection time.

In the same year, Ashmore and Duchowski published the idea of using a fisheye lens to support eye pointing [35].

In 2007, Kumar *et al.* presented an eye-gaze interface called *EyePoint* [36]. This interface technique uses expansions of the interaction targets, and it also uses a key as additionally needed input modality. When pressing down this key, the screen area that the gaze looks at becomes enlarged. Within this enlarged screen area, the user selects the target with the gaze and the action is triggered in the moment the user releases the key.

The inaccuracy of gaze pointing means that a pointing action has ambiguities on the target if the targets are close to each other. This gave Minotas *et al.* the idea to specify the target with an additional speech command [37]. They used targets with different colors and asked the users to speak out the color of the target loudly. They showed that this method allows addressing targets subtending 0.85 degrees in size with 0.3-degree gaps between them.

The method does not bring any speed benefit and it is not clear whether the better pointing accuracy is worth the extra effort of speaking, at least in the case of operating a standard GUI. However, the concept is interesting, because it is close to human-human interaction. Normally, we are aware of where other persons look, but with an accuracy much lower than that of an eye tracker. When we say, "Please give me the green book" and look on a table, we get the green book from the table and not the green book from the shelves. We assume that the other person knows where we are looking, and it is only necessary to specify the object within that scope.

8.6.3   Comparison of Mouse and Gaze Pointing

A good way to develop a deeper understanding of gaze pointing is a comparison with other pointing techniques. Besides accuracy and speed, pointing devices differ in space requirements, provision of feedback, support for multiple pointers and the modality to invoke a click. Table 8.1 gives an overview on these properties.

Fitt's law gives the relation of speed and accuracy for pointing devices; achieving a higher accuracy for a pointing operation demands more time. In the case of a touch screen or eye gaze, anatomic sizes limit the accuracy, and the question of how big the pointing targets should be is answered by the size of the finger or fovea, not by a speed-accuracy trade-off.

A mouse needs some space for its movements on the table. As sufficient space often is not available when sitting in a train or plane, mobile devices typically use a trackball, track point, or touchpad. A touch screen does not need additional space, but the precision of a finger is low because of the size of the fingertip, which hides the visual information. To achieve high precision in pointing on a touch screen, people use a pencil with a thin tip. The situation for gaze pointing regarding space demand and accuracy is similar to pointing with the finger on a

Table 8.1   Properties of pointing devices

	Mouse	Trackball	Track point	Touchpad	Touch screen	Eye gaze
Speed	fast	fast	medium	fast	fast	very fast
Accuracy	time	time	time	time	size of finger	size of fovea
Space demand	much	little	little	little	none	none
Feedback	yes	yes	yes	yes	no	no
Method	indirect	indirect	indirect	indirect	direct	direct
Multiple pointing	2 hands	2 hands	2 hands	10 fingers	10 fingers	1 pair of eyes
Intrinsic click	no	no	no	yes (no)	yes	No

touch screen. Eye gaze does not hide the visual information, but increasing the precision with a pencil is not an option.

The provision of feedback by a mouse pointer is mandatory for pointing devices which work indirectly. For direct pointing on a touch screen, feedback is not necessary. Gaze pointing is also a direct method and does not need feedback. The reason a feedback pointer for eye gaze is desirable is to make sure that the coordinates reported by the eye tracker are really at the gaze position. Section 8.6.5 discusses the feedback for gaze pointing and why introducing a gaze pointer can be counter-productive.

The use of multiple pointers is a topic of current research. There are many discussions on two-handed interaction and the use of all fingers for pointing. For the eyes, it is clear that both eyes move synchronously and we are not able to point with both eyes independently. Multiple gaze pointers only make sense for multiple people.

Many GUI operations use pointing in combination with a click on the mouse key, which means that there is a need of an extra modality. This is not the case for the touch screen, where the touch provides a position and a click event. The touchpad does not present the same possibility, as the indirect method does not allow touching a target directly. The touch happens before steering the feedback pointer to the target and, consequently, the touch event is not useful to trigger an action on the target. A touchpad can use increased pressure as a click event, but commercial devices with a touchpad normally provide an extra mouse key. From the traditional pointing devices, the touch screen is most similar to gaze pointing. The big difference is that a finger can be lifted to move to another location, while the gaze cannot do that. Consequently, the eye gaze cannot produce a click event like the finger.

8.6.4   Mouse and Gaze Coordination

Figures 8.10 and 8.11 show typical mouse and gaze movements for a click-the-target task. Interestingly, the gaze moves directly to the target and does not look to the position of the mouse pointer. Motion detection works well in the peripheral vision area, and there is no need to hit the mouse pointer by gaze.

Gaze pointing does not put an extra load on the eye muscles and does not cause more stress on the eyes than working with a traditional mouse. The simple reason for that is that we have problems hitting a target without looking at it. Under special conditions, like big targets which we can see with peripheral vision, it is possible to follow the motion of the mouse cursor and steer it to the target only with motion vision, but without the gaze hitting the target.

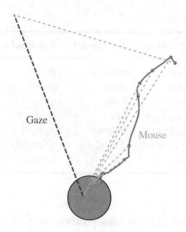

Figure 8.10 Gaze (dashed) and mouse movement (solid) for the classic mouse task without background. The dotted grey lines connect points of same time.

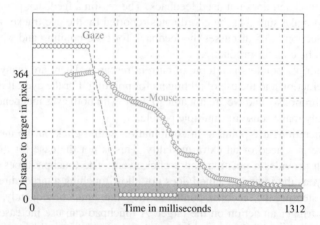

Figure 8.11 Gaze (dashed) and mouse movement (solid) for the classic mouse task without background as a plot of distance to target over time.

The fact that the mouse pointer moves directly towards the target means that the user was aware of the mouse pointer position before starting the movements. Conducting the click-the-target task on a complex background destroys the possibility of pre-attentive perception, and the user is not aware of the mouse pointer position. Typically, people start to stir the mouse to detect the mouse pointer by its motion. Figures 8.12 and 8.13 show the situation.

Figures 8.11 and 8.13 show that eye and hand have about the same reaction time, but the gaze arrives at the target much earlier. Therefore, gaze pointing is definitely faster than mouse pointing. With mouse pointing, it can happen that we are not aware of where the mouse pointer is and have to find it first, something that never happens for the gaze. Pointing with the eye and pressing a gaze key is the fastest pointing interaction known, and typically needs about 600 milliseconds, 300 milliseconds reaction time, 100 milliseconds to move the gaze to the target and 200 millisecond to press down the key.

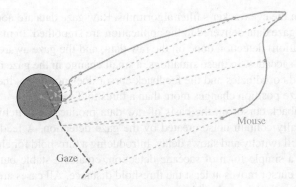

Figure 8.12   Gaze (dashed) and mouse trail (solid) for the classic mouse task with background. In the beginning, the user stirs the mouse to detect its position by movement.

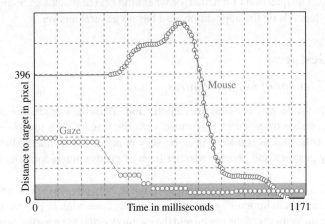

Figure 8.13   Gaze (dashed) and mouse trail (solid) for the classic mouse task on a complex background as a plot of distance to target over time.

8.6.5   Gaze Pointing Feedback

An interesting question connected with gaze pointing is the provision of feedback. Of course, the user knows where she or he is looking, but not with one-pixel accuracy and, additionally, there may be calibration errors, causing the gaze position to differ from the position reported by the eye tracker. Providing a gaze feedback cursor could result in chasing the cursor across the screen – or, as Jacobs expressed it:

> *"If there is any systematic calibration error, the cursor will be slightly offset from where the user is actually looking, causing the user's eye to be drawn to the cursor, which will further displace the cursor, creating a positive feedback loop."* [23]

However, such chasing of a gaze feedback cursor typically does not occur. It seems that the eyes do not care whether the cursor is exactly in the center of the area of clear vision and, therefore, the eyes are not drawn to the cursor. A further reason for the absence of the phenomenon lies in the eye tracker's filter algorithms. Raw gaze data are normally very noisy and, therefore, the gaze data delivered to the application are smoothed. In many cases, there is a saccade (and fixation) detection done on the raw data, and the gaze aware application only gets notified about saccades. In these situations, a small change in the gaze direction does not change the reported coordinates and the feedback cursor does not move; the feedback cursor will move if the gaze position changes more than a threshold value.

Providing a feedback cursor on the basis of raw data produces a twitchy cursor, because the raw data typically contain noise created by the gaze detection. A feedback cursor with smoothed data is still twitchy and shows delay. Introducing a threshold for changes in the gaze position, which is a simple form of saccade detection, creates a stable but jumpy feedback cursor because the cursor moves at least the threshold distance. All cases are more disturbing than helpful, and therefore there should be no gaze cursor. This does not mean that no feedback is necessary. Typically, systems which use gaze pointing highlight the object where the gaze is focused. If the system uses the dwell time method, it is a good idea to provide feedback on the time elapsed. The question of whether, or what kind of, feedback a gaze-aware application should provide depends on the application, and has no general answer.

8.7   Gaze Gestures

8.7.1   The Concept of Gaze Gestures

Gestures are a possible way for computer interaction. With the introduction of smartphones which are controllable by gestures performed with the fingers on a touch-sensitive display, this interaction concept has suddenly become very popular. 3D scanners allow the detection of hand or body gestures and provide another form of gesture interaction, mostly used in the gaming domain.

Of course, the idea of gestures performed with the eyes is close to being achieved. We certainly use eye gestures in human-human interaction (e.g., we wink or roll the eyes). Such eye gestures include movements of the eyelid and the eyebrows, and are a part of facial expressions. The gestures presented here are restricted to the movement of the eyeball or gaze direction, respectively. This kind of gestures is detectable in the data delivered by an eye tracker.

In 2000, Isokoski suggested the use of off-screen targets for text input [38]. The eye gaze has to visit off-screen targets in a certain order to enter characters. Although Isokoski did not use the term "gesture", the resulting eye movements are gaze gestures. However, off-screen targets force the gesture to be performed in a fixed location and in a fixed size. Such gestures still need a calibrated eye tracker.

Milekic used the term "gaze gesture" in 2003 [39]. Milekic outlined a conceptual framework for the development of a gaze-based interface for use in a museum context but, being from a Department of Art Education and Art Therapy, his approach is not strictly scientific – there is no algorithm given, and no user study was done.

In contrast to the gaze gestures of Isokoski, the gaze gestures presented by Wobbrock *et al.* [40] and Drewes and Schmidt [41] are scalable and can be performed in any location. The big advantage of such gestures is that they work even without eye tracker calibration.

8.7.2   Gesture Detection Algorithm

The popular mouse gesture plug-in for web browsers provided the inspiration for gaze gestures. This gesture plug-in traces the mouse movements and translates the movements into a string of characters or tokens representing strokes in eight directions. The eight directions are U, R, D, and L for up, right, down and left respectively, and 1, 3, 7, and 9 for the diagonal directions, according to the standard layout of the number pad on the keyboard. The mouse gesture detection algorithm receives x and y coordinates. Whenever one or both coordinates exceed a threshold distance (or grid size) from the start position, the algorithm outputs a character for the direction of the movement, but only when it is a character different from the last character. The current coordinates become the new start position, and the detection of a new stroke starts. The result is a translation of a stream of coordinates into a stream of characters (see Figure 8.14). A string of characters describes a gesture, and the algorithm signals the occurrence of a gesture when it finds the gesture string in the stream of characters.

The mouse gesture algorithm also works for gaze gestures. Interestingly, the gesture detection is better suited for gaze than to mouse movements. First, the natural motor space for hand movements is curved, while the saccadic movements of the eyes are straight lines. Second, the mouse trail is a continuous series of coordinates, and it is very unlikely that both coordinates cross the threshold in the same moment, which means diagonal strokes are difficult to detect. For detection of diagonal movements, it is good to have only the start- and end-points of a movement. Saccade detection on gaze movements delivers exactly this.

8.7.3   Human Ability to Perform Gaze Gestures

The most important question for gaze gestures as an active interaction method is whether people are able to perform them. Gaze gestures are definitely not very intuitive. Not all persons who try to perform a gaze gesture are successful immediately. Telling someone to move their gaze along a line confuses most subjects, while asking them to visit points with their gaze in a certain order brings better results. Therefore, it is a good idea to provide some supporting points (and not lines) for the users. The four corners of a display are good points for the gaze to perform a gaze gesture. Alternatively, the corners of a dialog window also work well.

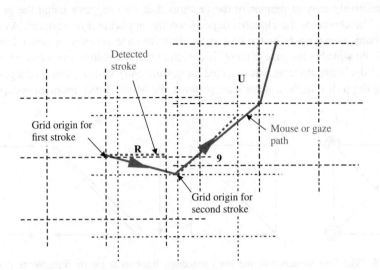

Figure 8.14   The figure shows how a mouse or gaze path translates into the string R9U. The end of a detected stroke is the origin for the grid to detect the next stroke.

The time to perform a gesture consists of the time for the strokes and the time for the stops, which are called fixation. Figure 8.4 shows saccade times over angle, which can be converted to the time for a stroke of certain length. From Figure 8.4, we learn that long saccades last around 100–150 milliseconds, and that there is only a little dependency on the length of a saccade. Therefore, the time to perform a gaze gesture does not depend very much on the size of the gesture, unless the gesture is very small in size. If n is the number of strokes, S the time for a saccade, and F the time for a fixation, the total time for a gesture T is:

$$T = nS + (n - 1)F$$

Theoretically, the fixation time could be zero and the minimum time to perform a gesture could be 120 milliseconds times the number of strokes in the gesture. Practically, untrained users in particular need some hundred milliseconds for the fixation.

8.7.4 Gaze Gesture Alphabets

The four corners of the display match perfectly with squared gestures, also called EdgeWrite gestures [42]. The EdgeWrite gestures use the order in which four points – the corners of a square – are reached (see Figure 8.15).

It is easy to describe all possible squared gestures with the notation introduced for gesture detection. The token string LD9DL, for example, represents the zero gesture in Figure 8.15. The token string URUR, on the other hand, is not a squared gesture and, therefore, the squared gestures are only a subset of the mouse gestures. Nevertheless, the squared gestures have the capability to define a large alphabet as shown by Wobbrock *et al.* [42], who assigned at least one gesture to each of the letters and digits in the Latin alphabet. The squared gestures seem to provide a suitable alphabet to start with. Figure 8.16 shows three examples of squared closed gaze gestures with four and six strokes.

The EdgeWrite alphabet uses gestures which resemble Latin characters and digits. This makes it relatively easy to memorize the gestures, but also suggests using the gestures for text entry. The choice of the alphabet depends on the application of gestures. As discussed later, text entry seems not to be the best application for gaze gestures. A general problem of the EdgeWrite alphabet for gaze gestures is detecting when a gesture starts and when it is finished. The EdgeWrite alphabet was invented for gesture entry with a stylus, and a gesture ends with lifting the stylus. In the case of gaze gestures, the Midas Touch problem occurs.

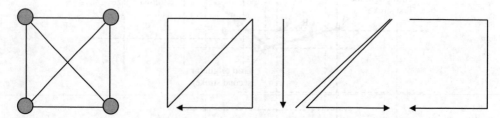

Figure 8.15 The four corners and the six connecting lines used for the EdgeWrite gestures and examples for EdgeWrite gestures (digits 0, 1, 2 and 3).

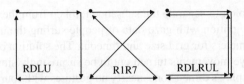

Figure 8.16   Three examples of squared closed gaze gestures with four and six strokes.

One possible application of gaze gestures could be remote control for a TV set. For such an application, there should be gaze gestures to switch channels up and down or to increase and decrease the volume. It is very likely that repetitions of the same gesture occur in this interaction (e.g., switching the channel up three times). In this situation, it is comfortable to use closed gestures where the end position of the gesture is also the starting point of the gesture. Otherwise, the gaze has to move from the end point of the gesture to the start point for the repetition, which means the entry of an extra stroke, which also creates the danger that the gesture detection recognizes another gesture.

8.7.5   Gesture Separation from Natural Eye Movement

The mouse gesture algorithm needs a gesture key, typically the right mouse key, to make the system detect gestures, otherwise the mouse movements for normal operations would conflict with gesture detection. Of course, it would be possible to use the same mechanism for gaze gestures, but it would make gaze gestures nearly useless. If a key has to be pressed while performing a gaze gesture invoking a command, a key could be pressed to trigger the action without the effort of a gaze gesture. Perhaps in very special situations (e.g., a mobile context with only a single key available), a gesture key makes sense but, in general, a gesture key destroys the benefits of gaze interaction. Therefore it is necessary to separate gaze gestures from natural eye movements. However, this seems not to be an easy task.

One possibility is to detect gestures only in a certain context. For example, if there is a gaze gesture to close a dialog box, gesture detection starts when the dialog appears and ends when the dialog closes. This means that gesture detection only takes place in the context of the open dialog, and typically only for a short period of time. The probability of an unintended gaze gesture from natural eye movements is small if the gesture detection is only active for a short time.

Another possibility to achieve a separation of intentional gaze gestures from natural eye movements lies in the selection of suitable gestures. The more strokes a gesture has, the less likely it is that it will occur in natural eye movements by chance. However, with an increasing number of strokes, the gesture takes more time to perform and is harder to memorize. An analysis of natural eye movements reveals that some gestures appear more frequently than other gestures. The eye movements of people sitting in front of a computer monitor contain many RLRLRL gestures which are generated by reading. The occurrence frequency of a gesture depends on the activity; many people generate many DUDUDU gestures when typing because they look down to the keyboard and back to the display.

Drewes *et al.* introduced a ninth token, the colon, to indicate a timeout condition. The idea behind a timeout is that a gesture should be performed within a short time. The detection algorithm does not report any token if the gaze stays within a grid cell. In this case, the modified algorithm produces a colon to separate the subsequent token from the previous. Drewes [43] tested gaze gesture recognition with gaze data of people surfing the internet and watching a video, varying the parameters for grid size and timeout. The solution of how to separate gaze gestures from natural eye movements turned out to be surprisingly simple. When using a big grid size close to the display's dimensions, nearly no gestures with four or more strokes occur. Long saccades across the screen seldom occur in natural eye movements, and four or more subsequent long saccades nearly never occur.

The timeout parameter is not critical in the value but crucial. With a big grid size, the gaze typically lingers for a long time inside a cell before moving into another cell. Without colons, it is possible that the algorithm detects a gesture performed over a long time, which is was not intended by the user. As long as the timeout interval is a bit smaller than the average stay of the gaze in a cell, the separation works. Further decreasing of the timeout does not improve the separation. Table 8.2 shows recorded eye movements translated into gesture strings using different values for the parameters.

8.7.6   Applications for Gaze Gestures

Text entry by gaze gestures is possible, but it is questionable whether it makes sense. The time to enter a character by a gaze gesture takes one to two seconds per character for an inexperienced user. Even a well-trained user may have problems to beat the standard eye-typing with dwell time, which typically needs 5000 milliseconds per character. Additionally, performing gaze gestures is less intuitive than looking at a key, so there is no reason to assume that users will prefer the gaze gesture input method.

Table 8.2   Translating the same eye movements (surfing the internet) into a gesture string using different values for the grid size s and timeout t

Parameters	Resulting Gesture String
$s = 80$ $t = 1000$	:3LUD::7R1L9:73LR:73LR:7379RL:U:D:U3:LU::RL::R13U::LR:R:73:73D:73: LRLRLR7373DU7LD:RUL13L:R:RL:RL:LRL:R7L3L9R1UR3DR::7: URLRLRLRLRDLR7U:R3RL:LR
$s = 80$ $t = 700$	:3LU:D::7R:1L9:7:3LR::73L:R::737:9:RL:U:D:U3:LU::RL::R13U::LR:R:73:73D:73: LR:LRLR7373DU7LD:RUL1:3L:R:RL:RL:LRL::R:7L:3:L9R:1UR3DR::7: URLRL:RLRDLR7U
$s = 250$ $t = 1000$	:3L:::7R:1U:U3::7RL:UDU:L:::::::::RD:R:::::73::73:::::L::L:::R::L:R:L:::R7:R:RL: DR:L:::::U:R:R::LR:L:RL:UD:R:::L::LR:L:3:L:::::::::D:RL:RLRL::DRL:: RLRLRLRLR:LRLRL:RLR:7:::
$s = 250$ $t = 700$	:3L:::::7R:1::U:U:3:::7RL:::UDU::L:::::::::RD::R:::::::73::73:::::L:::L:::::R::L:R::L:::::R7:: R::R:L:D:R:L:::::::::U:R:R:::LR:::L::L:RL::UD:R:::::L::LR:L::3::L:::::::::D::RL::RL: RL:::D:RL::RL:RL
$s = 400$ $t = 1000$	:3::::L:D::U3:::::U::L:::::::RD:::::73::73::::::L:::::RL:::::::U::RL:R:::::::7:R:::::::L:R::7: R::L::LR:::::L::R::::::RL::::::::::::::::::::::::::::::::::::R::L:RL:::::::RL:R::L:::RL:RL: RL::RL:RL:R::L::::RL:

One of the big advantages of gaze gestures is that they work without calibration. Therefore, an obvious idea is the use a gaze gesture to invoke the calibration procedure on systems for the disabled.

Gaze gestures provide an interface for instant use by different people, because of the absence of a calibration procedure. Additionally, gaze gestures work without touching anything and therefore could be used in highly hygienic environments, such as surgery rooms or labs. Gaze gestures provide more complex control than eye contact sensors or other touch-free techniques, such as capacity sensors or photoelectric barriers.

The question of whether gaze gestures could serve as a remote control for the TV set and become an input technology for the masses, and not only for special purposes, is interesting and still open. The big advantage of remote control by gaze gestures is the absence of a control device – nothing to mislay and no batteries to recharge. However, some TV manufacturers now sell TV sets which can be controlled by gestures performed with the hands. Hand gestures are more intuitive and bring the same benefit for the user.

Another field of applications for gaze gestures is mobile computing. In the mobile context, it is desirable to have the hands free for other tasks. Gaze interfaces can facilitate this. Gaze pointing requires objects to look at on the display and, therefore, portions of an augmented reality display would be needed for this purpose. However, such interaction objects obscure parts of the vision. Gaze gestures do not need any interaction objects and save space on the display. In the context of an augmented reality display with a graphical user interface, it is imaginable to close a dialog window by looking clockwise at the corners of this window. A RDLU gesture would close the dialog in the same way as clicking the OK button with the mouse. Looking at the corners counterclockwise could close the dialog with a NO and, if needed, a crosswise gesture like 3U1U could have the meaning of a cancel operation.

Bulling *et al.* showed that gaze gestures work for mobile applications [13]. They used eye tracking by electro-occulography for their research, and this shows that the concept of gaze gestures does not depend on the eye-tracking technology used.

8.8   Gaze as Context

Instead of using the gaze as an active input from the user who intentionally invokes commands, it is also possible to treat gaze data as context information or to utilize the user's eye gaze for implicit human computer interaction. The computer uses the information from the eye tracker to analyze the situation and activity of the user, and adapts its behavior according to the user's current state. The idea to take the situation of the user into account dates back to 1997 [44]. Since that time, consideration of the user's environment and situation has been a topic of HCI research called "context awareness".

8.8.1   Activity Recognition

A very important context for computer interaction is the current activity of the user. Therefore, it is an interesting question whether it is possible to estimate the user's activity, based on her or his eye movements.

Tables 8.3 and 8.4 show gaze statistics from people watching a video and surfing the internet, such as the average number of saccades performed per second, the average saccade time and length and the average fixation time.

Table 8.3 Mean values of gaze activity parameters for all participants (watching video)

Video	Sac./sec	Pixel/Sac.	Av. Sac. Time	Av. Fix. Time	Total Time
unit	1/sec	pixels	ms	ms	sec
P1	3.68	101.9	67.8	204.0	216.9
P2	3.43	87.2	69.3	221.9	219.2
P3	3.45	94.1	71.7	213.9	231.3
P4	2.74	107.5	76.3	282.0	228.3
P5	3.24	131.2	73.5	225.6	216.6
P6	2.79	134.1	99.1	259.3	225.1
P7	3.49	102.0	73.9	212.5	217.4
P8	2.76	111.3	91.1	270.6	219.9
Mean	**3.20**	**108.7**	**77.8**	**236.2**	**221.8**
Std. dev.	0.38	16.6	11.2	29.8	
Std. dev./ mean	11.8%	15.3%	14.4%	12.6%	

Table 8.4 Mean values of gaze activity parameters for all participants (surfing the internet)

Internet	Sac./sec	Pixel/Sac.	Av. Sac. Time	Av. Fix. Time	Total Time
unit	1/sec	Pixel	ms	ms	sec
P1	5.36	73.0	44.9	141.5	234.0
P2	5.10	73.2	43.9	137.2	229.4
P3	4.54	109.4	70.3	150.0	228.9
P4	5.51	69.0	41.5	140.0	229.8
P5	4.58	105.9	70.2	145.6	291.1
P6	4.59	106.7	54.7	156.0	264.0
P7	4.78	108.0	66.0	143.3	238.2
P8	3.17	123.3	104.5	177.1	374.7
Mean	**4.70**	**96.1**	**62.0**	**148.8**	**261.3**
Std. dev.	0.72	20.9	20.9	12.9	
Std. dev./ mean	15.4%	21.8%	33.7%	8.6%	

It is in the nature of statistics that measured values differ and, therefore, we have to answer the question whether different mean values in both tables are by chance or differ significantly. A standard method to answer this question is a t-test. The t-test offers the probability that the difference in the mean value are by chance. Table 8.5 shows the values for a paired student's t-test comparing both tasks. The values that differ significantly are the saccades per second and the average fixation time.

The results of the statistics are good news for attempts to use eye-gaze activity for context awareness. The strong significance proves that a gaze-aware system can guess the user's activity quite well. The low individual differences in the gaze activity parameters justify the hope that such activity recognition will work with universal threshold values and does not need an adaptation for individual users.

Table 8.5 t-test for "watching video" versus "surfing the internet"

	Sac./sec	Pixel/Sac.	Av. Sac. Time	Av. Fix. Time
t-test	0.00040	0.13136	0.05318	0.00003

At first it seems to be a little bit paradoxical that people perform fewer eye movements when watching a video full of action than surfing on static pages in the internet. The reason for that lies in the process of reading, which consists of saccades with short fixations in between. When reading, the eye moves as fast as possible but, when watching a movie, the eye waits for something to happen.

Most activities of humans involve eye movements related to the activity. Land [45] describes eye movements for daily life activities such as reading, typing, looking at pictures, drawing, driving, playing table tennis, and making tea. Retrieving context information means to go the other way round and conclude from eye movements to the activity. Using such an approach, Iqbal and Bailey measured eye-gaze patterns to identify user tasks [46]. Their aim was to develop an attention manager to mitigate disruptive effects in the user's task sequence by identifying the mental workload. The research showed that each task – reading, searching, object manipulation, and mathematical reasoning – has a unique signature of eye movement.

The approach to activity recognition with mean values is simple and more sophisticated math, which is beyond the scope of this text, can retrieve more information from gaze data. Bulling *et al.* described a recognition methodology that combines minimum redundancy maximum relevance (mRMR) feature selection with a support vector machine (SVM) classifier [47]. They tested their system with six different activities: copying a text, reading a printed paper, taking handwritten notes, watching a video, browsing the web, and no specific activity. They reported a detection precision of around 70%.

The approach of activity recognition is interesting, but the general problem is that the analysis of the eye movements can tell what the user's task was in the past, but not predict what the user is going to do. The period providing the data needed for the analysis causes latency. It is also not clear how reliable the task identification works and, finally, there is no concept for the social intelligence needed to make the right decisions.

8.8.2 Reading Detection

Reading is a very special activity which we do frequently, especially in the context of interacting with computer devices. Eye movements while reading were already a topic of research in the 19th century. Javal (1879) and Lamare (1892) observed the eye movements during reading and introduced the French originated word *saccade* for the abrupt movements of the eye. Psychologists intensively researched the reading process, which is understood in detail [48].

In many situations, especially when surfing the internet, people tend to read not very carefully and often do not read the text completely. Jakob Nielsen did a big survey with several hundred participants on reading behavior when looking at web pages. He visualized the eye gaze activity with heatmaps and found out that most heatmaps have an F-shape. This means that users tend to read only the first lines. As a consequence, web pages should state important things in the first lines.

Gaze analysis for reading is inspiring when it comes to finding design rules for interaction. However, if done in real-time, it can bring further benefits to the user. For example, if the system knows that the user is reading, it can stop distracting animations on the display and postpone disturbing notifications. There are several suggestions of algorithms for reading detection in the literature [49–51].

Reading detection, in general, is not really difficult. A series of forward saccades and a subsequent backward saccade is a strong indicator for reading activity. The gesture recognition algorithm introduced in the previous chapter outputs RLRLRL gestures when somebody is reading. The problems of reading detection are latency and reliability. The reading detector needs some saccades as input to know the user is reading, and therefore cannot detect that the user has just started reading. Therefore, reading detection has problems with detecting the reading of a short, single line of text. Reading detection can also signal reading activity although the user is doing something else, for example looking at people's heads on a group picture, which has a similar gaze pattern to reading.

Detecting reading as an activity is helpful, but it helps even more if the system also knows what was read. For this, it is worth to have a closer look at the gaze while reading. Figure 8.17 shows the gaze path while reading a text.

It is easy to see that the gaze moves in saccades with brief fixations. Revisits can occur, especially in difficult texts. An analysis of the data give saccade lengths of around 1° within a line, and backward saccades which are slightly smaller than the length of the line. The angle of the forward saccades is about the same as the angle of the fovea, which makes sense as this means that the coverage of the line is optimal – no gaps and no overlaps. It is interesting to note that, in consequence, the first fixation points in a line are about half a degree behind the beginning of the line and the last fixation is the same distance away from the end of the line. In the vertical direction, the low accuracy causes the problem that it is not possible to detect reliably which line the gaze is reading if the line height is within a normal range. The line height in Figure 8.17 is about 0.5°.

Figure 8.17 Gaze path reading a text.

In many professions, it is necessary to read many and long documents. A smart system could track how well documents were read and offer this information to the user. This raises the question of how to convert gaze activity into a number which tells how well a document was read.

One idea presented in [52] is to lay virtual cells over the text which cover the text completely. An algorithm can count the fixations in each cell. The total number of fixations on this document provides useful information, but it does not tell whether the document was read completely or whether one-third of the document was read three times. Therefore, a single number cannot provide a good indicator of how well a document was read. A second value is needed which indicates how the fixations are distributed over the text.

One possible definition for a second value is the variance of fixations in the cells. A low variance indicates that the fixations were equally distributed over the text. Another possibility would be the percentage of text cells hit by the gaze. This would be an easy understandable value which tells whether the document was read completely, but does not provide the information needed to tell whether the document was read several times.

The values for the quality of reading are helpful for queries finding unread documents. The document itself can provide feedback based on the gaze statistics. For example the document can display text which was already read on a different background color.

Reading detection is an example of eye-gaze context information which is not trivial, but also not too vague. When looking at the elaborate models on reading from the field of psychology [48], it becomes clear that there is further potential in reading detection. From the speed of reading, the number of backward saccades and the time needed to read difficult words, it should be possible to get information on the reading skills and formation of the reader. Just by displaying text in different languages or scripts, it is possible to find out which languages the user can read. An online bookshop could use such information for book recommendations in the future. There is definitely much space left for further research.

8.8.3 Attention Detection

The most obvious way to utilize the eye gaze as context information is the interpretation of the eye gaze as an indicator of attention. In most situations, we look at the object that we pay attention to. This may sound trivial, but attention is a very powerful context information which can provide real benefit for a user. A display of an electronic device, like a desktop computer or a mobile phone, provides information for the user. However, there is no need to display the information if the user does not look at it. Nowadays, systems which are normally not aware of the user's attention may display an important message while the user is not present and, when the user returns, the information may be overwritten by the next message. An attention-aware system, realized with an eye tracker, may give the user an overview on what has happened when the user's attention is back at the system.

Logging the user's attention also provides reliable statistics on how much time she or he spent on a particular document. The time the document was displayed is not reliable, as the user can open the document and leave the place to fetch a coffee. The statistics on how much time spent on which document can be very useful. In a working environment, where somebody works on several projects, it is necessary to calculate the costs of each project, which finally means to know how much time was spent on each project. Such statistics can also be useful for electronic learning.

8.8.4   Using Gaze Context

Eye gaze is very important for human communication. If we want good assistance from computers, the computers have to know where we are looking. However, the interpretation of gaze can be difficult, because even humans cannot always read wishes from the eyes of others. The correct interpretation of the way people look requires a kind of social intelligence, and it will not be easy to implement this on a computer.

In many cases, however, simple approaches can bring benefits to the users. A display could switch itself on if somebody is looking at it and switch off when nobody is looking any more. This would not only be convenient for the user, but could also save power, which is a limited resource for a mobile device. Another example is that messages displayed by the computer could disappear without a mouse click if the system recognizes that the message was read.

A further example is the Gazemark introduced by Kern et al. [54]. Gazemark is a substitute for the finger which we put at a location on a map or in a text, when we have to look somewhere else and intend to return our gaze to this location. Gazemark is a visual placeholder, highlighting the last position on the display where we looked at. Gazemark is helpful in situations with multiple monitor setups, or when we have to switch our attention between a display and a physical document. It may be even more helpful for user interfaces in cars, where our interaction with the navigation system can be interrupted by the traffic situation. Because the car moves, the content of the navigation system display may change during the interruption, and this means we need some time for a new orientation when looking at the display again. In this situation, the Gazemark has to move with the map and must not stay fixed on the display. As our attention should mainly focus on steering the car, it is a big advantage if we have a quicker orientation at the interface and shorten the time for interaction.

The examples above make clear that gaze interfaces may not only provide an alternative way of directing computers, but also have a big potential for new types of assistance systems.

8.9   Outlook

As mentioned in the introduction, interaction methods which use the eyes are desirable for many reasons. If we want to have computer devices with which we can interact in a similar way as we do with humans, especially for human-like robots, eye-tracking technology is even mandatory.

Looking at more than 20 years of eye-tracking research, it seems that the way how we expect to utilize the information which lies in our gaze changed over the years. While the early research mostly focused on operating graphical user interfaces with the eyes, the current focus seems to be gaze-aware applications.

Gaze pointing has some difficulties – the Midas Touch problem, the accuracy issue and the need for calibration. We do not need the speed benefit of eye interaction when working with a text processor or spreadsheet application, or operating an ATM. The few situations in which we appreciate speed are in playing shooter games or similar actions in a military context. Eye tracker hardware for the mass market will most probably arrive as an addition for a gaming console.

As soon as there are cheap eye trackers available, we can expect further applications beyond the gaming domain. However, it seems unlikely that we will operate graphical user interfaces mainly with the eye tracker as a full substitute for what we do with a mouse today.

It seems more likely that the eye tracker will give us additional functions which make the system smarter. An eye tracker can assist us by positioning the mouse pointer on multi-monitor setups; it can track and record our eye activity and tell us which documents or which part of a document we already read and how intense; animated figures in a digital encyclopedia may pause while we read text.

Interaction with mobile devices has increased drastically in recent years. There are two options for eye-tracking with a mobile device. One option is an eye tracker which resides inside a mobile device. The challenges of such eye trackers are compensation for hand movements and changing light conditions. The other option is a head-mounted eye tracker. A head-mounted eye tracker seems to be easier to realize, but is definitely obtrusive. However, it makes sense in combination with an augmented reality display in the form of glasses.

The fundamental problem of possible conflicts between vision and interface operations is especially severe in the domain of mobile computing. There could also be conflicts with our social protocols, which demand that we direct our gaze to the person's face to whom we are talking. Mobile computing has strong needs for controlling the device without hands. Voice commands are an option, but social protocols do not allow using the voice in many situations. Gaze control does not need the hands and is silent. Perhaps gaze gestures are useful for switching modes of an augmented reality display, such as switching the display on or off.

The step before the introduction of eye trackers as an interface technology for the masses seems to be the introduction of attention sensors. Attention sensors do not report the gaze's direction, but only whether someone is looking. Attention sensors are easier to build and do not need calibration. Some mobile phone manufacturers already have gaze-aware devices in the market which switch off the display for power saving when nobody is looking. Gaze-awareness is already present.

References

1. Fitts, P.M. (1954). The Information Capacity of the Human Motor System in Controlling the Amplitude of Movement. *Journal of Experimental Psychology* **47**, 381–391.
2. Ware, C., Mikaelian, H.H. (1987). An Evaluation of an Eye Tracker as a Device for Computer Input. *Proceedings of the SIGCHI/GI Conference on Human Factors in Computing Systems and Graphics interface* CHI '87, 183–188.
3. Miniotas, D. (2000). Application of Fitt's Law to Eye Gaze Interaction. *CHI '00 Extended Abstracts on Human Factors in Computing Systems*, 339–340.
4. Zhang, X., MacKenzie, I.S. (2007). Evaluating Eye Tracking with ISO 9241 – Part 9. *Proceedings of HCI International 2007*, 779–788.
5. Vertegaal, R.A. (2008). Fitt's Law comparison of eye tracking and manual input in the selection of visual targets. *Conference on Multimodal interfaces* IMCI '08, 241–248.
6. Carpenter, R.H.S. (1977). *Movement of the Eyes*. Pion, London.
7. Abrams, R., Meyer, D.E., Kornblum, S. (1989). Speed and accuracy of saccadic eye movements: characteristics of impulse variability in the occulomotor system. *Journal of Experimental Psychology: Human perception and performance* **15**, 529–543.
8. Tian, Y., Kanade, T., Cohn, J.F. (2000). Dual-State Parametric Eye Tracking. *Proceedings of the Fourth IEEE International Conference on Automatic Face and Gesture Recognition*, 110–115.
9. Ravyse, I., Sahli, H., Reinders MJT, Cornelis J. (2000). Eye Activity Detection and Recognition Using Morphological Scale-Space Decomposition. *Proceedings of the international Conference on Pattern Recognition* ICPR. IEEE Computer Society, **1**, 1080–1083.
10. von Romberg, G., Ohm, J. (1944). Ergebnisse der Spiegelnystagmographie. *GräfesArch Ophtalm* **146**, 388–402.
11. Robinson, D.A. (1963). A method of measuring eye movement using a scleral search coil in a magnetic field. *IEEE Trans Biomed Eng (BME)* **10**, 137–145.

12. Brown, M., Marmor, M., Zrenner, V., Brigell, E., Bach, M. (2006). ISCEV Standard for Clinical Electro-oculography (EOG). *Documenta Ophthalmologica* 2006 **113**(3), 205–212.

13. Bulling, A., Roggen, D., Tröster, G. (2008). It's in Your Eyes: Towards Context-Awareness and Mobile HCI Using Wearable EOG Goggles. *Proceedings of the 10th international Conference on Ubiquitous Computing, UbiComp '08*, **344**, 84–93.

14. Vertegaal, R., Mamuji, A., Sohn, C., Cheng, D. (2005). Media Eyepliances: Using Eye Tracking for Remote Control Focus Selection of Appliances. *CHI '05 Extended Abstracts on Human Factors in Computing Systems*, 1861–1864.

15. Kim, K.-N., Ramakrishna, R.S. (1999). Vision-based Eye-gaze Tracking for Human Computer Interface. *Proceedings of IEEE International Conference on Systems, Man and Cybernetics*, **2**, 324–329.

16. Dongheng, L., Winfield, D., Parkhurst, D.J. (2005). Starburst: A hybrid algorithm for video-based eye tracking combining feature-based and model-based approaches. *IEEE Computer Society Conference on Computer Vision and Pattern Recognition*, **3**, 79.

17. Nguyen, K., Wagner, C., Koons, D., Flickner, M. (2002). Differences in the infrared bright pupil response of human eyes. *Proc. Symposium on Eye Tracking Research & Applications (ETRA 2002)*, 133–138.

18. Ohno, T., Mukawa, N., Yoshikawa, A. (2002). FreeGaze: a gaze tracking system for everyday gaze interaction. *Proceedings of the symposium on ETRA 2002: eye tracking research & applications symposium*, 125–132.

19. Ohno, T., Mukawa, N. (2004). A Free-head, Simple Calibration, Gaze Tracking System That Enables Gaze-Based Interaction. *Proceedings of the symposium on ETRA 2004: eye tracking research & application symposium*, 115–122.

20. Hennessey, C., Noureddin, B., Lawrence, P. (2006). A single camera eye-gaze tracking system with free head motion. *Proceedings of the 2006 symposium on Eye tracking research & applications*, 87–94.

21. Zhai, S., Morimoto, C., Ihde, S. (1999). Manual And Gaze Input Cascaded (MAGIC) Pointing. *Proceedings of the SIGCHI Conference on Human Factors in Computing Systems CHI '99*, 246–253.

22. Morimoto, C.H., Koons, D., Amir, A., Flickner, M. (1999). Frame-Rate Pupil Detector and Gaze Tracker. *Proceedings of the IEEE ICCV'99 frame-rate workshop*.

23. Jacob, R.J.K. (1990). What You Look At is What You Get: Eye Movement-Based Interaction Techniques. *Proceedings of the SIGCHI Conference on Human Factors in Computing Systems CHI '90*, 11–18.

24. Majaranta, P., Räihä, K. (2002). Twenty Years of Eye Typing: Systems and Design Issues. *Proceedings of the 2002 Symposium on Eye Tracking Research & Applications ETRA '02*, 15–22.

25. Bolt, R.A. (1981). Gaze-orchestrated Dynamic Windows. *Proceedings of the 8th Annual Conference on Computer Graphics and interactive Techniques, SIGGRAPH '81*, 109–119.

26. Bolt, R.A. (1982). Eyes at the Interface. *Proceedings ACM Human Factors in Computer Systems Conference*, 360–362.

27. Card, S.K., Moran, T.P., Newell, A. (1983). *The Psychology of Human-Computer Interaction*. Lawrence Erlbaum.

28. Dickie, C., Vertegaal, R., Sohn, C., Cheng, D. (2005). eyeLook: using attention to facilitate mobile media consumption. *Proceedings of the 2005 ACM Symposium on User Interface Software and Technology 2005*, 103–106.

29. Drewes, H., Schmidt, A. (2009). The MAGIC Touch: Combining MAGIC-Pointing with a Touch-Sensitive Mouse. *Proceedings of Human-Computer Interaction – INTERACT 2009*, 415–428.

30. Salvucci, D.D., Anderson, J.R. (2000). Intelligent Gaze-Added Interfaces. *Proceedings of the SIGCHI Conference on Human Factors in Computing Systems CHI '00*, 273–280.

31. McGuffin, M., Balakrishnan, R. (2002). Acquisition of Expanding Targets. *Proceedings of the SIGCHI Conference on Human Factors in Computing Systems CHI '02*, 57–64.

32. Zhai, S., Conversy, S., Beaudouin-Lafon, M., Guiard, Y. (2003). Human On-line Response to Target Expansion. *Proceedings of the SIGCHI Conference on Human Factors in Computing Systems, CHI '03*, 177–184.

33. Miniotas, D., Špakov, O., MacKenzie, I.S. (2004). Eye Gaze Interaction with Expanding Targets. *CHI '04 Extended Abstracts on Human Factors in Computing Systems*, 1255–1258.

34. Špakov, O., Miniotas, D. (2005). Gaze-Based Selection of Standard-Size Menu Items. *Proceedings of the 7th international Conference on Multimodal interfaces, ICMI '05*, 124–128.

35. Ashmore, M., Duchowski, A.T., Shoemaker, G. (2005). Efficient eye pointing with a fisheye lens. *Proceedings of Graphics Interface 2005 (GI '05)*, 203–210.

36. Kumar, M., Paepcke, A., Winograd, T. (2007). EyePoint: Practical Pointing and Selection Using Gaze and Keyboard. *Proceedings of the SIGCHI Conference on Human Factors in Computing Systems CHI '07*, 421–430.

37. Miniotas, D., Špakov, O., Tugoy, I., MackKenzie, I.S. (2005). Extending the limits for gaze pointing through the use of speech. *Information Technology and Control* **34**, 225–230.

38. Isokoski, P. (2000). Text Input Methods for Eye Trackers Using Off-Screen Targets. *Proceedings of the 2000 Symposium on Eye Tracking Research & Applications, ETRA '00*, 15–21.

39. Milekic, S. (2003). The More You Look the More You Get: Intention-based Interface using Gaze-tracking. In: Bearman D, Trant J. (eds.). *Museums and the Web 2002*: Selected Papers from an International Conference. Archives & Museum Informatics, Pittsburgh, PA.

40. Wobbrock, J.O., Rubinstein, J., Sawyer, M., Duchowski, A.T. (2007). Not Typing but Writing: Eye-based Text Entry Using Letter-like Gestures. *Proceedings of COGAIN*, 61–64.

41. Drewes, H., Schmidt, A. (2007). Interacting with the Computer using Gaze Gestures. *Proceedings of Human-Computer Interaction – INTERACT 2007*, 475–488.

42. Wobbrock, J.O., Myers, B.A., Kembel, J.A. (2003). EdgeWrite: A Stylus-Based Text Entry Method Designed for High Accuracy and Stability of Motion. *Proceedings of the 16th Annual ACM Symposium on User interface Software and Technology, UIST '03*, 61–70.

43. Drewes, H. (2010). *Eye Gaze Tracking for Human Computer Interaction*. Dissertation an der Ludwig-Maximilians-Universität München.

44. Hull, R., Neaves, P., Bedford-Roberts, J. (1997). Towards Situated Computing. *Tech Reports*: HPL-97-66, HP Labs Bristol.

45. Land, M.F. (2006). Eye movements and the control of actions in everyday life. *Prog Retinal & Eye Res* **25**, 296–324.

46. Iqbal, B., Bailey, P. (2004). Using Eye Gaze Patterns to Identify User Tasks. *Proceedings of The Grace Hopper Celebration of Women in Computing*.

47. Bulling, A., Ward, J.A., Gellersen, H., Tröster, G. (2009). Eye movement analysis for activity recognition. *Proceedings of the 11th international conference on Ubiquitous computing*, 41–50.

48. Reichle, E.D., Pollatsek, A., Fisher, D.L., Rayner, K. (1998). Toward a Model of Eye Movement Control in Reading. *Psychological Review* **105**, 125–157.

49. Campbell, C.S., Maglio, P.P. (2001). A Robust Algorithm for Reading Detection. *Proceedings of the 2001 Workshop on Perceptive User interfaces, PUI '01*, **15**, 1–7.

50. Keat, F.-T., Ranganath, S., Venkatesh, Y.V. (2003). *Eye Gaze Based Reading Detection. Conference on Convergent Technologies for Asia-Pacific Region, TENCON '03*, **2**, 825–828.

51. Bulling, A., Ward, J.A., Gellersen, H., Tröster, G. (2008). Robust Recognition of Reading Activity in Transit Using Wearable Electrooculography. *Proc. of the 6th International Conference on Pervasive Computing (Pervasive 2008)*, 19–37.

52. Drewes, H., Atterer, R., Schmidt, A. (2007). Detailed Monitoring of User's Gaze and Interaction to Improve Future E-Learning. *Proceedings of the 12th International Conference on Human-Computer Interaction HCII '07*, 802–811.

53. Kern, D., Marshall, P., Schmidt, A. (2010). Gazemarks – Gaze-Based Visual Placeholders to Ease Attention Switching. *Proceedings of the 28th ACM Conference on Human Factors in Computing Systems (CHI '10)*. ACM, Atlanta (GA), USA.

54. Wobbrock, J.O., Rubinstein, J., Sawyer, M.W., Duchowski, A.T. (2008). Longitudinal Evaluation of Discrete Consecutive Gaze Gestures for Text Entry. *Proceedings of the 2008 Symposium on Eye Tracking Research & Applications, ETRA '08*, 11–18.

9

Multimodal Input for Perceptual User Interfaces

Joseph J. LaViola Jr., Sarah Buchanan and Corey Pittman
University of Central Florida, Orlando, Florida

9.1 Introduction

Ever since Bolt's seminal paper, "Put that there: Voice and Gesture at the Graphics Interface", the notion that multiple modes of input could be used to interact with computer applications has been an active area of human computer interaction research [1]. This combination of different forms of input (e.g., speech, gesture, touch, eye gaze) is known as multimodal interaction, and its goal is to support natural user experiences by providing the user with choices in how they can interact with a computer. These choices can help to simplify the interface, to provide more robust input when recognition technology is used, and to support more realistic interaction scenarios, because the interface can be more finely tuned to the human communication system. More formally, multimodal interfaces process two or more input modes in a coordinated manner which aim to recognize natural forms of human language and behavior and, typically, incorporate more than one recognition-based technology [2].

With the advent of more powerful perceptual computing technologies, multimodal interfaces that can passively sense what the user is doing are becoming more prominent. These interfaces, also called perceptual user interfaces [3], provide mechanisms that support unobtrusive interaction where sensors are placed in the physical environment and not on the user. Previous chapters in this book have focused on various input technologies and associated interaction modalities. In this chapter, we will examine how these different technologies and their input modalities – specifically speech, gesture, touch, eye gaze, facial expressions, and brain input – can be combined, and the types of interactions they afford. We will also examine the strategies for combining these input modes together, otherwise known as multimodal integration or fusion. Finally, we will examine some usability issues with multimodal interfaces and methods for handling them. Research in multimodal interfaces spans many fields, including psychology, cognitive science, software engineering, and human computer interaction [4].

Interactive Displays: Natural Human-Interface Technologies, First Edition. Edited by Achintya K. Bhowmik.
© 2015 John Wiley & Sons, Ltd. Published 2015 by John Wiley & Sons, Ltd.

Our focus in this chapter will be on the types of interfaces that have been created using multimodal input. More comprehensive surveys can be found in [5] and [6].

9.2 Multimodal Interaction Types

Multimodal interaction can be defined as the combination of multiple input modalities to provide the user with a richer set of interactions, compared to traditional unimodal interfaces. The combination of input modalities can be divided into six basic types: complementarity; redundancy; equivalence; specialization; concurrency; and transfer [7]. In this section, we briefly define each:

- **Complementarity**. Two or more input modalities complement each other when they combine to issue a single command. For example, to instantiate a virtual object, a user makes a pointing gesture and then speaks. Speech and gesture complement each other, since the gesture provides the information on where to place the object and the speech command provides the information on what type of object to place.
- **Redundancy**. Two or more input modalities are redundant when they simultaneously send information to the application. By having each modality issue the same command, redundant information can help resolve recognition errors and reinforce what operation the system needs to perform [8]. For example, a user issues a speech command to create a visualization tool, while also making a hand gesture which signifies the creation of that tool. When more than one input stream is provided, the system has a better chance of recognizing the user's intended action.
- **Equivalence**. Two or more input modalities are equivalent when the user has a choice of which modality to use. For example, the user can create a virtual object by either issuing a voice command or by picking the object from a virtual palette. The two modalities present equivalent interactions, in that the end result is the same. The user can choose which modality to use, based on preference (they simply like speech input over the virtual palette) or on circumvention (the speech recognition is not accurate enough, thus they move to the palette).
- **Specialization**. A particular modality is specialized when it is always used for a specific task because it is more appropriate and/or natural for that task. For example, a user wants to create and place an object in a virtual environment. For this particular task, it makes sense to have a "pointing" gesture determine the object's location, since the number of possible voice commands for placing the object is too large, and a voice command cannot achieve the specificity of the object placement task.
- **Concurrency**. Two or more input modalities are concurrent when they issue dissimilar commands that overlap in time. For example, a user is navigating by gesture through a virtual environment and, while doing so, uses voice commands to ask questions about objects in the environment. Concurrency enables the user to issue commands in parallel, reflecting such real-world tasks as talking on the phone while making dinner.
- **Transfer**. Two input modalities transfer information when one receives information from another and uses this information to complete a given task. One of the best examples of transfer in multimodal interaction is the push-to-talk interface [9]; the speech modality receives information from a hand gesture, telling it that speech should be activated.

9.3 Multimodal Interfaces

In this section, we examine how the different technologies and input modalities discussed in this book have been used as part of multimodal interaction systems. Note that, although speech input is a predominant modality in multimodal interfaces, we do not have a dedicated section for it in this chapter. Rather, uses of speech are found as part of each modality's subsection.

9.3.1 Touch Input

Multi-touch devices have become more prevalent in recent years with the growing popularity of multi-touch phones, tablets, laptops, table-top surfaces and displays. As a result, multitouch gestures are becoming part of users' everyday vocabulary, such as swipe to unlock or pinch to zoom. However, complex tasks, such as 3D modeling or image editing, can be difficult when using multi-touch input alone. Multimodal interaction techniques are being designed to incorporate multi-touch interfaces with other inputs, such as speech, to create more intuitive interactions for complex tasks.

9.3.1.1 3D Modeling and Design

Large multitouch displays and table-top surfaces are marketed as natural interfaces that foster collaboration. However, these products often target commercial customers in public settings, and are more of a novelty item. Thus, the question remains as to whether they provide utility as well as unique experiences. Since the mouse and keyboard are no longer available, speech can provide context to operations where WIMP paradigms were previously used, such as in complex engineering applications (e.g., Auto-CAD). For instance, MozArt [10] combines a tiltable multi-touch table with speech commands to provide an easier interface to create 3D models, as shown in Figure 9.1. A study was conducted, evaluating MozArt versus a multi-touch CAD program with novice users. The majority of users preferred the multimodal interface, although a larger number of users would need to be tested in order to evaluate efficiency and accuracy. Similar interfaces could be improved upon by using speech and touch, as stated in work on a multitouch-only interface for performing 3D CAD operations [11].

Figure 9.1 MozArt Table hardware prototype. Source: Sharma A, Madhvanath S, Shekhawat A and Billinghurst M 2011. Reproduced with permission.

9.3.1.2 Collaboration

Large multi-touch displays and tabletop surfaces are ideal for collaboration, since they have a 360-degree touch interface, a large display surface, and allow many input sources. For instance, Tse *et al.* (2008) [12] created a multimodal multi-touch system that lets users gesture and speak commands to control a design application called The Designer's Environment. The Designer's Environment is based on the KJ creativity method that industrial designers use for brainstorming. The four steps of the KJ creativity method are:

1. create notes;
2. group notes;
3. label groups; and
4. relate groups.

In The Designer's Environment, multiple users can complete these tasks by using a combination of touch, gesture, and speech input, as shown in Figure 9.3. However, there are some obstacles as explored by Tse *et al.* (2008) [12]: parallel work, mode switching, personal and group territories, and joint multimodal commands. Tse *et al.* proposed solutions to these issues, such as allowing for parallel work by creating personal work areas on the surface.

Tse *et al.* (2006) [13] also created the GSI Demo (Gesture and Speech Infrastructure created by Demonstration). This system demonstrates multimodal interaction by creating a multiuser speech and gesture input wrapper around existing mouse/keyboard applications. The GSI Demo can effectively convert a single-user desktop application to a multi-touch table application, such as maps, command and control simulations, simulation and training, and games. Tse *et al.* (2007) [14] specifically discuss how playing Blizzard's *Warcraft III* and *The Sims* can become a collaborative effort on a multi-touch tabletop system. Their proposed interface allows players to use gesture and speech input to create a new and engaging experience that is more similar to the social aspect provided by arcade gaming, shown in Figure 9.2.

Another interesting aspect of collaborative environments is how to track who did or said what during a collaborative effort. Collaboration data can then shed light on the learning or collaboration process. This type of data can also act as input to machine learning or data mining algorithms, providing adapted feedback or personalized content.

Figure 9.2 Two people interacting with Warcraft III (left) and The Sims game (right). Source: Tse E, Greenberg S, Shen C, Forlines C and Kodama R 2008. Reproduced with permission.

Figure 9.3 A two person grouping hand gesture in the Designer's Environment. Source: Tse E, Greenberg S, Shen C and Forlines C 2007. Reproduced with permission.

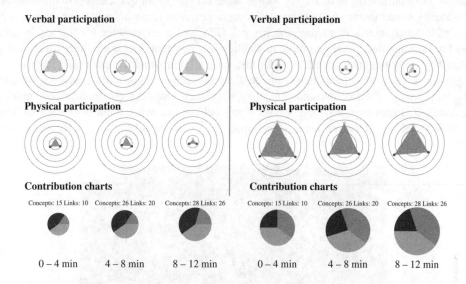

Figure 9.4 A collaborative visualization corresponding to 12 minutes of activity from a communicative group (left) and a less collaborative group (right). Source: Martínez-Maldonado R, Collins A, Kay J and Yacef K 2011. Reproduced with permission.

Collaid (Collaborative Learning Aid) is an environment that captures multimodal data about collaboration in a tabletop learning activity [15]. Data is collected using a microphone array and a depth sensor, integrated with other parts of the learning system, and finally transformed into visualizations showing the collaboration processes that occurred at the table. An example visualization of data collected from a collaborative group versus a less collaborative group is shown in Figure 9.4. Other work on multimodal collaboration using distributed whiteboards can be found in [16].

9.3.1.3   Communication with Disabled or Elderly Patients

Multimodal surface applications have also been shown to support medical communication with hearing impaired patients [17] and is being explored as way to enhance medical communication with the elderly by transcribing speech [18]. The Shared Speech Interface (SSI) is an application for an interactive multitouch tabletop display designed to facilitate medical communication between a deaf patient and a hearing, non-signing physician. The Deaf patient types on a keyboard, and the hearing doctor speaks into a headset microphone. As the two people communicate, their speech is transcribed and appears on the display in the form of movable speech bubbles, shown in Figure 9.5.

Multimodal surface technology can also benefit other populations with different communication needs. For example, a student learning a foreign language could access text-based representations of speech, along with audio clips of the instructor speaking the phrases.

9.3.1.4   Mobile Search

Mobile users are becoming more tech-savvy with their devices and expect to be able to use them while multitasking or on the go, such as when driving, or for quick access of information. In addition, contemporary mobile devices incorporate a wide range of input technologies, such as a multi-touch interface, microphone, camera, GPS, and accelerometer. Mobile applications need to leverage these alternate input modalities in a way that does not require users to stop what they are doing but, instead, provides fast, on-the-go input and output capabilities.

Mobile devices also face their own challenges, such as slower data transfer, smaller displays, smaller keyboards, and thus they have their own design implications, making desktop paradigms even less appropriate. Some believe that voice input is an easy solution to these problems. However, using voice input alone is not feasible, since voice recognition

Figure 9.5   A doctor (left) and patient (right) communicate using the Shared Speech Interface (SSI). Meanwhile, movable speech bubbles appear on the multi-touch surface. Source: Piper AM 2010. Reproduced with permission from Anne-Marie Piper.

is error-prone, especially in noisy environments, and does not provide fine-grained control. Many multimodal mobile interfaces are emerging that use voice input combined with other interactions in a clever way. For example, voice input can be used to bring context to the desired operation, while leaving fine-grained manipulation to direct touch manipulation. Or, voice input can be used at the same time as text entry, to ensure that text is entered correctly.

Voice input is ideal for mobile search, since it is quick and convenient. However, since voice input is prone to errors, error correction should be quick and convenient as well. The Search Vox system uses multimodal error correction for mobile search [19]. After a query is spoken, the user is presented with an N-best list of recognition results for a given query. The N-best results comprise a word palette that allows the user to conveniently rearrange and construct new queries based on the results, by using touch input.

Another subset of mobile search, local search, is desirable in mobile contexts, allowing searches to be constrained to the current location. Speak4it is a mobile app that takes voice search a step further by allowing users to sketch with their finger the exact area they wish to search in [20]. Speak4it supports multimodal input by allowing users to speak or type search criteria and sketch the desired area with touch input. An example scenario of Speak4it would be a biker searching the closest repair shop along a trail, by using speech and gesture to get a more refined search. For instance, the query by voice, "bike repair shops in the Stuyvesant Town", combined with a route drawn on the display, will return the results along the specified route traced on the display (Figure 9.6).

Research prototypes with these capabilities have existed for many years, such as QuickSet [21]. However, they have not been available to everyday users until the widespread adoption of mobile devices that come equipped with touchscreen input, speech recognition, and mobile internet access. Other work that has explored multimodal interaction for mobile search is the Tilt and Go system by Ramsay *et al.* (2010) [22]. A detailed analysis of speech and multimodal interaction in mobile search is presented in Feng *et al.* (2011) [23].

9.3.1.5 Mobile Text Entry

Typing on a touchscreen display using a soft keyboard remains the most common, but time consuming, text input method for many users. Two resolutions for faster text entry

Figure 9.6 Speak4it gesture inputs. Reproduced by permission of Patrick Ehlen.

are gesture keyboards and voice input. Gesture keyboards circumvent typing by allowing the user to swipe over the word path quickly on a familiar QWERTY keyboard; however, gestures can be ambiguous for prediction. Voice input offers an attractive alternative that completely eliminates the need for typing. However, voice input relies on automatic speech recognition technology, which has poor performance in noisy environments or for non-native users. Speak-As-You-Swipe (SAYS) [24] is a multimodal interface that integrates a gesture keyboard with speech recognition to improve the efficiency and accuracy of text entry, as shown in Figure 9.7. The swipe gesture and voice inputs provide complementary information for word prediction, allowing the SAYS system to extract useful cues intelligently from the ambient sound to improve the word prediction accuracy. In addition, SAYS builds on previous work [25] by enabling continuous and synchronous input.

A similar interface created by Shinoda *et al.* (2011) [26], allows semi-synchronous speech and pen input for mobile environments, as shown in Figure 9.8. There is an inherent time lag between speech and writing, making it difficult to apply conventional multimodal recognition algorithms. To handle this time-lag, they developed a multimodal recognition algorithm that used a segment-based unification scheme and a method of adapting to the time-lag characteristics of individual users. The interface also supports keyboard input and was tested in a variety of different manners:

1. the user writes the initial character of each phrase in a saying;
2. the user writes the initial stroke of the initial character as in (1);
3. the user inputs a pen touch to cue the beginning of each phrase;
4. the user taps the character table to which the first character of each phrase belongs to; and
5. the user inputs the initial character of each phrase using a QWERTY keyboard.

These five different pen-input interfaces were evaluated using noisy speech data, and the recognition accuracy of the system was higher than that of speech alone in all five interfaces.

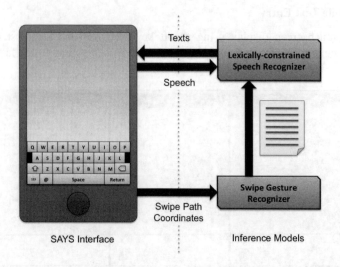

Figure 9.7 Speak As You Swipe (SAYS) interface. Source: Reproduced with permission of Khe Chai Sim.

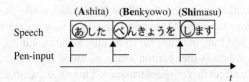

Figure 9.8   Relationship between speech and pen input. Source: Shinoda K, Watanabe Y, Iwata K, Liang Y, Nakagawa R and Furui S 2011. Reproduced with permission.

They also conducted a usability test for each interface, finding a trade-off between usability and improvement in recognition performance.

Other work that compares different multimodal interaction strategies using touch input for mobile text entry can be found in [27].

9.3.1.6   Mobile Image Editing

Another mobile app that incorporates multimodal inputs in novel way is PixelTone [28]. Pixel-Tone is a multimodal photo editing interface that combines speech and direct manipulation to make photo editing easier for novices and when using mobile devices. It uses natural language for expressing desired changes to an image, and direct manipulation to localize these changes to specific regions, as shown in Figure 9.9. PixelTone does more than just provide a convenient interface for editing photos. The interface allows for ambiguous commands that a novice user would make such as "Make it look good", as well as advanced commands such as "Sharpen the midtones at the top". Although users performed the same using the multimodal interface versus just a touch interface, they preferred the multimodal interface overall and were able to use it effectively for a realistic workload.

9.3.1.7   Automobile Control

Although a majority of US states ban text messaging for all drivers, there are still a variety of activities that drivers attempt to complete during their commute. Research is being conducted to help drivers be able to achieve the primary task of driving effectively, while still completing tertiary tasks such as navigating, communicating, adjusting music and controlling climate [29]. The Society of Automotive Engineers recommends that any tertiary task taking more than 15

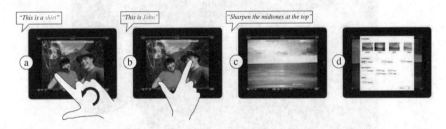

Figure 9.9   PixelTone. Source: Laput GP, Dontcheva M, Wilensky G, Chang W, Agarwala A, Linder J and Adar E 2013. Reproduced with permission.

seconds to carry out while stationary should be prohibited while the vehicle is in motion. Voice controls are exempt from the 15 second rule, since they do not require the user to take their eyes off the road, and they may appear to be an obvious solution.

However, some data suggests that certain kinds of voice interfaces impose high cognitive loads and can negatively affect driving performance. This negative affect is due to the technical limitations of the speech recognition, or to usability flaws such as confusing or inconsistent command sets and unnecessarily deep and complex dialog structures. Multimodal interfaces may be a way to address these problems, by combining the best modes of input depending on the driving situation.

Incorporating speech, touch, gesture, and haptic touchpad input have been explored separately as input to a driving interface. However, none of these inputs alone create an ideal solution. A multimodal interface created by Pfleging *et al.* (2012) [30], combines speech with gestures on the steering wheel to minimize driver distraction, shown in Figure 9.10. Pfleging *et al.* point out that speech input alone does not have fine-grained control, while touch input alone requires too much visual interaction, and gesture input alone does not scale well [31]. They propose a multimodal interaction style that combines speech and gesture, where voice commands select visible objects or functions (mirror, window, etc.) and simple touch gestures are used to control these functions. With this approach, recalling voice commands is simpler, as the users see what they need to say. By the use of a simple touch gesture, the interaction style lowers the visual demand and provides, at the same time, immediate feedback and easy means for undoing actions. Other work that looks at multimodal input for automobile control can be found in Gruenstein *et al.* (2009) [32].

9.3.2 3D Gesture

Devices like Microsoft's *Kinect* and depth cameras driven by Intel's *Perceptual Computing* SDK, such as Creative's Senz3D camera, have steadily become more widespread and have provided for new interaction techniques based on 3D gestures. Using a combination of depth

Figure 9.10 Multimodal automobile interface combining speech with gestures on the steering wheel. Source: Pfleging B, Kienast M, Schmidt A and Doring T 2011. Reproduced with permission.

cameras and standard RGB cameras, these devices allow for accurate skeletal tracking and gesture detection. Such 3D gestures can be used to accomplish a number of tasks in a more natural way than WIMP interfaces. In order to enrich the user's experiences, the gestures can be combined with different modalities, such as speech and face tracking. Interestingly enough, the *Kinect* and *Senz3D* camera have built in microphones, making them ideal for designing multimodal interfaces.

Another common technique for gesture detection is to use stereo cameras to detect gestures, while using machine learning and filtering to properly classify them. Prior to the development of these technologies, the use of 3D gesture in multimodal interfaces was limited to detecting deictic gestures for selection or similar tasks using simple cameras. Speech is one of the more common modalities to be combined with gesture, as modern 3D gestures tend to involve a large portion of the body and limit what other modalities could realistically be used simultaneously [5].

Numerous applications of the *Kinect* sensor have been realized, with a number of games developed that have used both speech and gesture developed by Microsoft's first party developers. Outside of the game industry, sensors like the *Kinect* have been used for simulation and as a means to interface with technology using more natural movements. Some work has been done in the fields of human-robot interaction (HRI) and medicine for the utility of hands-free, gesture-controlled applications.

9.3.2.1   Games and Simulation

Interacting with a simulation using body gestures is commonly used in virtual environments when combined with speech to allow for simultaneous inputs. Williamson *et al.* (2011) [33] developed a system combining the *Kinect*, speech commands, and a Sony Playstation Move controller to make a full-body training simulation for soldiers (Figure 9.11). The RealEdge prototype allows users to march through an environment using a marching gesture, and to make small movements of their virtual avatar by leaning. The user can also look around the

Figure 9.11   RealEdge prototype with *Kinect* and PS Move Controller. Source: Reproduced by permission of Brian Williamson.

environment using the Move controller, which is attached to a weapon-like apparatus. The user may also use voice commands to give orders to virtual characters.

One shortcoming of the depth camera-based gesture recognition devices currently available is that they generally require the user to be facing them to accurately track the skeleton of the user. The RealEdge Fusion System, an extension of the RealEdge prototype, allows for accurate 360 degree turning by adding multiple *Kinect*s around the user and fusing retrieved skeletal information at the data level to allow for robust tracking of the user within the range of the sensors regardless of the users orientation [34]. The skeletal tracking information is passed from a Kinect depth sensor through a client laptop to a server, where the data is then fused. The system only requires the addition of more *Kinect* sensors, laptops, and a head mounted display to give the user the proper view of their environment.

A large amount of research has been done emphasizing the addition of speech to 3D interfaces for segmentation and selection. Budhiraja *et al.* specify that a problem with deictic gesture alone is that large numbers of densely packed or occluded objects make selection difficult [35]. In order to solve this problem, the authors add speech as a modality to help specify defining attributes of the desired object, such as spatial location, location relative to other objects, or physical properties of the object. This allows for specific descriptions, such as "the blue one to the left", to be used to clarify what object the user aims to select. Physical attributes and locations must be clearly defined for proper identification.

There are instances where 3D gesture is not the primary method of control in an interface. The SpeeG input system is a gesture-based keyboard replacement that uses speech augmented with a 3D gesture interface [36]. The interface is based on the Dasher interface [37] with the addition of speech and hand gestures, which replace the mouse. The system uses intermediate speech recognition to allow the user to guide the software to the correct phrase using their hand.

Figure 9.12 shows an example of the virtual environment and the pointing gesture. Though the prototype did not allow for real-time input, due to problems with speech recognition latency, users found that SpeeG was "the most efficient interface" when compared to an onscreen keyboard controlled with a Microsoft Xbox 360 controller, speech control only, and Microsoft *Kinect* keyboard only.

3D gestures are not limited to full body movements. Bohus and Horvitz (2009) [38] developed a system to detect head pose, facial gestures, and a limited amount of natural language in a conversational environment. Head position tracking and gaze estimation were done using

Figure 9.12   SpeeG interface and example environment. Source: Reproduced by permission of Lode Hoste.

a basic wide-angle camera and commercially available software. A linear microphone was used to determine the sources of user voices. These modalities were fused and analyzed, and a suitable response was given to the users.

This multiparty system was used for observational studies of dialogues, in which the system asks quiz questions and waits for responses from the users. Upon receiving a response from the user, the system will vocally ask for confirmation from the users. Depending on their responses, the system will respond appropriately and either continue on, or ask for another answer. The system's behavior is based on a turn-taking conversational model, for which the system has four behaviors: Hold, Release, Take, and Null. An example of this system functioning can be seen in Figure 9.13.

Hrúz *et al.* (2011) [39] designed a system for multimodal gesture and speech recognition for communication between two disabled users – one hearing impaired and one vision impaired. The system recognizes signed language from one of the users, using a pre-trained sign recognizer. The recognizer uses a single camera for capture of the hand signs, so the user must wear dark clothes to create contrast between the hands and the background and to

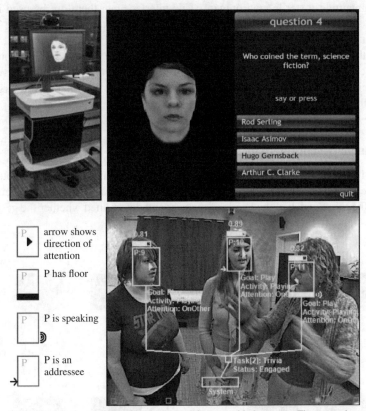

P_9, P_{10}, and P_{11} are all engaged in playing a trivia game with the system. The system is currently looking towards P_{11}, as shown by the red dot. The participants' focus of attention is directed towards each other. P_{11} has the floor and is currently speaking to P_9 and P_{10}.

Figure 9.13 Example of system for turn-taking model for multiparty dialogues. Source: Reproduced with permission from Microsoft Corporation.

allow for more accurate detection of the signs The signs are converted to text and then played for the other user using a language-specific text-to-speech system. The other user speaks to the system, which uses automatic speech recognition to translate the spoken words to text, and this is then displayed for the other user. Each one of the users represents a separate input modality for the system architecture, which is illustrated in Figure 9.14. The system allows for communication between two users who otherwise would be unable to find a medium to speak to one another.

9.3.2.2   Medical Applications

Gallo *et al.* (2011) [40] developed a system for navigating medical imaging data using a *Kinect*. The user is able to traverse the data as though it were a virtual environment, using hand gestures for operations like zooming, translating, rotating, and pointing. The user is also able to select and extract regions of interest from the environment. The system supports a number of commonly used imaging techniques, including CT, MRI, and PET images.

The benefit of having a 3D gesture interface in medicine is that there is no contact with any devices, maintaining a sterile environment. This interface can therefore be used in surgical environments to reference images without having to repeat a sterilization process. The sterile environment problem arises often in computerized mechanical systems. Typically, there is an assistant in the operating room who controls the terminals that contain information about the surgical procedure and images of the patient. These assistants typically do not have the same level of training as the surgeon in charge, and may misinterpret information the surgeon could have correctly deciphered.

9.3.2.3   Human-Robot Interaction

Perzanowski *et al.* (2001) [41] designed a method of communicating with robots in a way that was already natural for humans – using natural language and gestures. A stereo camera attached to the robot watched the hands of the user and determined whether the gestures were

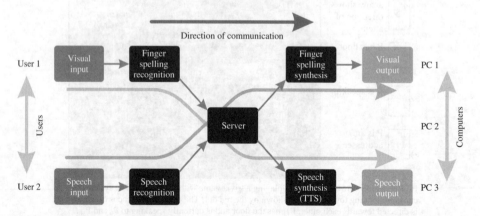

Figure 9.14   Flow of system for communication between two disabled users. Source: Reproduced with permission from Marek Hruz.

meaningful. Combining these gestures with the voice input from the user, the robot would carry out the desired commands from the user. The speech commands could be as simple as "Go over there" or "Go this far".

Along with the information the gesture gives, which could be something like a position in the environment from a pointing gesture or a distance from holding hands spread apart, the robot is able to make intelligent decisions on how far to move. The robot was also capable of being controlled using a PDA that mapped out the environment. The user could combine the PDA commands with speech commands or gesture commands or be used in place of speech and/or gesture. The tasks given to robots could be interrupted and the robot would eventually return to its original task to complete it. Similar interfaces that combine 3D gestures and speech for controlling robots include those described in [42] and [43].

9.3.2.4   Consumer Electronics

Multimodal input technologies have matured to the point where they are becoming interfaces to consumer electronics devices, particularly large screen displays and televisions. One commercial example is Samsung's Smart TV series, which includes 3D gestures, speech input, and facial recognition. In the research community, Lee *et al.* (2013) [44] combined 3D hand gestures and face recognition to interact with Smart TVs. Facial recognition was used as a viewer authentication system and the 3D hand gestures are used to control volume and change channels. Takahashi *et al.* (2013) [45] also explored 3D hand gestures coupled with face tracking using a ToF depth camera to assist users in TV viewing.

Krahnstoever *et al.* (2002) [46] constructed a similar system for large screen displays that combined 3D hand gestures, head tracking, speech input, and facial recognition. This system was deployed in a shopping mall to assist shoppers in finding appropriate stores. As the technology continues to improve and becomes smaller and less expensive, these types of multimodal interfaces will be integrated into more consumer electronic devices, such as PCs and laptops, as their main user interface.

9.3.3   Eye Tracking and Gaze

The ability to track a user's eyes in order to know their gaze location has been an important area of research over the last several years [47], as there are a variety of commercial eye tracking and gaze determination devices used in many different applications, from video games to advertising analysis. One of the more common uses of eye gaze in a multimodal input system has been for object selection by either coupling eye gaze with keyboard button presses [48, 49] or using eye gaze as a coarse selection technique, coupled with a mouse for finer pointing [50, 51].

Eye gaze in multimodal interfaces include integration with hand gestures for multi-display environments with large area screens [52]. Coupling eye gaze with a variety of different modalities including a mouse scroll wheel, tilting a handheld device, and touch input on a smart phone have also be used for pan and zoom operations for large information spaces [53]. Text entry has been explored using eye gaze by combining it with speech input [54]. In this interface, users would focus on a character of interest and issue a voice command to type the character in a document, although in usability testing, a traditional keyboard was found to be faster and more accurate. Multimodal interfaces making use of eye gaze have also be developed in

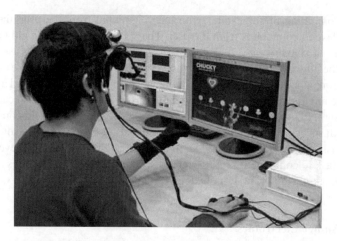

Figure 9.15   A video game that combines eye gaze, hand gestures, and bio-feedback. Reproduced by permission of Hwan Heo.

the entertainment domain. For example, Heo *et al.* (2010) [55] developed a video game (see Figure 9.15) that combined eye gaze, hand gestures, and bio-signal analysis and showed the multimodal interface was more compelling than traditional keyboard and mouse controls.

Finally, eye gaze has recently been explored in combination with brain-computer input (BCI) to support disabled users [56, 57]. In this interface, eye gaze was used to point to objects with the BCI component used to simulate dwell time for selection. Integrating eye gaze with BCI for multimodal interaction makes intuitive sense, given that movement is often limited when using BCI (see Section 9.3.5).

9.3.4   Facial Expressions

Recognizing facial expressions is a challenging and still open research problem that can be an important component of perceptual computing applications. There has been a significant amount of research on facial expression recognition in the computer vision community [58]. In terms of multimodal interaction, facial expressions are commonly used in two ways.

In the first way, other modalities are integrated with facial expression recognizers to better improve facial expression recognition accuracy, which ultimately supports human emotion recognition. For example, De Silva *et al.* (1997) [59] combined both video and audio information to determine what emotions are better detected with one type of information or another. They found that anger, happiness, surprise and dislike were better recognized by humans with visual information, while sadness and fear tended to be easier to recognize using audio.

Busso *et al.* (2004) [60] found similar results by integrating speech and facial expressions together in a emotion recognition detection system. Kessous *et al.* (2010) [61] used facial expressions and speech, as well as body gestures, to detect human emotions in a multimodal-based recognizer. Another example of an emotion detection system that couples facial expressions and speech input can be found in [62].

The second way that facial expression recognition has been used in the context of multimodal interfaces is to build affective computing systems which determine emotional state or mood in an attempt to adapt an application's interface, level of difficulty, and other parameters to improve user experience. For example, Lisetti and Nasoz (2002) [63] developed the MAUI system, a multimodal affective user interface that perceives a user's emotional state combining facial images, speech and biometric feedback.

In another example, Caridakis *et al.* (2010) [64] developed an affective state recognizer using both audio and visual cues using a recurrent neural network. They were able to achieve recognition rates as high as 98%, which shows promise for multimodal perceptual computing systems that can observe and understand the user not only when they are actively issuing commands, but when they are being passively monitored as well.

9.3.5 Brain-computer Input

Modern brain-computer interfaces (BCI) are capable of monitoring our mental state using nothing more than an electroencephalograph (EEG) connected to a computer. In order for signals to be tracked using today's technologies, a number of electrodes must be connected to the user's head in specific locations. These connections limit the possibilities for the addition of certain modalities to systems that the user can interact with. If the user moves while wearing a BCI, there is typically some noise in the signal, decreasing the accuracy of the signal. This is often not a large issue, as BCIs have typically been used to facilitate communication and movement from disabled individuals.

A number of companies, including Emotiv (Figure 9.16) and Neurosky, have begun developing low-cost BCIs for previously unconventional applications, one of which is video games. With the decrease in the cost of consumer BCIs, a number of multimodal applications using them have been proposed. Currently, the most commonly used modalities with BCIs are speech and eye gaze, as these do not require significant movement of the body. Gürkök and Nijholt (2012) [65] cite numerous examples of BCIs being used to improve user experience and task performance by using a brain-control interface as a modality within multimodal interfaces.

Electroencephalography (EEG) is often combined with additional neuroimaging methods, such as electromyography (EMG), which measures muscle activity. Leeb *et al.* (2010) [66] showed significant improvement in recognition performance when EEG is used in conjunction with EMG when compared to either EEG or EMG alone. Bayesian fusion of the two signals

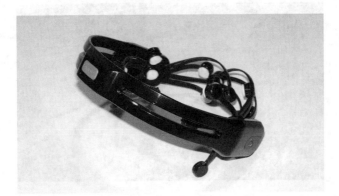

Figure 9.16   Emotiv EEG neuroheadset. Source: Corey Pittman.

was used to generate the combined signal. EEG, combined with NIRS (Near-infrared spectroscopy), has also been shown to significantly improve classification accuracy of signals [67]. NIRS creates a problem for real-time BCIs because of its significant latency.

Gürkök *et al.* (2011) [68] studied the effects of redundant modalities on a user's performance in a video game of their own design. The game, titled "Mind the Sheep!", required the player to move a group of dogs around to herd sheep into a pen. The system setup is shown in Figure 9.17. The game used a mouse in conjunction with one or two other modalities. The participant controlled which dog was selected using either speech or a BCI. To select a dog with speech, participants needed only to say the name of the dog. To control the dog with the BCI, the participant was required to focus on the desired dog's location, and then release the mouse button on the location where the dog was to move. Participants were asked to play the game under multiple conditions: with automatic speech recognition (ASR); with BCI; and lastly with both in a multimodal configuration. The study found that being given the opportunity to select the current modality did not give a significant performance increase when compared to either of the single modality modes, with some participants not even changing modalities once.

One further application for BCIs is modeling interfaces. Sree *et al.* (2013) [69] designed a framework for using BCI as an additional modality to assist in 3D modeling. EEG and EMG are again combined, this time in conjunction with a keyboard and mouse to control the modeling operations, with the Emotiv EEG neuroheadset as the primary device. The software accompanying the Emotiv is used to set some parameters for the device's signals and to calibrate the device for each particular user. Facial movements are detected using the EMG component, with gestures like looking left, controlling drawing an arc or blinking to left click on the mouse. Mouse movement is controlled using the EEG component, which detects the intent of the user.

The Emotiv API allows for interpretation of twelve movements – six directional movements and six rotational movements – which can all be used in the CAD environment. Participant fatigue was a common problem with the system, along with some problems with EEG signal strength. Additional input modalities, such as speech, could be added to this system to allow for improved usability.

Zander *et al.* (2010b) [70] allowed users the freedom to control a BCI using either imagined movement, visual focus, or a combination of both. The authors proposed that since BCI does not always work for all users when there is only one method of control, providing alternatives

Figure 9.17 "Mind the Sheep!" system interface. Source: Reproduced with permission from Hayrettin Gürkök.

or hybrid techniques allows for significant improvement in accuracy. Maye *et al.* (2011) [71] present a method for increasing the number of distinct stimuli (different tactile and visual stimuli) when using a BCI increases the number of the dimensions the user can control, while maintaining similar mental effort. Brain activity can be more easily classified by asking the user to shift focus between different stimuli. Zander *et al.* (2010b) [70] separate BCIs that are used in HCI into three categories: active, reactive, and passive.

9.4   Multimodal Integration Strategies

One of the most critical parts of a multimodal interface is the integration component, often called the fusion engine, that combines the various input modalities together to create a cohesive interface that supports meaningful commands [72]. There are many technical challenges with building a multimodal integration engine seems extraneous.

- First, the different input modalities may have very different characteristics, including data formats, frequencies, and semantic meaning, making it difficult to combine them together.
- Second, timing plays a significant role in integration since different modalities may come at different times during an interaction sequence, requiring the integration engine to be flexible in its ability to process the input streams.
- Third, and related to timing, resolving ambiguities is a significant challenge when the integration engine is under-constrained (the engine does not have enough information to make a fusion decision) or over-constrained (the engine has conflicting information and has more than one possible fusion decision to make).
- Finally, the input modalities used in multimodal interfaces often stem form natural communication channels (e.g., 3D gesture, speech, facial expressions) where recognition technologies are required to segment and classify the incoming data. Thus, levels of probabilistic uncertainty will exist with these modalities, making the integration engine's job more complex.

There are two basic approaches to performing multimodal integration in the fusion engine. The first is early integration and the second is late integration and, within these two approaches, there are several different integration strategies [73]. The premise behind early integration is that the data is first integrated before any major processing is done (aside for any low-level processing). In contrast to early integration, late integration processes the data from each mode separately and interprets it as if it is unimodal before integration occurs. The advantage of late integration is that, since the input modalities can be analyzed individually, time synchronization is not an issue and software development tends to be simpler.

However, late integration suffers from a potential loss of information, in that it can miss potential cross-modal interaction. For example, in early integration, the results from a gesture recognizer could be improved, based on the results from a speech recognizer or vice versa. However, with late integration, each individual recognizer must make a decision about its modality independently. The choice of whether to use early or late integration is still an open research question, and depends on the types of modalities used, as well as the multimodal interaction styles supported by an application. Note that, in some cases, a compromise between early and late integration can be used to perform the multimodal integration. For example, taking an early integration of 3D gestures and eye gaze with late integration with speech.

Within the context of early and late integration, there are three different integration levels for any incoming data stream [74]. Data-level and feature-level integration both fit into the early integration approach. Data-level integration focuses on low-level processing and is often used when the input modalities are similar, such as lip and facial expressions. This type of processing is also used when there is minimal integration required. Since it is closest to the raw data source, it can provide the highest level of detail, but is susceptible to noise.

Feature-level integration is used when the modalities are closely coupled or time-synchronized. Example modalities would be speech recognition from voice and lip movements, and example strategies include neural networks and hidden Markov models. Feature-level integration is also less sensitive to noise, but does not provide as much detail as low-level integration.

Decision-level integration (i.e., dialog level fusion [72]) is a late integration approach and is the most common type of multimodal integration. Its main strength is its ability to handle loosely coupled modalities (e.g., touch input and speech) but it relies on the accuracy of the processing already done on each individual modality.

Frame-based, unification-based, procedural, and symbolic/statistical integration are the most common integration strategies under decision-level integration.

9.4.1   Frame-based Integration

Frame-based integration focuses on data structures that support attribute-value pairs. Such frames collect these pairs from the various modalities to make an overall interpretation. For example, for speech input, an attribute-value pair could be "operation", with possible values of "delete", "add entry", "modify entry", etc. Each frame supports an individual input modality, and the integration occurs as the union of the sets of values in the individual frames. Scores are assigned to each attribute, and the overall score from the integration determines the best course of action to take. Koons *et al.* (1993) [75] were one of the first groups to explore feature-based integration that combined 3D gestures, eye gaze, and speech. More recently, Dumas *et al.* (2008) [76] developed the HephaisTK multimodal interface toolkit, which includes the frame-based approach to multimodal integration. Other multimodal interfaces that make use of frame-based integration using a variety of different input modalities include [77–79].

9.4.2   Unification-based Integration

The main idea behind unification-based integration is the use of the unification operator. Taken from natural language processing [80], it determines the consistency of two pieces of partial information, and if consistent, combines them into a single result [81]. For example, Cohen *et al.* (1997b) [82] was the first to use unification coupled with typed feature structures to integrate pen gesture and speech input in the QuickSet system. More recently, Taylor *et al.* (2012) [83] chose a unification integration scheme for combining speech and 3D pointing gestures and speech with touch gestures to support interaction with an unmanned robotic vehicle, while Sun *et al.* (2006) [84] also used unification, coupled with a multimodal grammar syntax in a traffic management tool that used 3D gestures and speech. Unification tends to work well when fusing only two modalities at any one time, and most unification-based integration research tends to focus on input pairs. Other examples of multimodal integration based on unification can be found in [85, 86].

9.4.3   Procedural Integration

Procedural integration techniques explicitly represent the multimodal state space through algorithmic management [72]. Common example representations using procedural integration include augmented transition networks and finite state machines. For example, both Neal *et al.* (1989) [87] and Latoschik (2002) [88] made use of augmented transition networks, and Johnston and Bangalore (2005) [89] and Bourguet (2002) [90] used finite state automata for procedural integration. Other approaches to procedural integration include Petri nets [91] as well as guided propagation networks [7]. In these systems, speech input was combined with either the mouse, keyboard, pen input, touch input, or 3D gestures.

9.4.4   Symbolic/Statistical Integration

Symbolic/statistical integration takes more traditional unification-based approaches and combines them with statistical processing to form hybrid multimodal integration strategies [92]. These strategies also bring in concepts from machine learning. Although machine learning has been primarily used with feature level integration, it also have been explored at the decision level [4]. As an example, Pan *et al.* (1999) [93] used Bayesian inference to derive a formula for estimating joint probabilities of multisensory signals that uses appropriate mapping functions to reflect signal dependencies. Mapping selection is guided by maximum mutual information.

Another early example of a symbolic/statistical integration technique is the Member Team Committee (MTC) architecture [94] used in the QuickSet application. In this approach, modalities are integrated on the basis of their posterior probabilities. Recognizers from individual modes become members of an MTC statistical integrator, and multiple teams are then trained to coordinate and weight the output from the different modes. Each team establishes a posterior estimate for a multimodal command, given the current input received. The committee of the MTC integrator analyzes the empirical distribution of the posteriors and then establishes an n-best ranking for each possible command. Flippo *et al.* (2003) [95] used a similar approach to MTC as part of a multimodal interaction framework, where they used parallel agents to estimate a posterior probability for each modal recognition result and then weighed them to come to an overall decision on the appropriate multimodal command.

More recently, Dumas et al. (2012) [96] developed a statistical multimodal integration scheme that uses hidden Markov models to select relevant modalities at the semantic level via temporal relationship properties. More information on using machine learning in multimodal integration can be found in [97, 98]. Although these methods can be very powerful at modeling the uncertainty in multimodal integration, their main drawback is the need for an adequate amount of training data.

9.5   Usability Issues with Multimodal Interaction

Given the very nature of multimodal interaction in the context of perceptual computing, usability becomes a critical part of multimodal interface design because of the attempt to couple different input modalities tightly together in a way that provides an intuitive and powerful user experience [99]. To discuss some of the usability issues with multimodal input, we use Oviatt's ten myths of multimodal interaction as starting points for discussion [100]. Although written several years ago, they are still applicable today.

If you build a multimodal system, users will interact multimodally. Just because an application supports multiple modes of input does not mean that users will make use of them for all commands. Thus, flexibility in the command structure is critical to support natural forms of communication between the human and the computer. In other words, multimodal interfaces should be designed with flexibility for the user in mind, so that they can issue commands in different ways. For example, a user should be able to use both speech and 3D gesture simultaneously to issue a command, and support should also be provided for using speech and eye gaze, or using 3D gesture and eye gaze, or using speech in isolation. This design choice makes the overall multimodal user interface more complex in terms of how the input modes are integrated, but provides the most generality.

Speech and pointing is the dominant multimodal integration pattern. From a usability perspective, speech and pointing make for an intuitive multimodal input combination, especially when the user wants to select some virtual object and then perform some operation on it (e.g., paint [this] cylinder blue). However, as we have seen in this chapter, there are a variety of different multimodal input combinations that are possible. The key question from a usability perspective is, does a particular input combination make sense for a particular task?

As a general guideline, it is important to provide multimodal input combinations that will support a simple and natural interaction metaphor for any given task. For example, speech and pointing may not be the best input combination in mobile settings, where touch input or 3D gestures may be more appropriate.

Multimodal input involves simultaneous signals. Not all multimodal input strategies require users to perform the individual inputs at the same time. Certain modalities extraneous afford such temporal integration but, in many cases, individual input modes are used consecutively (e.g., saying something, then performing a 3D gesture, and vise versa) but they still act as complementary modes of input. In fact, multimodal input strategies can also use one mode for certain tasks and a second or third mode for other tasks. Thus, from a usability perspective, it is important to recognize that there are a number of different ways that individual inputs can be combined, and not all of them need to support simultaneous input.

Speech is the primary input mode in any multimodal system that includes it. While speech is a dominant input channel that humans use to communicate, in many cases it does not need to be the primary modality in a multimodal interface. Unfortunately, speech recognition can be compromised in noisy environments, making it a less robust input mechanism. In addition, it may be the case that users do not want to use voice input due to privacy concerns. In other cases, speech may simply be a backup modality, or other modalities may be better suited to combine for a given task. Thus, when designing multimodal interaction scenarios, it is not a requirement to make speech the primary input mode, and it should only be used when it makes the most sense.

Multimodal language does not differ linguistically from unimodal language. One of the benefits of multimodal interaction is that it tends to simplify input patterns. As an example, consider a user who wants to move an object from one location to another. Using speech in isolation would require the user not only to describe the object of interest, but also to describe the location where the objects needs to be placed. However, with a combination of speech and gesture, the user can simplify the description of the object, because they are also using a second modality (in this case, pointing) both to identify the object and to place it in a different location. This input combination implies that one can use simpler input commands for each individual mode of input when performing concurrent multimodal interaction. In terms of usability, it is important to understand that unimodal language can be more complex than multimodal language, and that multimodal input can take away this complexity, making for an easier-to-use interface.

Multimodal integration involves redundancy of content between modes. One of the key ideas behind multimodal integration is that providing redundant modes of input can help to support a better user experience, because each individual mode can be used to reinforce the others. This is certainly true from a computational point of view, and it does have its place in multimodal integration. However, the complementary nature of multimodal input should not be ignored for its benefits from a usability perspective. Thus, ensuring proper multimodal integration to support complementarity is important from the user's perspective.

Individual error-prone recognition technologies combine multimodally to produce even greater unreliability. One of the interesting challenges with multimodal input, especially with perceptual computing, is that the input modalities used (e.g., speech, 3D gesture, 2D gesture) require recognition technologies designed to understand the input. Unfortunately, recognition is error-prone, based on the accuracy of the individual recognizers. However, combining multiple recognition-based input together actually can help to improve accuracy for the overall command, producing a more reliable interface. Multimodal integration strategies are key to this improvement. In addition, users, if given a choice, will tend to work with modalities that they believe have higher accuracies. Thus, from a usability perspective, this usage pattern is another reason to ensure the multimodal interface is flexible.

All users' multimodal commands are integrated in a uniform way. Users of multimodal interfaces tend to identify an integration pattern for how they will use the interface fairly early on, and they will stick with this pattern. However, as we have seen, there are many different ways that people can use a multimodal interface. Thus, it is important for the multimodal integration scheme to be flexible and to try to identify the dominant integration pattern on a per-user basis. This approach then can support improved recognition rates, because the fusion engine would be aware of how the user is interacting with the different modalities.

Different input modes are capable of transmitting comparable content – but not all input modalities are created equal. In other words, different modalities have strengths and weaknesses, depending on the type of information that the user wants to convey. For example, eye gaze will produce very different types of information than speech. Thus, from a usability perspective, it is important to understand what modalities are available, and what each one is ideal for. Trying to use a given modality as input for tasks that it is not suited for will only make the interface difficult to use.

Enhanced efficiency is the main advantage of multimodal systems. Finally, speed and efficiency are not the only advantages of multimodal interfaces. Reducing errors from individual recognition systems, as well as providing more flexibility to interact with an application in the way that users want to, are also important advantages of multimodal interaction. In addition, a properly designed multimodal interface will provide a level of generality to the user to support a variety of different tasks, applications, and environments.

9.6 Conclusion

In this chapter, we have explored how combining different input modalities can form natural and expressive multimodal interfaces. We have examined the types of multimodal input strategies and presented a variety of different multimodal interfaces that offer different combinations of touch input, speech, 3D gesture, eye gaze and tracking, facial expressions, and brain-computer input. We have also examined multimodal integration or fusion, a critical component of a multimodal architecture that integrates different modalities together to form a cohesive interface by examining the different approaches and levels of integration. Finally, we have presented a number of usability issues as they relate to multimodal input. Clearly, multimodal interfaces have come a long way from Bolt's "put-that-there" system [1].

However, more work is still needed in a variety of areas, including multimodal integration, recognition technology and usability, to fully support perceptual computing applications that provide powerful, efficient, and expressive human computer interaction.

References

1. Bolt, R.A. (1980). *"Put-that-there": voice and gesture at the graphics interface.* Proceedings of the 7th annual conference on computer graphics and interactive techniques, SIGGRAPH '80, 262–270. ACM, New York, NY, USA.
2. Oviatt, S. (2003). Advances in robust multimodal interface design. *IEEE Computer Graphics and Applications* **23**(5), 62–68.
3. Turk, M., Robertson, G. (2000). Perceptual user interfaces (introduction). *Communications of the ACM* **43**(3), 32–34.
4. Dumas, B., Lalanne, D., Oviatt, S. (2009). Multimodal interfaces: A survey of principles, models and frameworks. In: Lalanne, D., Kohlas, J. (ed.). *Human Machine Interaction,* 3–26. vol. 5440 of Lecture Notes in Computer Science. Springer Berlin, Heidelberg.
5. Jaimes, A., Sebe, N. (2007). Multimodal human-computer interaction: A survey. *Computer Vision and Image Understanding* **108**(12), 116–134. Special Issue on Vision for Human-Computer Interaction.
6. Oviatt, S. (2007). Multimodal interfaces. In: Sears, A., Jack, J. (eds.). *The Human-Computer Interaction Handbook: Fundamentals, Evolving Technologies and Emerging Applications,* Second Edition, 413–432. CRC Press.
7. Martin, J.C. (1998). Tycoon: Theoretical framework and software tools for multimodal interfaces In: Lee, J, (Ed.). *Intelligence and Multimodality in Multimedia Interfaces.* AAAI Press.
8. Oviatt, S., Van gent, R. (1996). *Error resolution during multimodal human-computer interaction,* 204–207.
9. Bowman, D.A., Riff, E., Lavolta, J.J., Papyri, I. (2004). *3D User Interfaces: Theory and Practice.* Addison Wesley Longman Publishing Co., Inc., Redwood City, CA, USA.
10. Sharma, A., Madhvanath, S., Shekhawat, A., Billinghurst, M. (2011). *Mozart: a multimodal interface for conceptual 3D modelling.* Proceedings of the 13th international conference on multimodal interfaces, ICMI '11. ACM, New York, NY, USA, 307–310.
11. Radhakrishnan, S., Lin, Y., Zeid, I., Kamarthi, S. (2013). Finger-based multitouch interface for performing 3D CAD operations. *International Journal of Human-Computer Studies* **71**(3), 261–275.
12. Tse, E., Greenberg, S., Shen, C., Forlines, C., Kodama, R. (2008). *Exploring true multi-user multimodal interaction over a digital table.* Proceedings of the 7th ACM conference on Designing interactive systems, DIS '08. ACM, New York, NY, USA, 109–118.
13. Tse, E., Greenberg, S., Shen, C. (2006). *GSI demo: multiuser gesture/speech interaction over digital tables by wrapping single user applications.* Proceedings of the 8th international conference on Multimodal interfaces, 76–83.
14. Tse, E., Greenberg, S., Shen, C., Forlines, C. (2007). Multimodal multiplayer tabletop gaming. *Computers in Entertainment (CIE)* **5**(2), 12.
15. Martínez, R., Collins, A., Kay, J., Yacef, K. (2011). *Who did what? Who said that? Collaid: an environment for capturing traces of collaborative learning at the tabletop.* Proceedings of the ACM International Conference on Interactive Tabletops and Surfaces, 172–181.
16. Barthelmess, P., Kaiser, E., Huang, X., Demirdjian, D. (2005). *Distributed pointing for multimodal collaboration over sketched diagrams.* Proceedings of the 7th international conference on Multimodal interfaces, ICMI '05. ACM, New York, NY, USA, 10–17
17. Piper, A.M., Hollan, J.D. (2008). *Supporting medical conversations between deaf and hearing individuals with tabletop displays.* Proceedings of the 2008 ACM conference on Computer supported cooperative work, CSCW '08. ACM, New York, NY, USA, 147–156.

18. Piper, A.M. (2010). *Supporting medical communication with a multimodal surface computer.* CHI '10 Extended Abstracts on Human Factors in Computing Systems, CHI EA '10. ACM, New York, NY, USA, 2899–2902.

19. Paek, T., Thiesson, B., Ju, Y.C., Lee, B. (2008). *Search VOX: Leveraging multimodal refinement and partial knowledge for mobile voice search.* Proceedings of the 21st annual ACM symposium on User interface software and technology, 141–150.

20. Ehlen, P., Johnston, M. (2011). *Multimodal local search in speak4it.* Proceedings of the 16th international conference on Intelligent user interfaces, 435–436 ACM.

21. Cohen, P.R., Johnston, M., McGee, D., Oviatt, S., Pittman, J., Smith, I., Chen, L., Clow, J. (1997a). *Quickset: Multimodal interaction for distributed application.* Proceedings of the fifth ACM international conference on Multimedia, 31–40.

22. Ramsay, A., McGee-Lennon, M., Wilson, G.A., Gray, S.J., Gray, P., De Turenne, F. (2010). Tilt and go: exploring multimodal mobile maps in the field. *Journal on Multimodal User Interfaces* **3**(3), 167–177.

23. Feng, J., Johnston, M., Bangalore, S. (2011). Speech and multimodal interaction in mobile search. *IEEE Signal Processing Magazine* **28**(4), 40–49.

24. Sim, K.C. (2012). *Speak-as-you-swipe (says): a multimodal interface combining speech and gesture keyboard synchronously for continuous mobile text entry.* Proceedings of the 14th ACM international conference on Multimodal interaction, 555–560.

25. Kristensson, P.O., Vertanen, K. (2011). *Asynchronous multimodal text entry using speech and gesture keyboards.* Proceedings of the 12th Annual Conference of the International Speech Communication Association (Interspeech 2011), 581–584.

26. Shinoda, K., Watanabe, Y., Iwata, K., Liang, Y., Nakagawa, R., Furui, S. (2011). Semi-synchronous speech and pen input for mobile user interfaces. *Speech Communication* **53**(3), 283–291.

27. Dearman, D., Karlson, A., Meyers, B., Bederson, B. (2010). *Multi-modal text entry and selection on a mobile device.* Proceedings of Graphics Interface 2010, 19–26 Canadian Information Processing Society.

28. Laput, G.P., Dontcheva, M., Wilensky, G., Chang, W., Agarwala, A., Linder, J., Adar, E. (2013). *Pixeltone: a multimodal interface for image editing.* Proceedings of the SIGCHI Conference on Human Factors in Computing Systems, CHI '13. ACM, New York, NY, USA, 2185–2194.

29. Muller, C., Weinberg, G. (2011). Multimodal input in the car, today and tomorrow. *IEEE Multimedia* **18**(1), 98–103.

30. Pfleging, B., Schneegass, S., Schmidt, A. (2012). *Multimodal interaction in the car: combining speech and gestures on the steering wheel.* Proceedings of the 4th International Conference on Automotive User Interfaces and Interactive Vehicular Applications, AutomotiveUI '12. ACM, New York, NY, USA, 155–162.

31. Pfleging, B., Kienast, M., Schmidt, A., Döring, T. (2011). *Speet: A multimodal interaction style combining speech and touch interaction in automotive environments.* Adjunct Proceedings of the 3rd International Conference on Automotive User Interfaces and Interactive Vehicular Applications, AutomotiveUI, 65–66.

32. Gruenstein, A., Orszulak, J., Liu, S., Roberts, S., Zabel, J., Reimer, B., Mehler, B., Seneff, S., Glass, J., Coughlin, J. (2009). *City browser: developing a conversational automotive HMI.* CHI '09 Extended Abstracts on Human Factors in Computing Systems, CHI EA '09. ACM, New York, NY, USA, 4291–4296.

33. Williamson, B.M., Wingrave, C., LaViola, J.J., Roberts, T., Garrity, P. (2011). *Natural full body interaction for navigation in dismounted soldier training.* The Interservice/Industry Training, Simulation & Education Conference (I/ITSEC), NTSA.

34. Williamson, B.M., LaViola, J.J., Roberts, T., Garrity, P. (2012). *Multi-kinect tracking for dismounted soldier training.* The Interservice/Industry Training, Simulation & Education Conference (I/ITSEC), NTSA.

35. Budhiraja, P., Madhvanath, S. (2012). *The blue one to the left: enabling expressive user interaction in a multimodal interface for object selection in virtual 3D environments.* Proceedings of the 14th ACM international conference on Multimodal interaction, 57–58.

36. Hoste, L., Dumas, B., Signer, B. (2012). *Speeg: a multimodal speech-and gesture-based text input solution.* Proceedings of the International Working Conference on Advanced Visual Interfaces, 156–163.

37. Ward, D.J., Blackwell, A.F., MacKay, D.J. (2000). *Dashera data entry interface using continuous gestures and language models.* Proceedings of the 13th annual ACM symposium on User interface software and technology, 129–137.

38. Bohus, D., Horvitz, E. (2009). Dialog in the open world: platform and applications Proceedings of the 2009 international conference on Multimodal interfaces, 31–38, ACM.

39. Hrúz, M., Campr, P., Dikici, E., Kındıroğlu, A.A., Krňoul, Z., Ronzhin, A., Sak, H., Schorno, D., Yalçın, H., Akarun, L. *et al.* (2011). Automatic fingersign-to-speech translation system. *Journal on Multimodal User Interfaces* **4**(2), 61–79.

40. Gallo, L., Placitelli, A.P., Ciampi, M. (2011). *Controller-free exploration of medical image data: Experiencing the Kinect.* IEEE International Symposium on Computer-Based Medical Systems (CBMS), 1–6.

41. Perzanowski, D., Schultz, A.C., Adams, W., Marsh, E., Bugajska, M. (2001). Building a multimodal human-robot interface. *IEEE Intelligent Systems* 16(1), 16–21.

42. Burger, B., Ferrané, I., Lerasle, F., Infantes, G. (2012). Two-handed gesture recognition and fusion with speech to command a robot. *Autonomous Robots* 32(2), 129–147.

43. Stiefelhagen, R., Fugen, C., Gieselmann, R., Holzapfel, H., Nickel, K., Waibel, A. (2004). *Natural human-robot interaction using speech, head pose and gestures.* Proceedings of International Conference on Intelligent Robots and Systems, IROS, 3, 2422–2427.

44. Lee, S.H., Sohn, M.K., Kim, D.J., Kim, B., Kim, H. (2013). *Smart TV interaction system using face and hand gesture recognition.* IEEE International Conference on Consumer Electronics (ICCE), 173–174.

45. Takahashi, M., Fujii, M., Naemura, M., Satoh, S. (2013). Human gesture recognition system for TV viewing using time-of-flight camera. *Multimedia Tools and Applications* 62(3), 761–783.

46. Krahnstoever, N., Kettebekov, S., Yeasin, M., Sharma, R. (2002). *A real-time framework for natural multimodal interaction with large screen displays.* Proceedings of the 4th IEEE International Conference on Multimodal Interfaces, 349.

47. Duchowski, A.T. (2007). *Eye Tracking Methodology: Theory and Practice.* Springer-Verlag New York, Inc., Secaucus, NJ, USA.

48. Kumar, M., Paepcke, A., Winograd, T. (2007). *Eyepoint: practical pointing and selection using gaze and keyboard.* Proceedings of the SIGCHI Conference on Human Factors in Computing Systems, CHI '07. ACM, New York, NY, USA, 421–430

49. Zhang, X., MacKenzie, I. (2007). *Evaluating eye tracking with ISO 9241– part 9.* In: Jacko, J. (ed.). *Human-Computer Interaction. HCI Intelligent Multimodal Interaction Environments.* vol. 4552 of Lecture Notes in Computer Science. Springer Berlin, Heidelberg, 779–788.

50. Hild, J., Muller, E., Klaus, E., Peinsipp-Byma, E., Beyerer, J. (2013). *Evaluating multi-modal eye gaze interaction for moving object selection.* The Sixth International Conference on Advances in Computer-Human Interactions, ACHI, 454–459.

51. Zhai, S., Morimoto, C., Ihde, S. (1999). *Manual and gaze input cascaded (magic) pointing.* Proceedings of the SIGCHI conference on Human Factors in Computing Systems, CHI '99. ACM, New York, NY, USA, 246–253.

52. Cha, T., Maier, S. (2012). *Eye gaze assisted human-computer interaction in a hand gesture controlled multi-display environment.* Proceedings of the 4th Workshop on Eye Gaze in Intelligent Human Machine Interaction, Gaze-In '12. ACM, New York, NY, USA, 13, 1–13:3.

53. Stellmach, S., Dachselt, R. (2012). *Investigating gaze-supported multimodal pan and zoom.* Proceedings of the Symposium on Eye Tracking Research and Applications, ETRA '12. ACM, New York, NY, USA, 357–360

54. Beelders, T.R., Blignaut, P.J. (2012). *Measuring the performance of gaze and speech for text input.* Proceedings of the Symposium on Eye Tracking Research and Applications, ETRA '12. ACM, New York, NY, USA, 337–340.

55. Heo, H., Lee, E.C., Park, K.R., Kim, C.J., Whang, M. (2010). A realistic game system using multi-modal user interfaces. *IEEE Transactions on Consumer Electronics* 56(3), 1364–1372.

56. Vilimek, R., Zander, T. (2009). Bc(eye): Combining eye-gaze input with brain-computer interaction. In: Stephanidis, C. (ed.). *Universal Access in Human-Computer Interaction. Intelligent and Ubiquitous Interaction Environments.* Vol. 5615 of Lecture Notes in Computer Science. Springer Berlin, Heidelberg, 593–602.

57. Zander, T.O., Gaertner, M., Kothe, C., Vilimek, R. (2010a). Combining eye gaze input with a brain-computer interface for touchless human-computer interaction. *Intl. Journal of Human-Computer Interaction* 27(1), 38–51.

58. Sandbach, G., Zafeiriou, S., Pantic, M., Yin, L. (2012). Static and dynamic 3D facial expression recognition: A comprehensive survey. *Image and Vision Computing* 30(10), 683–697.

59. De Silva, L., Miyasato, T., Nakatsu, R. (1997). *Facial emotion recognition using multi-modal information.* Proceedings of 1997 International Conference on Information, Communications and Signal Processing (ICICS), 1, 397–401.

60. Busso, C., Deng, Z., Yildirim, S., Bulut, M., Lee, C.M., Kazemzadeh, A., Lee, S., Neumann, U., Narayanan, S. (2004). *Analysis of emotion recognition using facial expressions, speech and multimodal information.* Proceedings of the 6th international conference on Multimodal interfaces, ICMI '04. ACM, New York, NY, USA, 205–211.

61. Kessous, L., Castellano, G., Caridakis, G. (2010). Multimodal emotion recognition in speech-based interaction using facial expression, body gesture and acoustic analysis. *Journal on Multimodal User Interfaces* 3(1–2), 33–48.

62. Wöllmer, M., Metallinou, A., Eyben, F., Schuller, B., Narayanan, S.S. (2010). Context-sensitive multimodal emotion recognition from speech and facial expression using bidirectional LSTM modeling. Proceedings of Interspeech, Japan, 2362–2365.

63. Lisetti, C.L., Nasoz, F. (2002). *Maui: a multimodal affective user interface.* Proceedings of the tenth ACM international conference on Multimedia, MULTIMEDIA '02. ACM, New York, NY, USA, 161–170.

64. Caridakis, G., Karpouzis, K., Wallace, M., Kessous, L., Amir, N. (2010). Multimodal users affective state analysis in naturalistic interaction. *Journal on Multimodal User Interfaces* **3**(1–2), 49–66.

65. Gürkök, H., Nijholt, A. (2012). Brain-computer interfaces for multimodal interaction: a survey and principles. *International Journal of Human-Computer Interaction* **28**(5), 292–307.

66. Leeb, R., Sagha, H., Chavarriaga, R., del R Millan, J. (2010). *Multimodal fusion of muscle and brain signals for a hybrid-BCI.* 2010 Annual International Conference of the IEEE Engineering in Medicine and Biology Society (EMBC), 4343–4346.

67. Fazli, S., Mehnert, J., Steinbrink, J., Curio, G., Villringer, A., Müller, K.R., Blankertz, B. (2012). Enhanced performance by a hybrid NIRS-EEG brain computer interface. *Neuroimage* **59**(1), 519–529.

68. Gürkök, H., Hakvoort, G., Poel, M. (2011). *Modality switching and performance in a thought and speech controlled computer game.* Proceedings of the 13th international conference on multimodal interfaces, ICMI '11. ACM, New York, NY, USA, 41–48.

69. Sree, S., Verma, A., Rai, R. (2013). *Creating by imaging: Use of natural and intuitive BCI in 3D CAD modelling.* ASME International Design Engineering Technical Conference ASME/DETC/CIE ASME.

70. Zander, T.O., Kothe, C., Jatzev, S., Gaertner, M. (2010b). Enhancing human-computer interaction with input from active and passive brain-computer interfaces. *Brain-Computer Interfaces*, Springer, 181–199.

71. Maye, A., Zhang, D., Wang, Y., Gao, S., Engel, A.K. (2011). Multimodal brain-computer interfaces. *Tsinghua Science & Technology* **16**(2), 133–139.

72. Lalanne, D., Nigay, L., Palanque, P., Robinson, P., Vanderdonckt, J., Ladry, J.F. (2009). *Fusion engines for multimodal input: a survey.* Proceedings of the 2009 international conference on Multimodal interfaces, ICMIMLMI '09. ACM, New York, NY, USA, 153–160.

73. Turk, M. (2014). Multimodal interaction: A review. *Pattern Recognition Letters* **36**, 189–195.

74. Sharma, R., Pavlovic, V., Huang, T. (1998). Toward multimodal human-computer interface. *Proceedings of the IEEE* **86**(5), 853–869.

75. Koons, D.B., Sparrell, C.J., Thorisson, K.R. (1993). Integrating simultaneous input from speech, gaze, and hand gestures. In: Maybury, M.T. (Ed.). *Intelligent multimedia interfaces.* American Association for Artificial Intelligence, Menlo Park, CA, USA, 257–276.

76. Dumas, B., Lalanne, D., Guinard, D., Koenig, R., Ingold, R. (2008). *Strengths and weaknesses of software architectures for the rapid creation of tangible and multimodal interfaces.* Proceedings of the 2nd international conference on Tangible and embedded interaction, TEI '08. ACM, New York, NY, USA, 47–54.

77. Bouchet, J., Nigay, L., Ganille, T. (2004). *Icare software components for rapidly developing multimodal interfaces.* Proceedings of the 6th international conference on Multimodal interfaces, ICMI '04. ACM, New York, NY, USA, 251–258.

78. Nigay, L., Coutaz, J. (1995). *A generic platform for addressing the multimodal challenge.* Proceedings of the SIGCHI Conference on Human Factors in Computing Systems, CHI '95. ACM Press/Addison-Wesley Publishing Co., New York, NY, USA, 98–105.

79. Vo, M.T., Wood, C. (1996). *Building an application framework for speech and pen input integration in multimodal learning interfaces.* Proceedings of IEEE International Conference on Acoustics, Speech, and Signal Processing, ICASSP-96, **6**, 3545–3548.

80. Calder J. (1987). Typed unification for natural language processing. In: Kahn, G., MacQueen, D., Plotkin, G. (eds.). *Categories, Polymorphism, and Unification.* Centre for Cognitive Science University of Edinburgh, Edinburgh, Scotland.

81. Johnston, M., Cohen, P.R., McGee, D., Oviatt, S.L., Pittman, J.A., Smith, I. (1997). *Unification-based multimodal integration.* Proceedings of the 35th Annual Meeting of the Association for Computational Linguistics and Eighth Conference of the European Chapter of the Association for Computational Linguistics, ACL '98. Association for Computational Linguistics, Stroudsburg, PA, USA, 281–288.

82. Cohen, P.R., Johnston, M., McGee, D., Oviatt, S., Pittman, J., Smith, I., Chen, L., Clow, J. (1997b). *Quickset: multimodal interaction for distributed applications.* Proceedings of the fifth ACM international conference on Multimedia, MULTIMEDIA '97. New York, NY, USA, 31–40.

83. Taylor, G., Frederiksen, R., Crossman, J., Quist, M., Theisen, P. (2012). *A multi-modal intelligent user interface for supervisory control of unmanned platforms.* International Conference on Collaboration Technologies and Systems (CTS), 117–124.

84. Sun, Y., Chen, F., Shi, Y.D., Chung, V. (2006). *A novel method for multi-sensory data fusion in multimodal human computer interaction.* Proceedings of the 18th Australia conference on Computer-Human Interaction: Design: Activities, Artefacts and Environments, OZCHI '06. ACM, New York, NY, USA, 401–404.

85. Holzapfel, H., Nickel, K., Stiefelhagen, R. (2004). *Implementation and evaluation of a constraint-based multimodal fusion system for speech and 3D pointing gestures.* Proceedings of the 6th international conference on Multimodal interfaces, ICMI '04. ACM, New York, NY, USA, 175–182.

86. Pfleger, N. (2004). *Context-based multimodal fusion.* Proceedings of the 6th international conference on Multimodal interfaces, ICMI '04. ACM, New York, NY, USA, 265–272.

87. Neal, J.G., Thielman, C.Y., Dobes, Z., Haller, S.M., Shapiro, S.C. (1989). *Natural language with integrated deictic and graphic gestures.* Proceedings of the workshop on Speech and Natural Language, HLT '89. Association for Computational Linguistics, Stroudsburg, PA, USA, 410–423.

88. Latoschik, M. (2002). *Designing transition networks for multimodal VR-interactions using a markup language.* Proceedings of Fourth IEEE International Conference on Multimodal Interfaces, 411–416.

89. Johnston, M., Bangalore, S. (2005). Finite-state multimodal integration and understanding. *Natural Language Engineering* **11**(2), 159–187.

90. Bourguet, M. (2002). *A toolkit for creating and testing multimodal interface designs.* Proceedings of User Interface Software and Technology (UIST 2002) Companion proceedings, 29–30.

91. Navarre, D., Palanque, P., Bastide, R., Schyn, A.,Winckler, M., Nedel, L.P., Freitas, C.M.D.S. (2005). *A formal description of multimodal interaction techniques for immersive virtual reality applications.* Proceedings of the 2005 IFIP TC13 international conference on Human-Computer Interaction, INTERACT'05. Springer-Verlag, Berlin, Heidelberg, 170–183.

92. Wu, L., Oviatt, S.L., Cohen, P.R. (1999). Multimodal integration – a statistical view. *IEEE Transactions on Multimedia* **1**, 334–341.

93. Pan, H., Liang, Z.P., Huang, T. (1999). *Exploiting the dependencies in information fusion.* IEEE Computer Society Conference on Computer Vision and Pattern Recognition, **2**, 412.

94. Wu, L., Oviatt, S.L., Cohen, P.R. (2002). From members to teams to committee – a robust approach to gestural and multimodal recognition. *IEEE Transactions on Neural Networks* **13**(4), 972–982.

95. Flippo, F., Krebs, A., Marsic, I. (2003). *A framework for rapid development of multimodal interfaces.* Proceedings of the 5th international conference on Multimodal interfaces, ICMI '03. ACM, New York, NY, USA, 109–116.

96. Dumas, B., Signer, B., Lalanne, D. (2012). *Fusion in multimodal interactive systems: an hmm-based algorithm for user-induced adaptation.* Proceedings of the 4th ACM SIGCHI symposium on Engineering interactive computing systems, EICS '12. ACM, New York, NY, USA, 15–24.

97. Damousis, I.G., Argyropoulos, S. (2012). Four machine learning algorithms for biometrics fusion: a comparative study. *Appl. Comp. Intell. Soft Comput.* 242401-1-7.

98. Huang, X., Oviatt, S., Lunsford, R. (2006). Combining user modeling and machine learning to predict users multimodal integration patterns. In: Renals, S., Bengio, S., Fiscus, J. (eds.). *Machine Learning for Multimodal Interaction,* vol. 4299 of Lecture Notes in Computer Science Springer Berlin, Heidelberg, 50–62.

99. Reeves, L.M., Lai, J., Larson, J.A., Oviatt, S., Balaji, T.S., Buisine, S., Collings, P., Cohen, P., Kraal, B., Martin, J.C., McTear, M., Raman, T., Stanney, K.M., Su, H., Wang, Q.Y. (2004). Guidelines for multimodal user interface design. *Communications of the ACM* **47**(1), 57–59.

100. Oviatt, S. (1999). Ten myths of multimodal interaction. *Communications of the ACM* **42**(11), 74–81.

10

Multimodal Interaction in Biometrics: Technological and Usability Challenges

Norman Poh[1], Phillip A. Tresadern[2] and Rita Wong[3]

[1] Department of Computing, University of Surrey, UK
[2] University of Manchester, UK
[3] University of Surrey, UK

10.1 Introduction

In our increasingly interconnected society, the need to establish a person's identity is becoming critical. Traditional authentication, based on what one possesses (e.g., token/access card) or what one knows (password), no longer suffices. Biometrics, described as the science of recognizing an individual based on his/her physiological or behavioral traits, which is effectively based on what one *is*, offers an important solution to identity security. Since unimodal biometric systems have their own limitations, a multimodal biometric system is often needed. This provides an excellent case study for multimodal interaction.

In this chapter, we report the technological design and usability issues related to the use of multimodal biometrics in various applications, ranging from their use in large-scale multi-site border controls to securing personal portable devices.

In the context of the Mobile Biometric project (MOBIO), we combine face and voice biometrics for secure, yet rapid, user verification, to ensure that someone who requests access to data is entitled to it. In addition, we also report our experience of empowering blind users to use the face as biometrics on their mobile device. In order to help them capture good quality face images, we designed an audio feedback mechanism that is driven by an estimate of their facial quality and usability of the technology by blind users.

Interactive Displays: Natural Human-Interface Technologies, First Edition. Edited by Achintya K. Bhowmik.
© 2015 John Wiley & Sons, Ltd. Published 2015 by John Wiley & Sons, Ltd.

10.1.1   Motivations for Identity Assurance

Questions of the following sorts are posed everyday and everywhere: "Is (s)he really whom (s)he claims to be?"; "Is (s)he authorized to access this premise/resources/information?"; and "Is (s)he a person that the authority seeks?" Traditionally, a person's identity is routinely verified using a driver's license, passport, or national identity document. Access to guarded resources is granted if the person knows his/her password or PIN number. A token- or password-based means of authentication can be easily compromised through increasingly sophisticated techniques deployed by criminals. This has caused significant economic loss, as well as loss of trust in our modern society. For instance, according to the Javelin Strategy & Research report[1] in the US, identity fraud caused a lost of USD 21 billion in 2012, while in the UK, this is estimated to be GBP 1.3 billion. Business across the world could lose USD 221 billion per year due to identity fraud. The need for reliable user authentication techniques is paramount.

10.1.2   Biometrics

Increasingly, biometrics has gained widespread acceptance as a legitimate method for determining an individual's identity. Today, the security of border controls are enhanced by using iris, face, and/or fingerprint alongside travel documents. However, using one biometric system alone is often not sufficient. Unimodal biometric systems have to contend with a variety of problems such as noisy data, intra-subject variations, restricted degrees of freedom imposed during the acquisition process, non-universality (i.e., not everyone can provide fingerprints with clear ridges), spoof attacks, and unacceptable error rates for some users [1]. The noise is due to altered appearance of the biometric trait (e.g., voice modified by a cold), imperfection of the sensor (e.g., dirty sensor) or the environment in which the sensor acquires a biometric trait (e.g., face images affected by the lighting conditions).

One way to improve the robustness of the system is to use multimodal biometrics. Because different biometric modalities are affected by different sources of noise, the use of multimodal biometrics often result in a significant gain in performance, compared to any unimodal biometric systems.

10.1.3   Application Characteristics of Multimodal Biometrics

There are a number of application-related criteria that require a multimodal biometrics solution; these are listed below.

- **Enrolment requirements**: When a biometric technology is rolled out in a large scale – for instance, at the population level – one has to consider multiple biometric modalities. This is because a small fraction of the population of users may not be able to provide usable fingerprints, due to work or health-related reasons or handicap. For example, people who do not have a right arm cannot provide the fingerprint of any of the right fingers. For this reason, large-scale biometric projects such as United States Visitor and Immigrant

---

[1] https://www.javelinstrategy.com/brochure/276

Status Indicator Technology (US-VISIT) and Unique Identity (UID) project have to consider multiple biometric modalities to ensure that the technology can be used by all users in the target population. The US-VISIT program requires that every visitor to the United States must provide fingerprint images of left and right index fingers and a facial image at their port of entry, whereas the UID project uses fingerprint, iris, and face biometrics [2].

- **Risk and viability of spoofing**: In applications involving critical infrastructures, where the risk of intrusion is significant, multimodal biometrics may be used. Because several biometric traits are involved, spoofing or forging all the biometric modalities of a victim in order to gain illegitimate access to the facilities will be very difficult. Biometric modalities such as finger vein and iris are difficult to collect and, hence, harder to spoof. Incorporating these modalities, along with other biometric modalities, could deter spoof attacks aiming at the biometric sensors. However, recent studies [3, 4] show that, even though the risk of spoof attack is reduced, a multimodal biometric system can still be vulnerable to spoof attacks. This is because if one or more biometric sub-systems are compromised, then the performance of a multimodal system will be adversely affected.

- **Integrity requirements**: Once a user has been successfully authenticated and granted access to a secure resource, it is often desirable to ensure that the same user actually uses the system. For this reason, a continuous, non-intrusive authentication solution is often required after the initial authentication. For instance, in logical access control, where the end-user logs on to a secure resource, the terminal which the user uses will attempt to continuously authenticate the user. This prevents an attacker from accessing the system while the genuine user is momentarily absent. This line of research was reported in Altinok and Turk (2003) [5] using face, voice, and fingerprint; and in Azzini *et al.* (2008) [6] and Sim *et al.* (2007) [7] using face and fingerprint modalities. Niinuma and Jain (2010) [8] used the face modality as well as the color of the clothing to enable continuous authentication. The latter information is most useful when the face is not observable.

- **Accuracy requirements**: One of the most demanding applications that justifies the use of a multimodal biometric system is negative identification or record deduplication. The objective here is to prevent duplicate entries of identities. Unlike positive identification, negative identification ensures that a person does not exist in a database. Examples of the application are to prevent a person claiming social benefits twice under different two different identities; or to prevent blacklisted workers from entering into a country. While ten finger biometric and photographs can ensure de-duplication accuracy higher than 95%, depending upon quality of data collection, adding the iris modality could boost the accuracy to as high as 99% [2].

- **Uncontrolled environment**: While most biometric applications require user cooperation, there is a niche application where the cooperation of the users is not required nor desirable. The line of biometric research addressing this problem is called non-cooperative biometrics. Typically, the subjects can be several meters away from the sensor and may not be aware that they are under surveillance. The biometric sensor continuously tracks and recognizes all subjects passing by a strategic location. For this application, several biometric modalities or visual cues are often required, but the actual recognition mechanism may rely only on one or two biometric modalities that are usable. Li *et al.* (2008) [9] used a wide field-of-view (FOV) camera coupled with two narrow FOV cameras to track a subject for maritime surveillance. If the wide FOV camera detects a human silhouette, one narrow FOV will be activated to zoom into the person's face and the another will attempt to acquire the person's iris.

In a forensic application, Nixon *et al.* (2010) [10] proposed to combine gait and ear to recognize a criminal. Because criminals often attempt to evade their identity by disguise or concealment, gait is often the natural biometric candidate that is available. At the same time, the ear shape hardly changes over time, thus making the combination of gait and ear a potentially useful forensic tool for identity recognition under a non-cooperative scenario such as forensics surveillance.

In our own research, we addressed the issue of biometric authentication using general-purpose mobile devices equipped with a microphone and a camera. This field, referred to as "mobile biometry", has a significant practical application, because the biometric system can prevent other people from accessing potentially highly private and highly sensitive data such as that stored in a mobile phone. Furthermore, the biometric system can also be used as an authentication mechanism for electronic transactions. This would add significantly value and trust to transaction services.Our own study [11] shows that combining talking face images with speech recorded simultaneously lead to a better authentication performance than using any biometric modality alone. All the three examples above show that multimodal biometrics is instrumental in enabling identity recognition in uncontrolled environments.

10.1.4 2D and 3D Face Recognition

One of the most ubiquitous biometric modalities is arguably the face. Even before photography became common, drawn facial portraits had already been used for wanted criminals. Alphonse Bertillon invented a criminal identification system in 1882, now, known as the Bertillion system, that takes anthropometric measures, as well as a mugshot of face images.

Today, cameras are already equipped with face detection capability. Facebook provides a face-tagging service which automatically recognizes the face of a person in an uploaded photograph. Its face recognition engine, formally provided by Face.com but now is part of Facebook, can recognize images in highly unconstrained environments [12]. For this reason, and the fact that most mobile devices are installed with face recognition software, we also provide a brief discussion of automatic 2D and 3D face recognition in this section.

The process of face recognition normally involves three main stages: detection, feature extraction (dimensionality reduction) and classification (storage). The detection stage typically involves facial localization and normalization in order to handle viewpoint and lighting changes.

The face recognition technology can be categorized by the type of sensor used, which can be based on image sensors, video cameras, and depth sensors. 2D face recognition is by far the most common type of face recognition. Image sensors such as digital charge-coupled device (CCD) or complementary metal-oxide-semiconductor (CMOS) active pixel sensors can be produced cheaply and are small enough to fit inside every personal device.

Early 2D facial recognition algorithms are based on holistic approaches such as principal component analysis (PCA) – the image representation of which is known as Eigenfaces; or, linear discriminant analysis (LDA) – the resultant facial representation of which is known as Fisherfaces. Although the results are satisfactory, they lack robustness, since they all suffer from the common 2D drawbacks due to high variation in images.

Consequently, parts-based methods are consequently used [13]. These methods break up images into local regions or components, so that recognition is done by parts. The component-wise results are then subsequently combined to form the final output hypothesis.

Video-based face recognition research [14, 15] extends 2D face recognition by considering multiple images taken consecutively in time, in order to eliminate the uncertainty that is present in a single 2D face image. Although frame-based face recognition methods using temporal voting schemes is common, more powerful methods aim either at combining hypothesis derived from the image frame level, or at obtaining images with a higher resolution than any of the single images. The latter technique is called superresolution for face recognition [16].

Another attempt to overcome geometric distortion due to head pose is to warp an off-frontal image to a frontal view via image warping. An advanced approach to do so is the Active Appearance Model (AAM). An AAM composes an image of interest into a shape and textual (appearance) model. When applied to facial images, interest points are often manually marked around a face image, so that the facial features are tracked over time. Facial images that are not frontal can be warped to a frontal one, thanks to the tracked points. The resultant warped image often produces better recognition performance than an original off-frontal one.

Inspired by AAM, this line of research exploits 3D models for 2D face recognition. The pioneering work of Blanz and Vetter (1999) [17] proposes a 3D morphable model that enables a 3D model to be fitted on a 2D image, so that a face image can be re-rendered in any viewpoint. This has vastly improved unconstrained 2D face recognition. A variant of this method was reported by Face.comTM (now part of Facebook^{TM}) which makes it possible to recognize faces from any arbitrary viewpoint.

The recent introduction on the market of the '2.5D' Kinect^{TM} sensor opens a new era for human-computer interaction and, implicitly, for face recognition. Its software development kit, known as *Kinect Identity*, enables instantaneous face recognition of the players in real time. The sensor provides both depth information and also a visible image that are always aligned and synchronized. This makes it possible to implement 4D unconstrained face recognition [18] that involves 3D data along with time.

10.1.5 A Multimodal Case Study

For the rest of this chapter, we will present an in-depth case study on mobile biometry. A mobile device is ubiquitous; it is user-friendly for personal communication, anytime and anywhere.

There are several potential use cases of biometrics when implemented on a mobile device. Firstly, biometrics can prevent a mobile device from being used illegitimately if lost or stolen. Secondly, it can be used to digitally sign audio, text or image files, providing proof of their origin and authenticity [19]. We shall refer to biometric authentication on mobile device as "mobile biometry", or MoBio.

The MoBio project provides a software verification layer that uses your face and voice, captured by a mobile device, to ensure that you are who you claim to be (Figure 10.1). The software layer not only authenticates the face and voice, but also combines them to make the system more robust. It also updates system models in order to allow for changing conditions over time – all within the hardware limitations of a consumer-grade mobile platform. Though other studies have investigated face and voice authentication [20, 21], MoBio is the

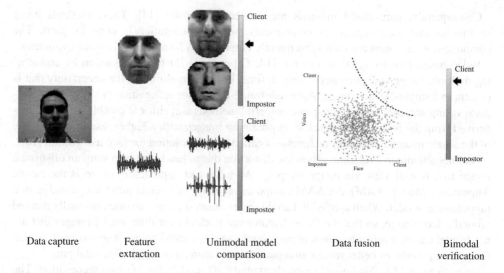

| Data capture | Feature extraction | Unimodal model comparison | Data fusion | Bimodal verification |

Figure 10.1 The MoBio identity verification system computes feature vectors from the captured and normalized face and voice, compares the features to stored models, fuses the scores for improved robustness, and performs a bimodal verification.

first to assess bimodal authentication under the challenging conditions imposed by a mobile architecture (e.g., limited processing power, shaking of the handheld camera).

The key-enabling technological components of the project include:

- Rapid face detection with an exponential reduction in false positives [22].
- Highly efficient facial feature localization algorithm implemented on the Nokia N900 mobile device, operating at frame-rate performance [23].
- Image descriptors for improved face verification [24, 25].
- Efficient speaker verification using a novel features based on the i-vector approach [26], combined with a method to decouple core speaker recognition from session variability with limited training data [27].
- A score-level classifier fusion algorithm EER by accommodating a wider range of capture based on [28].

10.1.6 Adaptation to Blind Subjects

There are several engineering, as well as user-interaction, challenges related to mobile biometry. From the engineering viewpoint, the realization of mobile biometry is made difficult mainly due to reduced computation capability (i.e., smaller memory, computation power, limited support for floating-point calculation), compared to that of a desktop computer. Due to the portability of the device and the way it is being used – any time and anywhere, the captured biometric data can, potentially, be of poor quality. For instance, it is well known that the performance of speech recognition can be severely degraded when recorded under a noisy environment [29].

From the user-interaction perspective, the mobile biometry problem is made difficult by the following factors:

- Dependency on the skill of the user: The entire process of capturing the face and speech biometrics relies on the skill of the user.
- Physical disability of the user: Users with visual impairment are likely to be excluded from the use of mobile devices for biometric identity verification.

Concerning the last point, according to the World Health Organization, there are over 161 million visually impaired globally and most of them are older people[2]. Considering that a potentially large population of users may be affected by visual impairment, we will address this issue in the second part of this chapter.

Therefore, in parallel to the development of MoBio, we also explored how the platform might be used by blind users. From the outset, we identified that the facial biometrics would be extremely challenging, because blind users cannot make use of the visual cue about how well the camera has captured their face images.

Although advancements in face recognition suggest that bad illumination can be corrected [30, 31], and non-frontal pose can be rectified [32], the image restoration process is nevertheless not satisfactory.

Furthermore, variation in facial expression can also negatively impact on the recognition performance. Among these factors, head pose is arguably the most difficult to rectify since a perfect restoration procedure is very complex and computationally expensive [33].

Figure 10.2 gives an intuitive explanation of why head pose can severely impact on a face recognition system. Referring to this figure, under a perfect frontal pose, the system can give a

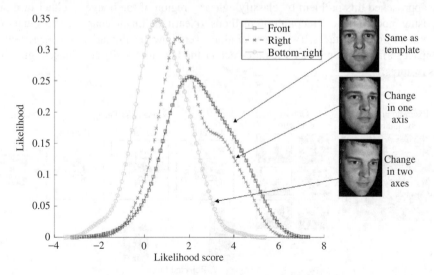

Figure 10.2   Effect of head pose on the face recognition system. Higher likelihood score implies more resemblance to the true identity. As the head pose changes, thus increasing in degrees of freedom, the likelihood score is likely to decrease, ultimately causing false rejection of the identity claim.

---

[2] Source: http://www.who.int/blindness/causes/magnitude/en

high likelihood that a processed face image belongs to the genuine person claiming to be that individual. However, as soon as there is a change in pose, the likelihood score values decrease and move closer to those of impostors (low likelihood scores imply that an image belongs to an impostor). As can be observed, the change of head pose in two axes (pan and tilt) affects the scores more severely than the head pose change in one axis (tilt, in this example).

10.1.7   Chapter Organization

We will present the design challenges of the Mobio platform in Section 10.2, and then address the usability issue of the platform for use by blind subjects through audio feedback in Section 10.3. This is followed by discussions and conclusions in Section 10.4.

10.2   Anatomy of the Mobile Biometry Platform

10.2.1   Face Analysis

10.2.1.1   Face Detection

To capture the user's appearance, we begin with an image that contains (somewhere) the user's face, and localize the face in the image to get a rough estimate of its position and size (Figure 10.3). This is difficult, because appearance varies in the image, and our system must detect faces regardless of shape and size, identity, skin color, expression and lighting. Ideally, it should handle different orientations and occlusion but, in mobile verification, we assume the person is looking almost directly into the camera most of the time.

We approached this problem by classifying every region of the image as either face or not face, using modern pattern recognition methods to learn the image characteristics that differentiate faces from non-faces. Two considerations were how to summarize the image region in a compact form (i.e., compute its feature vector) and how to classify the image region based on its features.

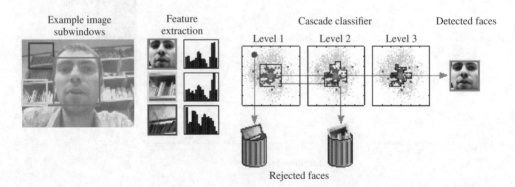

Figure 10.3 A "window" slides across the image, and the underlying region is sampled and reduced to a feature vector. This feature vector feeds into a simple classifier that rejects obvious non-faces; subwindows that are accepted then feed into a succession of more complex classifiers until all non-faces have been rejected, leaving only the true faces.

When searching an image, there are many thousands of possible locations for the face and it is important that every image region be summarized quickly. Using a variant of the Local Binary Pattern [34], we summarized local image statistics around each pixel with a binary code indicating the image gradient direction with respect to its eight neighbors (Figure 10.5). We then computed the histogram over transformed values for each patch, and fed this into a classifier to decide whether the patch was 'face' or 'not face'. In practice, we used a cascade of increasingly complex classifiers [35] to reject most image regions (ones that look nothing like a face) using simple but efficient classifiers in the early stages. The more accurate but computationally demanding classifiers were reserved for the more challenging image regions that look most face-like.

Our experiments on standard datasets (e.g., BANCA and XM2VTS) suggested that these methods detect over 97% of the true faces. In our application, however, we also prompted the user to keep their face in the centre of the image, so that we could restrict the search to a smaller region, thus reducing false positives and permitting more discriminative image representations that increase detection rates further.

To extend this baseline system, we developed a principled system that exponentially reduced false positives (background regions wrongly labeled as 'face') and clusters of several detections around the same true face, with little or no reduction in the true acceptance rate [21].

10.2.1.2 Face Normalization

Though we could try to recognize the user from the rectangular image region approximately surrounding the face in the image, performance will be impaired by factors such as background clutter, lighting, and facial expression. We therefore remove as many of these effects as possible, by normalizing the face so that it has similar properties (with respect to shape and texture) to the stored model of the user (Figure 10.4). First, we localize individual facial features such as the eyes, nose, mouth and jawline, and use these to remove any irrelevant background. Next, we stretch the face to fit a pre-defined shape, thus compensating for differences due to the direction in which the person is facing, their expression, and the shape of

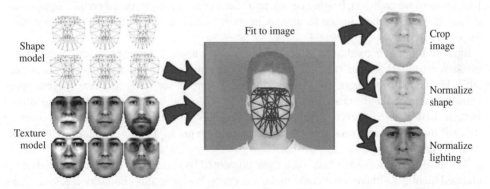

Figure 10.4   Statistical models of shape and texture are estimated from training data and fit to a new image using the Active Appearance Model. The underlying image can then be sampled to remove background information, warped to remove irrelevant shape information (e.g., due to expression), and normalized to standardized brightness and contrast levels.

their face (a weak cue for verification). Finally, we normalize lighting by adjusting brightness and contrast to some fixed values. The resulting image can then be directly compared with a similarly normalized model image for accurate verification.

To locate facial features, we fitted a deformable model to the image using a novel version of the Active Appearance Model (AAM) that we developed specifically for a mobile architecture using modern machine learning techniques [23]. The AAM uses statistical models of shape and texture variation – learned from a set of training images with hand-labeled feature locations – to describe the face using only a few parameters. It also learns to detect when the model is in the wrong place, and to correct parameters to align the model with the image. To predict these corrections, we trained a linear regressor to learn the relationship between sampled image data and true parameter values, using image samples with known misalignments.

When fitting the model to a new image, we initially aligned the model with the coarse face detection result, then sampled and normalized the corresponding part of the image (Figure 10.4). Our method then predicted and applied a correction to the shape and pose parameters to align the model more closely to the image. By repeating this sample-predict-correct cycle several times, we converged on the true feature locations, giving a normalized texture sample for verification.

Compared with the AAM, our approach achieved similar or better accuracy (typically within 6% of the distance between the eyes), while achieving a threefold speedup on a Nokia N900, reducing processing time from 44.6 ms to 13.8 ms and therefore reaching frame-rate performance [23]. Though this performance was achieved using a model that was trained from publicly available datasets, and which could be adapted to a specific user by retraining the predictor (online or offline), our results suggest that performance would improve little in return for the added computational cost.

10.2.1.3   Face Verification

Given a normalized image of the face, the final step is to assign a score describing how well it matches the stored model for the claimed identity, and to use that score to decide whether to accept or reject that person's claim (Figure 10.5). Again, we treat this is a classification problem, but here we want to label that person as a client or an impostor, based on their appearance as summarized by their image features. Clients are given access to the resource they need; impostors are not.

Since illumination conditions affect appearance, we applied gamma correction, difference of Gaussian filtering, and variance equalization to remove as many lighting effects as possible. For added robustness, we then subdivided the processed images into non-overlapping subwindows to make the descriptor more robust to occlusion, and computed the Local Binary Pattern (LBP) value for every pixel over three scales. We then summarized every window by its LBP histogram, and used the concatenated histograms as a feature vector for the whole image (Figure 10.5).

To classify an observed feature vector, we computed its distance to the stored model of the claimed identity. Although we could make a decision based on this similarity measure alone, we instead used a robust likelihood ratio, whereby the distance to a background model provided a reference that expressed how much more than average the observation matched the claimed

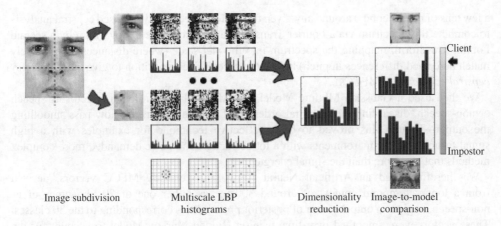

Image subdivision   Multiscale LBP   Dimensionality   Image-to-model
                            histograms   reduction   comparison

Figure 10.5 A cropped face window is subdivided into blocks, each of which is processed with a Local Binary Pattern operator at several scales. We then capture the distributions of LBP values in histograms that we concatenate and reduce in dimensionality (e.g., via Principal Component Analysis) before comparing with the stored model.

identity, thus indicating our confidence in the classification. Using this method, we were able to achieve half total error rates (where false acceptances are as likely as false rejections) of around 5% on the BANCA dataset.

We also developed a number of novel image descriptors that improved recognition performance. One based on Local Phase Quantization was designed for out-of-focus images and achieved a recognition rate of 93.5% (compared with 70.1% for Local Binary Patterns) on a face image that had been blurred [24]. Developing this descriptor further to include information at multiple scales, we improved recognition rates in some cases from 66% to 80% on a more challenging dataset with widely varying illumination [25].

10.2.2   Voice Analysis

Though face verification technology is maturing, we also exploit the fact that we have a microphone at our disposal by including voice-based speaker verification in our system.

10.2.2.1   Voice Activity Detection

Given a sound sample that was captured using the mobile's microphone, our first step is to separate speech (which is useful for speaker recognition) from background noise (which is not). As in face detection, however, speech detection is complicated by variation from speaker to speaker (e.g., due to characteristics of the vocal tract, learned habits and language) and from session to session for the same speaker (e.g., as a result of having a cold).

To summarize someone's voice, we represented the variation in the shape of the vocal tract by a feature vector summarizing frequency characteristics over a small window (on the order of a few tens of milliseconds) around any given time. More specifically, we used cepstral analysis to compute this spectrum via a Fourier Transform and decompose its logarithm by a second Fourier Transform, mapping the spectrum into the *mel* scale (where distances more closely match perceived differences in pitch) before the second decomposition to give *mel-frequency cepstral coefficients* (MFCCs).

We then used a Gaussian Mixture Model (GMM) to classify a feature vector as speech or non-speech, discarding the temporal ordering of feature vectors and low-pass smoothing the output. Though this proved to be an effective technique for examples with a high signal-to-noise ratio, environments with a lot of background noise demanded more complex methods that use more than the signal energies.

We therefore used an Artificial Neural Network to classify MFCC vectors, derived from a longer temporal context of around 300 ms, as either one of 29 phonemes or as non-speech, to give an output vector of posterior probabilities corresponding to the 30 classes. These vectors were smoothed over time using a Hidden Markov Model to account for the (language-dependent) known frequency of phoneme orderings learnt from training data, and the 29 phoneme classes were merged to form the 'speech' samples.

Because this approach was computationally demanding (and therefore not well suited to an embedded implementation) we also proposed a simpler feature set, denoted Boosted Binary Features [Roy et al. (2011b)], that was based on the relationship between pairs of filter responses and that achieved performance at least as good as existing methods ($\approx 65\%$ correct classification over 40 possible phonemes) while requiring only modest computation.

10.2.2.2 Speaker Verification

Having discarded background noise, we can then use the useful segments of speech to compute how well the person's voice matches that of the claimed identity, and decide whether to accept or reject their claim.

To describe the voice, we used 19 MFCCs (computed over a 20 ms window) plus an energy coefficient, each augmented with its first and second derivatives. After removing silence frames via voice activity detection, we applied a short-time cepstral mean and variance normalization over 300 frames.

As a baseline, to classify the claimant's feature vector, we used Joint Factor Analysis based on a parametric Gaussian Mixture Model, where the weights and covariances of the mixture components were optimized at the outset, but the centers were specified as a function of the data. These weights, covariances and means were learned using a large cohort of individuals, and the subject-subspace was learned using a database of known speakers, pooling over sessions to reduce inter-session variability. The session-subspace was then learned from what was left.

When testing, we used every training example to estimate the speaker and session, and adapted a generic model to be user-specific. We then discarded the session estimate (since we were not interested in whether the sessions matched, only the speaker) and computed the likelihood of the test example, given the speaker-specific model. Score normalization then gave a measure to use for classification.

On the BANCA dataset, this baseline system achieved equal error rates of around 3–4% for speaker verification. We showed, however, that we could improve the related *i -vector* estimation approach (the current state of the art in speaker recognition) to make speaker modeling 25–50 times faster using only 10–15% of the memory, with only a small penalty in performance (typically increasing Equal Error Rate from $\approx 3\%$ to $\approx 4\%$ [26]).

We also demonstrated that decoupling the core speaker recognition model from the session variability model – thus allowing us to optimize the two models independently and giving a more stable system with limited training data – resulted in little or no penalty in performance [27]. Finally, we showed that using pairwise features achieved a Half Total Error Rate (HTER) of 17.2%, compared with a mean HTER of 15.4% across 17 alternative systems, despite being 100–1000 times more efficient [37].

10.2.3 Model Adaptation

One challenge in biometric verification is accommodating factors that change someone's appearance over time – either intentionally (e.g., makeup) or unintentionally (e.g., wrinkles) – as well as external influences in the environment (e.g., lighting, background noise) that affect performance. Therefore, the model of the user that was created when they enrolled cannot remain fixed – it must adapt to current conditions and adjust its criteria for accepting or rejecting a claim accordingly.

In experiments with face verification, we began by building a generic model of appearance from training data that included many individuals, enabling us to model effects such as lighting and head pose that were not present in every individual's enrolment data. We then adapted this generic model to each specific user, by adjusting model parameters based on user-specific training data. In our case, we used a Gaussian Mixture Model to represent facial appearance, because of its tolerance to localization errors. We also adapted the model a second time to account for any expected variation in capture conditions.

To account for changes in the capture environment (e.g., the BANCA dataset contains examples captured under controlled, adverse and degraded conditions), we computed parameters of error distributions for each condition, q, independently during training, and used score normalization such as the Z-norm,

$$z_q(y) = \frac{y - \mu_q}{\sigma_q} \tag{10.1}$$

or Bayesian-based normalization (implemented via logistic regression):

$$P(q|y) = \frac{1}{1 + \exp(\alpha_q y - \beta_q)} \tag{10.2}$$

to reduce the effect of capture conditions (where μ_q, σ_q, α_q and β_q are parameters estimated by learning). During testing, we computed measures of signal quality that identified which of the known conditions most closely matched the current environment, and adapted the classifier score accordingly.

In our experiments [28], normalizing the score reduced the equal error rate in some tests by 20–30% (from 19.59% to 15.31% for the face; from 4.80% to 3.38% for speech), whereas adapting the model to capture conditions had an even greater effect on performance, reducing equal error rates by over 50% in some trials (from 19.37% to 9.69% for the face; from 4.80% to 2.29% for speech).

10.2.4   Data Fusion

At this point, every sample in a video sequence has a score that tells us how much the person looks like their claimed identity and another score for how much they sound like their claimed identity. To give a system that performs better than either biometric on its own, we fuse these two modalities, either by classifying each modality independently and feeding the resulting pair of scores into a third classifier (score-level fusion), or by fusing the features and passing the result to a single classifier (feature-level fusion). Since we are concerned with video sequences, it is also beneficial to fuse scores (or features) over time.

A naïve approach is to score-level fusion pools data over time by averaging scores over the sequence. More principled methods model the distribution of scores over the observed sequence and compare this to distributions, learned from training data, that correspond to true and false matches. As a baseline, we computed non-parametric statistics (such as mean, variance and inter-quartile range) of the score distributions, and separated true and false matches using a classifier trained via logistic regression. Again, score normalization can be used to ensure that the outputs from different sensing modalities are comparable, while also taking into account the quality of the signal [38].

Though score-level fusion is popular when using proprietary software (where the internal classifier workings are hidden), feature-level fusion can capture relationships between the two modalities. Feature-level fusion may, however, result in a large joint feature space where the "curse of dimensionality" becomes problematic, and we must also take care when fusing sources with different sampling rates (e.g., video and audio).

We therefore developed a novel feature-level fusion technique, dubbed the "boosted slice classifier", that searched over the space of feature pairs (one face and one speech) to find the pair for which quadratic discriminant analysis (QDA) minimized misclassification rate, iteratively reweighting training samples in the process. Although this approach had only a small effect under controlled conditions, it outperformed the baseline score-level fusion system when one modality was corrupted, confirming that fusion does, indeed, make the system more robust.

In a different experiment, the benefit of fusing modalities was more pronounced, as indicated by the Detection Error Tradeoff curves (Figure 10.6a).[3] This illustrates the tradeoff between false rejections and false acceptances for varying thresholds of the classifier score – accepting more claimants reduces false rejections but increases false acceptances (and vice versa).

10.2.5   Mobile Platform Implementation

To run the system on a mobile device, we need to consider the limitations of the available hardware such as low-power processing, a fixed-point architecture and limited memory. We therefore carried out experiments that looked at the effect on accuracy when making approximations that would make the system more efficient.

One very effective modification was to implement as many methods as possible using fixed-point (rather than floating-point) arithmetic; although some modern devices are equipped with floating-point units, they are far from common and are less efficient. Other ways to

---

[3] The DET curve shows the same variables as the Receiver Operating Characteristic (ROC) curve but on a logarithmic scale; this makes the curve almost linear and gives a more uniform distribution of points, making interpretation easier.

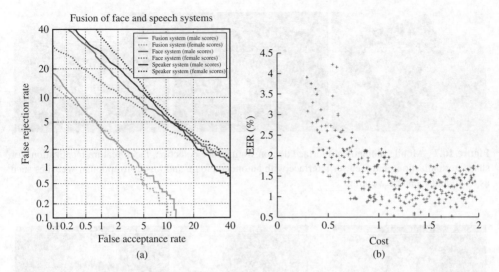

Figure 10.6   (a) Detection Error Tradeoff curves for unimodal and fused bimodal systems tested on the MoBio database; plots show false acceptance and false rejection rates on a logarithmic scale for a range of decision thresholds, where the lower left point is optimal. The equal error rate (EER) for a given curve occurs at the point where the curve intersects the line $y = x$. (See color figure in color plate section). (b) EER vs. efficiency for various scaled systems, confirming that better accuracy comes at a cost, defined as the lower of two proportions (memory consumption and time taken) with respect to a baseline system.

reduce computation included applying an early stopping criterion for the face detection and reducing the number of iterations used in facial feature localization. Because reducing memory consumption also has performance benefits, we made further gains by reducing parameters such as the number of LBP scales, the dimensionality of feature vectors and the number of Gaussian mixture components for speech verification. As a quantitative evaluation of these approximations, we rated 1296 scaled systems (48 face × 27 speech) by two criteria: an abstract cost reflecting both memory consumption and speed, and resultant generalization performance measured by equal error rate. As expected, increasing efficiency came at a cost in accuracy, whereas increasing complexity resulted in much smaller gains (Figure 10.6b).

To test the system under real conditions, we developed a prototype application (Figure 10.7) for the Nokia N900 that has a front-facing VGA camera for video capture, a Texas Instruments OMAP3 microprocessor with a 600 MHz ARM Cortex-A8 core, and 256 Mb RAM. Using GTK for the user interface and gstreamer to handle video capture, we achieved near frame-rate operation for the identity verification system.

10.2.6   *MoBio Database and Protocol*

One major difference between the MoBio project and other related projects is that the MoBio system is a *bimodal* system that uses the face and the voice and, therefore, needs a bimodal dataset on which to evaluate performance. Many publicly available datasets, however, contain either face data or voice data but not both. Even those few that do include both video and audio [20, 21] captured the data using high-quality cameras and microphones under controlled

Figure 10.7 Mobile biometrics interface demonstrating face detection, facial feature localization (for shape normalization) and the user interface with automated login and logout for popular websites such as email and social networking.

Figure 10.8 Screenshots from database, showing the unconstrained nature of the indoor environments and uncontrolled lighting conditions.

conditions, and are therefore not realistic for our application; we are limited to a low-quality, hand-held camera. The few that come close (e.g., the BANCA dataset) use a static camera and so do not have the image 'jitter', caused by small hand movements, that we have to deal with.

Since we anticipate other mobile recognition and verification applications in the future, we used a handheld mobile device (the Nokia N93i) to collect a new database that is realistic and is publicly available[4] for research purposes (Figure 10.8). This database was collected over a period of 18 months from six sites across Europe, contains 150 subjects and was collected in two phases for each subject. The first phase includes 21 videos per session for six sessions; the second contains 11 videos per session for six sessions. A testing protocol is also supplied with the data, defining how the database should be split into training, development and test sets, and how evaluation scores should be computed. This protocol was subsequently applied in a competition entered by 14 sites: nine for face verification and five for speaker verification [39].

10.3   Case Study: Usability Study for the Visually Impaired

This section explores the usability of face recognition system for visually impaired users. Inspired by the success of audio feedback in assisting the visually impaired in various human-computer interface applications, we attempted to improve the quality of the images

---

[4] http://www.idiap.ch/dataset/mobio

captured by the visually impaired using audio feedback. The problem will be addressed in several stages. In the first stage, we extensively evaluated the impact of head pose on image quality and on the face verification performance. In the second stage, a prototype system was developed to integrate the head pose scores with increasing frequencies and tempos to provide a user-interaction mechanism and feedback. Finally, the last stage consists of conducting experiments with visually impaired subjects, interacting with a face authentication system enhanced by head-pose-driven audio feedback.

10.3.1   Impact of Head Pose Variations on Performance

In order to quantify how head pose can impact the performance of a face recognition system, we need a database annotated with the ground-truth pose information. The database should contain one degradation factor only, e.g., head pose variations, and should not contain other factors such as illumination, facial expression and background variations. For this purpose, we used a database of 3D models consisting of 168 subjects collected at the University of Surrey [40]. For each subject, we rendered their 2D images at different tilt and pan angles, such that the angles are more densely sampled around a frontal face image and sparser towards the extreme poses. The mean images of each of the 81 poses used in this study are shown in Figure 10.9.

Let θ be a vector describing the pan and tilt angles, and $P(error|\theta)$ be the system error that is dependent on θ. Formally, we aspire to find a set of allowable θ_* such that an acceptable level of recognition error, δ, can be tolerated:

$$\theta_* \in \{\theta|P(error|\theta) < \delta\}$$

where δ is a small number.

In the sequel, $P(error|\theta)$ is approximated by Equal Error Rate (EER). EER is a point at which the probability of a false accept is equal to the probability of a false reject. For a perfect face authentication module, the error is zero; for a poorly performing system, its error can be at most 50% (beyond which the system would have accepted an impostor and rejected a genuine user). Such an approximation implies that the error estimate enforces equal prior class probability. This is desirable, considering that there are many more non-match (impostor) accesses than match (genuine) ones in a typical biometric experiment.

In order to estimate EER, all the accesses of each of the 150 legitimate users are matched against the remaining 18 users (who serve as impostors), and this is done for each of the 81 possible head poses.

We expect that EER would varies as a function of tilt and pan. This conjecture is confirmed by Figure 10.10, which serves as a basis for the specification of our prototype system. For instance, based on the above results, in order to attain EER below 5% (hence setting $\delta = 0.05$), the head pose variation should be within five degrees in both the pan and tilt directions (the range of values for θ_*). On the other hand, if the accuracy is relaxed to, say less than 15% EER, a greater head pose variation would be tolerated.

10.3.2   User Interaction Module: Head Pose Quality Assessment

While the previous section provides a specification of how head pose would affect the system performance, this section examines the mechanism that can be used to drive the user feedback.

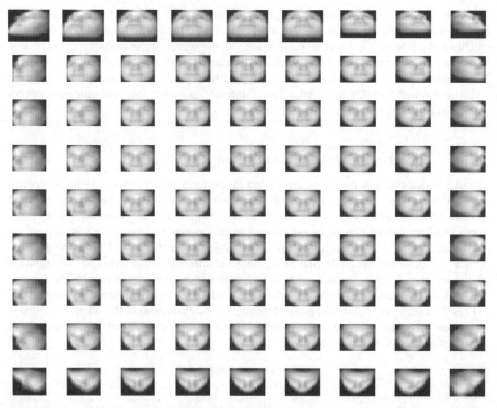

Figure 10.9 The 81 head poses generated from a single 3D model. Each image shown here is an average image of the training data at a given pose. The sampling angles in each of the pan and tilt directions are sampled in logarithmic scale as follows: $\{-45, -16.7, -5.8, -1.6, 0, 1.6, 5.8, 16.7, 45\}$.

In this work, we explored two approaches: (i) using the face detection confidence, and (ii) estimating the head pose.

10.3.2.1 Face Detection Approach

Face detection has been intensively studied for the last decade and many approaches have been proposed [41]. Among these are feature dimension reduction methods (e.g., Principle Component Analysis, Linear Discriminant Analysis), skin color analysis, filtering techniques and image-based methods (e.g., Adaboost, Neural Networks). In this study, an image-based face detection module using a cascade of classifiers is used [42], called WaldBoost. This is a variant of Adaboost, a state-of-the-art approach proposed by [43]. This detector is appealing because it can operate in real time, with images of varying resolutions, and it is not affected by cluttered background. In addition, can we use the output of the face detector as a quality assessment. The output is a log-likelihood ratio indicating how likely it is that a face is detected.

Let f denote the face detection output. Then, using the same database as before, we estimate $p(f|\theta)$, where θ is a vector of pan and tilt angles. The median value of this distribution is shown

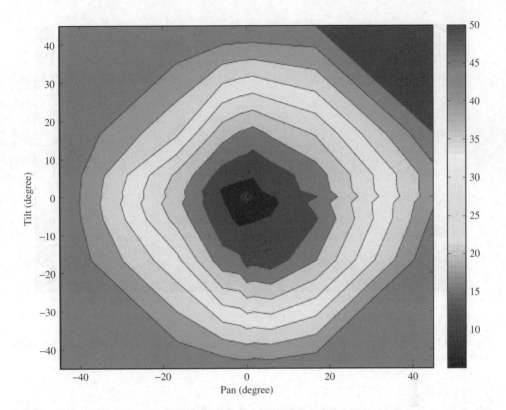

Figure 10.10 EER as a function of pan and tilt directions in degrees. (See color figure in color plate section).

in Figure 10.11. We note that this figure is correlated with the EER contour plot to some extent. This implies that it is feasible to drive the feedback based on the face detection output.

10.3.2.2 Head Pose Estimation Approach

While there exist many algorithms that can be used for head pose estimation [44], our choice is restricted by the application requirements on handheld devices: real time, lightweight computation, small memory consumption and coarse head pose estimation. The real-time requirement is important in order to feed the pose information back to the user.

In this study, we employed a dimensionality reduction method which meets these requirements – in particular, the Learning Discriminative Projections and Prototypes (LDPP) algorithm [45]. This algorithm simultaneously learns a linear projection base and a reduced set of prototypes for Nearest-Neighbor classification. However, since our task is regression rather than classification, the algorithm is modified slightly to cater for our needs. The description here explains briefly the algorithm and the introduced modification.

Let $x_{N \times 1}$ be a cropped image for an arbitrary pose, represented as a column vector having N image pixels (in gray level). The projected image (of size $b \times 1$) can be written as:

$$\widetilde{x} = B^{\mathrm{T}} x$$

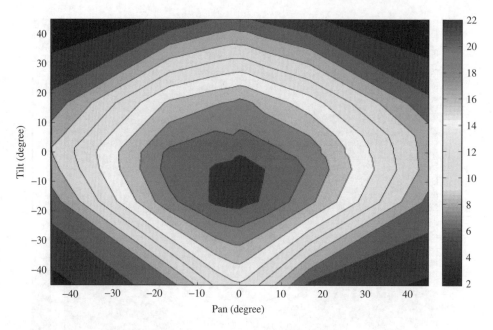

Figure 10.11 A contour plot of the median of face detection output as a function of pan and tilt directions in degrees. (See color figure in color plate section).

where $B_{N \times b}$ is a projection basis matrix and T is a matrix transpose operation. The number of bases to use, b, is determined by trading off the generalization performance for speed of computation requirement (smaller b implies smaller computation). Note that the projection basis B is not necessarily orthogonal, since it is obtained by LDPP via gradient descent.

Let θ be a bi-variate vector consisting of the tilt and pan angles of a head pose. Using the 81 discrete poses as defined by angle vectors θ_i, for $i = 1, 2, \dots, 81$, we can effectively cover the continuum of the entire view-based head pose range. Furthermore, let p_i be the prototype (the mean image) of one of the 81 head poses (as shown in Figure 10.9) and $\widetilde{p}_i = B^T p_i$ be its corresponding projected vector.

The original formulation of LDPP solves the classification problem via the nearest-neighbor rule, i.e., a query sample $\widetilde{x}$ is assigned the class label whose prototype $\widetilde{p}_i$ is closest to the query sample. However, since our problem here is regression, the nearest-neighbor rule is not applicable here. Instead, we need a function that quantifies how similar a query sample $\widetilde{x}$ is to a given prototype $\widetilde{p}_i$, for all possible poses spanned by i. The similarity measure should give a high response when $\widetilde{x}$ is near $\widetilde{p}_i$, eventually producing a peak value when $\widetilde{x} = \widetilde{p}_i$. Conversely, when $\widetilde{x}$ is far from $\widetilde{p}_i$, the measure should be small, eventually attaining zero.

A possible measure exhibiting the above characteristics is the radial basis function (RBF), also commonly referred to as the Gaussian kernel, having the form $\exp - \frac{\|\widetilde{x} - \widetilde{p}_i\|^2}{2\sigma^2}$, where σ is the kernel width, a parameter that controls how drastic the measure will drop as the sample $\widetilde{x}$ is located further away from the centroid $\widetilde{p}_i$. The optimal value of σ is data- and problem-dependent (since it is defined on the manifold spanned by $\widetilde{p}_i, \forall_i$) and is determined by experiments. We found that $\sigma = 1$ is appropriate for our task at hand. The RBF, when used in the context of other poses, can be interpreted as the posterior probability of a head pose, i.e.,

$$P(\theta_i | \widetilde{x}) = \frac{1}{Z} \exp \frac{(-\|\widetilde{x} - \widetilde{p}_i\|^2)}{2\sigma^2}$$

where Z is a normalizing factor such that axiom of probability is respected, i.e., $\sum_i P(\theta_i | \widetilde{x}) = 1$. From this, it follows that $Z = \sum_i \exp(-\frac{\|\widetilde{x} - \widetilde{p}_i\|^2}{2\sigma^2})$. Then, the expected head pose is:

$$\widehat{\theta} = \sum_i \theta_i P(\theta_i | \widetilde{x}) \forall_i \text{s.t.} P(\theta_i | \widetilde{x}) > \eta$$

which is, in essence, an expectation operator (in the usual statistic sense) under the posterior distribution of head poses, given the observation x in the reduced dimension, $P(\theta_i | \widetilde{x})$.

The condition $P(\theta_i | \widetilde{x}) > \eta$, where η is a small value, is in order here because in case the original image x is not a face image, the response of RBF, hence that of $P(\theta_i | \widetilde{x})$, is likely to be random to the extent that $P(\theta_i | \widetilde{x})$, for all i's, will be small. The consequence of this is that $\widehat{\theta}$ will converge to the mean value. By setting η, one effectively rules out the head pose whose corresponding RBF response is too small.

Another sanity check that we have considered is to ensure that x is, indeed, a face prior to head pose estimation. This is achieved by using the face detection confidence, via the quality(f) function, as discussed in Section 10.3.2. Since the details of head pose estimation are not extremely crucial in this paper, the validation of this approach is not further discussed.

Prior to concluding this section, it is instructive to visualize the distribution of samples spanned by $\widetilde{x}$. For this purpose, we chose five distinctive head poses, covering essentially frontal, upper left, upper right, lower left and lower right poses. The scatter plots of these poses on the test data sets are shown in Figure 10.12. As can be observed, the poses are somewhat well separated.

10.3.3 User-Interaction Module: Audio Feedback Mechanism

A generic way to feed the head pose quality information back to the user is to continuously assess the probability of error and instantaneously control a feedback mechanism (see Algorithm 1).

Let us define the quality information to be $q = [f, \widehat{\theta}]$, which is composed of the face detection confidence and an estimated head pose. The quality-conditioned probability of error to be characterized by $P(error|q)$. In the literature, several methods exist that could be used to estimate $P(error|q)$, such as generalized linear mixed model (GLMM) [46] and logistic regression (noting that logistic regression is a special case of the former). The advantage of using GLMM is the possibility of identifying several factors or covariates simultaneously, e.g., gender, presence of classes, ethnics, etc.

Although the vector q is rich in information (i.e., containing the estimated head pose and face detection confidence), it is not clear at this point how this rich information can be conveyed to the user in a meaningful way, for instance using 3D sound to guide the user, or giving explicit instructions. In either case, conveying this information may incur some mental workload upon users. Another observation is that the face detection confidence alone correlates strongly with the head pose, i.e., $P(f|q)$ and $P(error|q)$ happen to be closely related (compare Figure 10.10

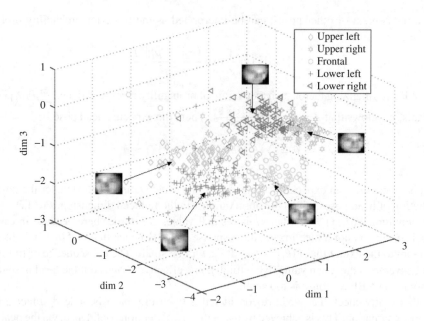

Figure 10.12   A scatter plot of five out of 81 head poses in three dimensions.

with 10.11). Based on the above reasoning and observation, instead of estimating $P(error|q)$ using a separate database, we opted for the following simpler deterministic function:

$$quality(f) = \begin{cases} \text{unknown} & \text{if } f \leq \Delta_{lower} \\ \text{non-frontal face} & \text{if } \Delta_{lower} < f < \Delta_{upper} \\ \text{frontal face} & \text{if } \Delta_{upper} \leq f \end{cases} \qquad (10.3)$$

which implies that the quality of a detected face is defined by a lower threshold (_lower) and an upper threshold (_upper) of the face detection output.

Algorithm 1   Head-Pose Driven Audio Feedback Module

```
δ ∈ ℜ: an error-tolerant threshold
  while true do
    Acquire a sample
    Estimate the quality, q
    if P(error|q) < δ then
      Perform matching
      exit
    else
      Produce feedback
    end if
  end while
```

An immediate utility of the above function is to drive the user feedback mechanism differently, depending on the state of the facial quality. In our knowledge, although there exist already biometric systems designed with feedback, they remain very basic. For instance, the feedback information consists of two states, each denoting the start and end of the data acquisition process. In our proposed feedback mechanism, a richer information (head pose) is conveyed to the user. However, this information is not as rich as using the estimated head pose, as discussed in the previous section. This is because we have no means of conveying this even richer information in a meaningful way to the user. This part of the exercise is left for a future research study.

Having determined the level of quality information to be fed to the user, the next issue is the actual type of feedback modality, which can be transmitted to the user via a screen display – hence visual feedback; or sound – audio feedback; and vibration – in tactile/haptic mode.

In this study, the audio feedback will be used. We created three different sinusoidal wave sounds at increasing frequencies, and the tempos were used to indicate three different qualitative stages (from unknown and non-frontal to frontal face).

In our study, the "unknown" qualitative stage is associated with a low-frequency sound played in a slow tempo; the "non-frontal face" stage is associated with a moderate frequency sound played at a faster tempo; and, finally, the "frontal face" stage has the highest frequency sound played at the fastest tempo. The frequencies used are respectively, 400, 800 and 1200 Hz.

The feedback is provided instantaneously and continuously during the acquisition process. Figure 10.13 shows the proposed system architecture when head pose quality assessment and feedback mechanism (the dotted lines square box) are integrated in a biometric system. When a user acquires a biometric data, the quality is checked in the quality assessment module. If the quality is considered high, the biometric data will be passed on to the feature extraction module, otherwise the system will feedback the quality to the user, and another interaction is necessary to acquire new biometric data. The process will continue until the timeout, or when a head pose with sufficient quality is acquired.

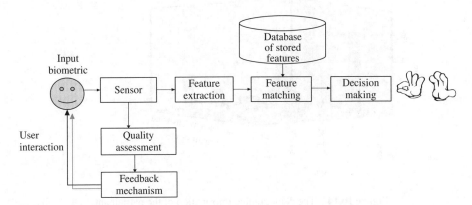

Figure 10.13   The system architecture of the proposed quality assessment and feedback mechanism in a biometric system.

10.3.4   Usability Testing with the Visually Impaired

This section explores the usability issue specifically related to how a visually impaired user interacts with a mobile biometry system. For this purpose, we recruited 40 visually impaired subjects from different age groups, genders, and level of impairment from St. Nicholas's Home, Penang, Malaysia. The demographic is depicted in Figure 10.14.

Although these participants can use mobile phones, they are not familiar with the camera functionality. Some of them have had undergone training to use computers. They are cooperative and sensitive to sound changes.

During the testing session, the participants were arranged in a quite room and asked to perform an image acquisition task using our prototype system. Each subject was told to capture his/her own face image as frontal as possible in a video clip. They had to perform it under three conditions: no feedback, with audio feedback, and with oral instruction followed by audio feedback. By audio feedback, we understand that sounds with different frequency tunes are played in order to provide an indication of their head pose, as in Equation 10.3. The mode of instruction, followed by audio feedback – audio + instruction – refers to the combination of audio feedback and oral explanation given to the volunteers about the association of the frequency and tempo to the image quality. In particular, the participants were instructed to hold their camera at arm's length just before starting the image acquisition process.

There are two possible outcomes in a session, i.e., either no face is detected or that a face is detected in the sequence. The first case is considered a failure, while the second case is considered a success. However, for the latter case, the level of success can still be further differentiated, depending on the confidence of the detected face. There are, therefore, two types of statistics that we can derive from each session of trial: success or failure of the image acquisition, and the value of confidence produced by the face detector in the case of a successful image acquisition.

We report the success/failure rate on visually impaired subjects for a given condition in Figure 10.15. This figure shows that the success rate of the audio feedback alone, as well as

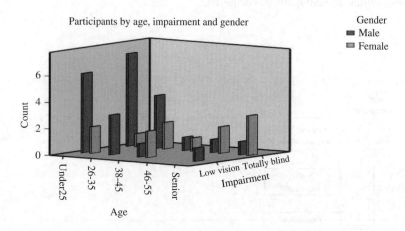

Figure 10.14   The demographic information of the participants.

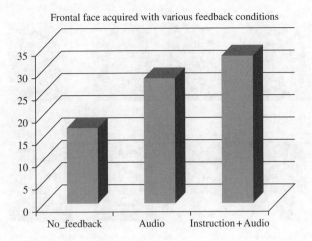

Figure 10.15   The confidence of the detected face under three experimental conditions.

that of instruction + audio, is significantly higher (both attaining 65% and 94%, respectively) than the baseline condition without feedback, which has a success rate of 49%. Therefore, the visually impaired can potentially use mobile biometry when the system is supplemented with audio feedback and instructions.

The procedure for obtaining the face detection confidence for each session of trial is as follows: when a sequence has no detected face, the facial detection confidence is set to the default zero value. However, when a face is detected, the highest value of facial detection confidence in the sequence is used. We show some of the detected face images in Figure 10.16.

We also performed a paired t-test on the continuous values of the detected facial confidence, in order to see if the mean value of the confidence of the detected faces of the three conditions are significantly different or not. The result is shown in Table 10.1. As can be seen, in each of the comparison under paired t-test, the results are all significant (under 5%). This further confirms the effectiveness of the proposed mechanism in improving the usability aspect.

Table 10.1   Paired significant t-test carried out under different conditions

Experiment	Paired differences					t	df	Sig. (2-tailed)
				95% CI				
	Mean	Std. Dev	SE Mean	Lower	Upper			
No FB vs. A	−4.64	8.42	1.40	−7.49	−1.79	−3.31	35	**.002**
No FB vs. A + I	−7.74	7.58	1.26	−10.31	−5.17	−6.124	35	**.000**
Audio vs. A + I	−3.01	4.86	.81	−4.74	−1.45	−3.824	35	**.001**

FB = Feedback, A = Audio, I = Instruction, CI = Confidence Interval

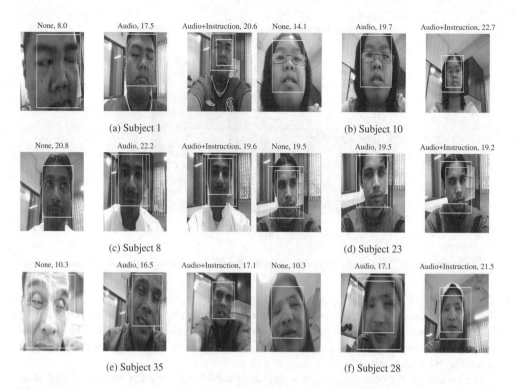

(a) Subject 1   (b) Subject 10

(c) Subject 8   (d) Subject 23

(e) Subject 35   (f) Subject 28

Figure 10.16   Example of detected facial images with the highest confidence in each of the four experimental configurations for six randomly subjects, all taken in session one. As can be observed, images taken with audio + instruction are likely to produce higher face detection confidence than audio or instruction alone.

10.4   Discussions and Conclusions

Identity assurance plays a central role in our daily lives. Biometrics, as an enabling technology that delivers this assurance, has to deal with a number of challenging issues. Although humans can recognize familiar people in very unconstrained environments, biometric systems can recognize millions of people, but under very controlled environments. The systems can be easily affected by sources of environmental noise, which humans can easily, and naturally, overcome.

In this chapter, we have explored multimodal biometrics as a viable means to take advantage of the diverse ways human interacts with the system. We have surveyed a number of scenarios where multimodal biometrics are required. In addition, we also outlined a new system for identity verification on a mobile platform that uses both face and voice characteristics. Specifically, state of the art video modules detect, normalize and verify the face, while audio modules segment speech and verify the speaker. To ensure that the system is robust, we adapt models to the estimated capture conditions and fuse signal modalities, all within the constraints of a consumer-grade mobile device.

In a separate case study, we also demonstrated that mobile biometry has a potential to be applied by visually impaired. Through providing appropriate feedback, the quality of the captured signals could be improved significantly.

Acknowledgements

This work was performed by the MOBIO project (http://www.mobioproject.org) 7th Framework Research Programme of the European Union (EU), grant agreement number 214324. The authors would like to thank the EU for the financial support and the consortium partners for a fruitful collaboration, especially Visidon Ltd., for their invaluable work in developing the mobile phone user interface. Rita Wong thanks the volunteers of St. Nicholas Home, Penang, Malaysia, for having participated in the usability case study.

References

1. Ross, A., Poh, N. (2009). *Fusion in Biometrics: An Overview of Multibiometric Systems*, chapter 8, 273–292. Springer, London.
2. Bhavan. Y., Marg, S. (2009). *Biometrics design standards for UID applications*.
3. Johnson, P.A., Tan, B., Schuckers, S. (2010). *Multimodal fusion vulnerability to non-zero effort (spoof) imposters*.
4. Rodrigues, R.N., Ling, L.L., Govindaraju, V. (2009). Robustness of multimodal biometric fusion methods against spoof attacks. *Journal of Visual Languages Computing* **20**, 169–179.
5. Altinok, A., Turk, M. (2003). *Temporal integration for continuous multimodal biometrics Multimodal User Authentication*, pp. 131–137.
6. Azzini, A., Marrara, S., Sassi, R., Scotti, F. (2008). A fuzzy approach to multimodal biometric continuous authentication. *Fuzzy Optimization and Decision Making* **7**(3), 243–256.
7. Sim, T., Zhang, S., Janakiraman, R., Kumar, S. (2007). Continuous verification using multimodal biometrics. *IEEE Trans. Pattern Anal. Mach. Intell.* **29**(4), 687–700.
8. Niinuma, K., Jain, A.K. (2010). *Continuous user authentication using temporal information*.
9. Li, X., Chen, G., Ji, Q., Blasch, E. (2008). *A non-cooperative long-range biometric system for maritime surveillance Pattern Recognition*. 19th International Conference on ICPR, 1–4.
10. Nixon, M.S., Bouchrika, I., Arbab-Zavar, B., Carter, J.N. (2010). *On use of biometrics in forensics: gait and ear*, 1655–1659.
11. Tresadern, P., Cootes, T., Poh, N., Matejka, P., Hadid, A., Levy, C., McCool, C., Marcel, S. (2013). Mobile biometrics: Combined face and voice verification for a mobile platform. *Pervasive Computing, IEEE* **12**(1), 79–87.
12. Taigman, Y., Wolf, L. (2011). Leveraging billions of faces to overcome performance barriers in unconstrained face recognition. *arXiv preprint arXiv: 1108.1122*.
13. Cardinaux, F., Sanderson, C., Bengio, S. (2006). User Authentication via Adapted Statistical Models of Face Images. *IEEE Trans. on Signal Processing* **54**(1), 361–373.
14. Gorodnichy, D. (2005). *Video-based framework for face recognition in video*. Proceedings of The 2nd Canadian Conference on Computer and Robot Vision, 330–338
15. Poh, N., Chan, C.H., Kittler, J., Marcel, S., McCool, C., Rua, E., Castro, J., Villegas, M., Paredes, R., Struc, V., Pavesic, N., Salah, A., Fang, H., Costen, N. (2010a). An evaluation of video-to-video face verification. *IEEE Transactions on Information Forensics and Security* **5**(4), 781–801.
16. Wheeler, F.W., Liu, X., Tu, P.H. (2007). *Multi-frame super-resolution for face recognition*. First IEEE International Conference on Biometrics: Theory, Applications, and Systems, BTAS 2007, 1–6.
17. Blanz, V., Vetter, T. (1999). *A morphable model for the synthesis of 3d faces*. Proceedings of the 26th annual conference on Computer graphics and interactive techniques, SIGGRAPH '99, 187–194. ACM Press/ Addison-Wesley Publishing Co., New York, NY, USA.
18. Schimbinschi, F., Wiering, M., Mohan, R., Sheba, J. (2012). 4D unconstrained real-time face recognition using a commodity depth camera. 7th IEEE Conference on Industrial Electronics and Applications (ICIEA), 166–173.

19. Ricci, R., Chollet, G., Crispino, M., Jassim, S., Koreman, J., Olivar-Dimas, M., Garcia-Salicetti, S., Soria-Rodriguez, P. (2006). *Securephone: a mobile phone with biometric authentication and e-signature support for dealing secure transactions on the fly.* Society of Photo-Optical Instrumentation Engineers (SPIE) Conference Series, vol. 6250 of Presented at the Society of Photo-Optical Instrumentation Engineers (SPIE) Conference.

20. Chetty, G., Wagner, M. (2006). Multi-level liveness verification for face-voice biometric authentication. *Biometric Symp.*

21. Teoh, A.B.J., Samad, S.A., Hussain, A. (2005). A face and speech biometric verification system using a simple Bayesian structure. *J. Inf. Sci. Eng.* **21**, 1121–1137.

22. Atanasoaei, C., McCool, C., Marcel, S. (2010). *A principled approach to remove false alarms by modelling the context of a face detector*, pp. 1–11, Aberystwyth, UK.

23. Tresadern, P.A., Ionita, M.C., Cootes, T.F. (2011). *Real-time facial feature tracking on a mobile device.*

24. Ahonen, T., Rahtu, E., Ojansivu, V., Heikkila, J. (2008). *Recognition of blurred faces using local phase quantization.* 19th International Conference on Pattern Recognition, ICPR 2008, 1–4.

25. Chan, C.H., Kittler, J. (2010). *Sparse representation of (multiscale) histograms for face recognition robust to registration and illumination problems.* 17th IEEE International Conference on Image Processing (ICIP), 2441–2444.

26. Glembek, O., Burget, L., Matějka, P., Karafiát, M., Kenny, P. (2011). *Simplification and optimization of i-vector extraction.*

27. Larcher, A., Lévy, C., Matrouf, D., Bonastre, J.F. (2010). *Decoupling session variability modelling and speaker characterisation.* Proceedings of the 11th Annual Conference of the International Speech Communication Association (INTERSPEECH).

28. Poh, N., Kittler, J., Marcel, S., Matrouf, D., Bonastre, J.F. (2010c). *Model and score adaptation for biometric systems: Coping with device interoperability and changing acquisition conditions.*

29. Furui, S. (1997). Recent advances in speaker recognition. *Pattern Recognition Letters* **18**(9), 859–872. Audio- and Video-Based Person Authentication.

30. Perronnin, F., Dugelay, J.L. (2003). *A Model of Illumination Variation for Robust Face Recognition Workshop on Multimodal User Authentication (MMUA 2003)*, pp. 157–164, Santa Barbara, CA.

31. Zou, X., Kittler, J., Messer, K. (2007). *Illumination invariant face recognition: A survey.* First IEEE International Conference on Biometrics: Theory, Applications, and Systems, BTAS 2007, 1–8.

32. Okada, K., Akamatsu, S., von der Malsburg, C. (2000). Analysis and synthesis of pose variations of human faces by a linear pcmap model and its application for pose-invariant face recognition system. Int'l Conf. on Automatic Face and Gesture Recognition, 142–149.

33. Salah, A.A., Nar, H., Akarun, L., Sankur, B. (2007). Robust facial landmarking for registration. *Annals of Telecommunications* **62**(1–2), 1608–1633.

34. Pietikainen, M., Hadid, A., Zhao, G., Ahonen, T. (2011). *Computer Vision Using Local Binary Patterns.* Springer.

35. Viola, P., Jones, M.J. (2004). Robust real-time face detection. *International Journal of Computer Vision* **57**(2), 137–154.

36. Roy, A., Magimai-Doss, M., Marcel, S. (2011b). Phoneme recognition using boosted binary features. Proc. IEEE Int. Conf. on Acoustics, Speech and Sig. Proc.

37. Roy, A., Magimai-Doss, M., Marcel, S. (2011a). *A fast parts-based approach to speaker verification using boosted slice classifiers.* DOI: 10.1109/TIFS.2011.2166387.

38. Poh, N., Kittler, J., Bourlai, T. (2010b). *Quality-based score normalization with device qualitative information for multimodal biometric fusion.* IEEE Trans. on Systems, Man, Cybernatics Part B : Systems and Humans **40**(3), 539–554.

39. Marcel, S., McCool, C., Matěcka, P., Ahonen, T., Černocky, J., Chakraborty, S., Balasubramanian, V., Panchanathan, S., Chan, C.H., Kittler, J., Poh, N., Fauve, B., Glembek, O., Plchot, O., Jančik, Z., Larcher, A., Lévy, C., Matrouf, D., Bonatre, J.F., Lee, P.H., Hung, J.Y., Wu, S.W., Hung, Y.P., Machlica, L.J.M., Mau, S., Sanderson, C., Monzo, D., Albiol, A., Albiol, A., Nguyen, H., Li, B., Wang, Y., Niskanen, M., Turtinen, M., Nolazco-Flores, J.A., Garcia-Perera, L.P., Aceves-Lopez, R., Villegas, M., Paredes, R. (2011). *On the results of the first mobile biometry (mobio) face and speaker verification evaluation ICPR (Contests).*

40. Tena, J. (2007). 3D *Face Modelling for 2D+3D Face Recognition.* PhD thesis, University of Surrey.

41. Hjelmas, E., Low, B. (2001). Face detection: A survey. *Computer Vision and Image Understanding* **83**, 236–274.

42. Kalal, Z., Matas, J., Mikolajczyk, K. (2008). *Weighted sampling for large-scale boosting.* Proc. BMVC.

43. Viola, P., Jones, M. (2001). *Rapid object detection using a boosted cascade of simple features.* IEEE Conference on Computer Vision and Pattern Recognition.

44. Murphy-Chutorian, E., Trivedi, M. (2009). Head pose estimation in computer vision: A survey. *IEEE Trans. Pattern Anal. Mach. Intell.* **31**(4), 607–626.

45. Villegas, M., Paredes, R. (2008). *Simultaneous learning of a discriminative projection and prototypes for nearestneighbor classification.* IEEE Computer Society Conference on Computer Vision and Pattern Recognition.

46. Givens, G.H., Beveridge, J.R., Draper, B.A., Phillips, P.J. (2005). *Repeated measures GLMM estimation of subject-related and false positive threshold effects on human face verification performance.* Proceedings of Computer Vision and Pattern Recognition (CVPR), 40.

11

Towards "True" 3D Interactive Displays

Jim Larimer[1], Philip J. Bos[2] and Achintya K. Bhowmik[3]

[1] *ImageMetrics, Half Moon Bay, California*
[2] *Kent State University, Kent, Ohio*
[3] *Intel Corporation, Santa Clara, California*

11.1 Introduction

We use all of our senses to interact with the environment, but there can be no doubt that vision provides the most immediate and important impressions of our surroundings. The signals that generate those experiences are in the light field. The light field is all of the information contained in light as it passes through a finite volume in space as we look in any direction from a vantage point within this small finite volume.

Displays today can reconstruct 2D and stereo-pair 3D (s3D) imagery, but there are visual signals in the natural environment that humans sense and use that are missing when images are reconstructed on these displays. This gap between a human's ability to extract information from the light field, and a similar ability to capture and reconstruct these signals, is being closed as improvements to cameras and displays replace current technologies used in computing and video entertainment systems. The missing signals are the subject of this section.

If you try to focus on an object that is out of focus in an image presented on a modern display, you will not be able to bring it into focus. This outcome can be irritating and uncomfortable. The information needed to bring the object into focus is absent in the signal as reconstructed on the display screen. 2D, s3D, and multi-view s3D displays do not reconstruct all of the signals that we routinely sample from the light field in the real world and, as a result, we are operating with an open control loop as we view images on these displays.

Of all the information in the light field, only a small fraction of it is sensed by human vision. There is information in frequencies outside of the narrow visual band of energies that we can see when we look about us. Additionally, there is information within the visual band of energies

Interactive Displays: Natural Human-Interface Technologies, First Edition. Edited by Achintya K. Bhowmik.
© 2015 John Wiley & Sons, Ltd. Published 2015 by John Wiley & Sons, Ltd.

we can see that we do not sample. We are insensitive to the polarization of light, we are unable to see large intensity differences or very small ones, small spatial details or rapid changes in time are invisible to us without special aids to vision, and we are not very good at identifying the precise spectral characteristics of the light we see.

In the vision research literature, the information in the light field that we cannot sense is said to be outside the window of visibility [1]. This window changes with viewing conditions – for example, the closer we are to a surface the more detail we can see on the surface and, as the level of illumination changes, so does our ability to see details and changes taking place during a brief temporal interval. Some information is always outside the window of visibility. Efficient engineering design requires knowing which signals the eye captures and which are outside the window of visibility. Including information that we cannot see is wasteful. Not including information that we *can* see and use means that the experience of viewing images on displays will be significantly different from viewing natural scenes with our unaided eyes.

Artists use all of the features and abilities of imaging technology to create unique visual experiences. Examples of this include the photographs of Ansel Adams, who manipulated image contrast to make images more dramatic, or the Joel and Ethan Coen, Roger Deakins movie, *O Brother, Where Art Thou?* that manipulated image color to give the impression of a drought to scenes captured during lush Mississippi summer months [2].

Not every manipulation of the signals captured from the light field and displayed on the screen will necessarily serve a desired purpose. Leaving out information needed for the visual system to close the loop can cause visual discomfort and fatigue [3]. The mismatch between where an object appears in space and where it is in focus in space in current s3D and multi-view s3D displays, and the visual discomfort that results, is an example of the kind of problem that can be created by not providing sufficient information for the visual task of viewing images on a display.

Ideally, imaging technology, cameras and displays would be capable of capturing or reconstructing all of the signals used by humans when interacting with the visual environment. For some tasks, capturing and reconstructing these signals faithfully is the appropriate goal. Artistic expression in visual media, however, implies that controlling and manipulating these signals is equally important. The image fidelity that matters to the artist is the reproduction of their artistic intent. For other applications – for example, medical imaging – it might be to detect disease. In this case, signal manipulation to enhance detection may be the engineering design goal. Nonetheless, these different intents all depend upon technologies capable of capturing or reconstructing a complete set of the visual signals sensed by humans. The choice to remove information from the reconstructed signal or to transform it should be by design, not by accident or due to technical limitations that could be overcome.

The process of sampling information from the natural light field with the unaided eye can produce distortions, artifacts and aliases, despite our belief that what we experience visually is an accurate representation of the environment. The Müller-Lyer illusion, shown in the upper left corner of Figure 11.1, is an example of how perception can distort the physical ground truth. These lines have exactly the same length, yet one appears shorter than the other. The colors we experience in photographs, in movies, and on displays are examples of a perceptual experience that can be generated by physically distinct input signals. The spectral energy distribution of the light imaged onto the retina when viewing a person's face is virtually never the same as the energy distribution when viewing their face in a photograph, despite the difference going unnoticed. The vision literature calls this phenomena metamerism; in communication theory, it

Figure 11.1 Three aspects of the visual system are illustrated in this figure. The Müller-Lyer illusion in the upper left, Mach Bands in the upper right, and a color appearance phenomenon related to spatial frequency and color contrasts in the bottom of the figure, all illustrate how the visual system interacts with the light field signal captured on the retina. There is a detailed discussion of these phenomena in the text. (See color figure in color plate section).

is called aliasing. Metamers, the technical name for two different spectral energy distributions that appear to have the same color, are rare in natural scenes but common in man-made ones.

The gray bars in the upper right of Figure 11.1 appear to be brighter on their left edge at the boundary with a darker bar, while at the same time appearing slightly darker on the right edge adjacent to a lighter bar. Despite the scalloped appearance of the bars the ground truth is that they are uniform in intensity within each bar. This phenomenon is called Mach Bands and it serves to enhance the visibility of the edges. It is an example of how evolution has evolved visual mechanisms that enhance perceptions to detect image details that might otherwise be less salient. This evolutionary enhancement is not unlike the goal in medical imaging to transform the image data to optimize the detection of pathologies.

The bottom of Figure 11.1 illustrates another feature of the human visual system. The words orange and pink are printed with the same color ink on the left and right sides of this illustration, but the colors appear very different on the right. Color perception is influenced by the surrounding imagery as much as by the signals created by the light at any particular location within the entire image. In this example, the spacing of the bars and whether or not a yellow or blue bar covers or goes under the letter determine the apparent color of the letters.

The spectral characteristic of the light reflected from surfaces depends upon the illuminant as well as the surface. The illuminant changes dramatically over the course of a day or in man-made illumination, yet most the time we are able to correctly identify the surface colors independent of these changes. In the example in the lower portion of Figure 11.1, however, a mechanism designed to discount the illuminant has produced a color perception different from the color of the surface. Magnifying the image in the lower right is all that is required to change the color percept. With increasing magnification, which changes the scale of the spatial relationships, the illusion of different colors for the letters vanishes.

The examples of artistic manipulations of imagery to achieve an artistic goal and of our visual system adding distortion, aliasing, and creating artifacts in the percepts corresponding to signals gathered from the light field, raise the issue of the purpose of vision. Why do we extract some information from the light field and not others, what is the evolutionary goal in seeing, and how closely related to the physical ground truth of objects are our perceptual experiences of them? To understand how missing signals from the light field in current imaging systems impact our perceptions and our ability to interact with machines, and to provide answers to some of these questions, it is useful to understand the evolutionary context of biological vision.

11.2 The Origins of Biological Vision

Biological sensory systems evolved shortly after the Cambrian explosion, 550 million years ago, when predation became a part of life. Vision evolved so that creatures could find food and avoid being eaten. The sense of vision plays a central role in cognition, the language we use to express ideas, and our understanding of the world about us. Vision provides the basic information we use to orient in our immediate environment, and it is the primary input data upon which we base many actions. Almost all ideas have images as correlates; a chair is a visual pattern, a tiger is a large cat. Even abstract concepts such as satisfaction can be imagined as a smile on someone's face. Visual cognition – understanding images – is not the mental equivalent of a photograph; our visual experience is more akin to Plato's concept of Ideals and Forms. We see people, objects, and actions, and not their images as projected onto our retinas. The process of seeing is dynamic and constructive; it is not a passive system.

Human vision is object-oriented. We use information extracted from the light field and neural signal processing based upon learning and memory to understand the environment we sense from the images projected onto our retinas. The image formed on the retina is the raw data for vision; it is not a sufficient signal for image understanding. To understand what we see, we change eye positions and focus to de-clutter or segment a scene into whole objects. Perception relies on experience as well as the immediately available data. This is obvious when you consider the common experience of finding and recognizing an object in a cluttered scene. Once an object is recognized and familiar, it is difficult, from that point onward, not to see it.

Images exist because we have a chambered eye with an entrance pupil similar to a pinhole camera such as the camera obscura shown in Figure 11.2. Understanding how our eyes extract useful information from the light field and the physics of light both began with the camera obscura. The camera obscura's connection to sight was described by Mozi and by Aristotle centuries ago and is featured in da Vinci's notes on light and imaging [4]. The idea that light from a point on any surface can be considered as rays emanating in all directions external to the surface, a central idea in geometric optics, is based upon the pinhole camera. Evolution discovered the pinhole camera shortly after the Cambrian explosion, and a chambered complex eye like ours evolved over 500 million years ago [5].

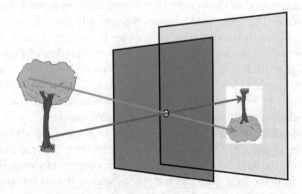

Figure 11.2 A camera obscura or pinhole camera is shown in this illustration. The discovery of the pinhole camera gave rise to geometric optics and the ray theory that postulates that light travels in straight lines.

Michael Faraday, in 1846 [6], was the first to describe light as a field similar to the field theory he developed for electricity and magnetism. Almost 100 years later, Gershun [7] defined the light field as the uncountable infinity of points in three-dimensional space where each point can be characterized as a radiance function that depends upon the location of the point in space and the radiance traversing through it in every direction (Figure 11.3).

Light traversing a point in the light field continues to move beyond the point until it encounters impedance that can alter its course by reflection, refraction, or extinction. Every line or ray passing through a point in the light field in terrestrial space will be terminated by two surfaces, one at each end. Every ray defined this way contains two packets of information, each going in opposite directions. If the ray is not very long and there is nothing to impede it, this information is almost entirely redundant at every point along the ray. The information carried on the ray is unique to the surfaces and light sources creating the packets, and this is the information biological vision has evolved to sample and use.

Adelson and Bergen [8] called Gershun's radiance function the 5-D plenoptic function, indicating that everything that is visible from a point in free space is contained in it. They described how our visual system extracts information from the light field to discover properties of objects and actions in our visual environment. The plenoptic function contains information about every unobstructed surface in the lines of sight to the point in space and, by adding time, the 6-D plenoptic function, how these surfaces change over time.

J. J. Gibson called the information we gather from the light field affordances [9], because it informs behavior; for example, when to duck as an object looms towards us. Information that is not useful in affording an action is not extracted from the light field. The Mach Bands illustrated earlier is evidence that our visual system sometimes enhances the information by signal processing to make features such as edges more salient and visible. Discounting the illuminant is an example of an affordance. It is more important for behavior to recognize the light-reflecting properties of a surface than it is to sense how the illuminant changes those signals. For example, it is important to detect ripe or spoiled food equally well in firelight or in sunlight. The illusionary colors in Figure 11.1 are actually examples of how the visual system attempts to remove changes in the illuminant that occur in shadowy environments, or as the illumination changes (e.g., when surface reflections alter the spectral content of illuminating a surface).

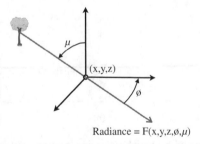

$$\text{Radiance} = F(x,y,z,\emptyset,\mu)$$

Figure 11.3 An arbitrary point in space is shown with a ray that originated on the trunk of a tree passing through it in a direction relative to the point at azimuth and elevation angles $\emptyset$ and μ, respectively. The radiance in this direction through the point is $F(x, y, z, \emptyset, \mu)$. This is Gershun's definition of a point in the light field; Adelson and Bergen called this the 5-D plenoptic function.

The physical ground truth of a perception can be considered in terms of the utility of the actions taken in response to a perceptual experience. Evolution has optimized percepts to generate behaviors that support survival. The percept of an object will correspond to the real physical properties of the object when those properties are critical to affording an appropriate action that assures the survival of the organism and, thereby, the genes it passes on to future generations [10].

A pinhole camera forms an image based upon half of the plenoptic function at the location of the pinhole. The pointing direction of the camera determines which half is selected. The ideal, or theoretical, pinhole camera consists of a vanishingly small hole or aperture that, in theory, would allow a vanishingly small amount of light through to form the image. In practice, the pinhole camera's aperture is always large relative to the wavelength of light and, therefore, consist of many ideal pinhole cameras, each slightly displaced spatially relative to their neighbors within the plane of the pinhole camera's aperture. Real pinhole cameras, as opposed to ideal ones, create an image that therefore is a composite of many ideal pinhole cameras, each projecting an image slightly displaced relative to its neighbor on the projection surface of the camera. The shifting and combination of images by light addition is illustrated in Figure 11.4; it is expressed as blur.

In the left panel, labeled A, in Figure 11.4 three points of the surface of three different objects within the field of view of a pinhole camera are shown. The arrowed lines indicate rays originating at these points as they pass through the pinhole and ending on the projection surface of the camera. In panel B, the pinhole has been moved slightly to the right and the corresponding locations of the projected points, again illustrated with the arrows, has shifted on the projection plane. The rightmost panel, labeled C, shows the composite of all of the idealized aperture projections when the pinhole is enlarged from its location in panel A to its location in panel B. Now a bundle of rays from each surface point send rays of light onto the projection surface. The resulting projections of each point on these three objects surfaces have grown larger on the projection surface in panel C. This is called blur, and it takes the shape of the pinhole aperture. Additionally, these regions of blur overlap, reducing image contrast in the overlapping regions. As the aperture is enlarged to let in more light, the resulting pinhole camera image is increasingly degraded by blur. The amount of blur depends upon the aperture's size and shape.

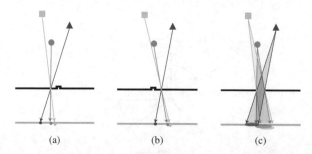

(a)   (b)   (c)

Figure 11.4 A pinhole camera with a variable size and location aperture is illustrated in this diagram. In panel A, a small aperture is located on the left and rays from points on three objects are traced to the projection surface of the camera. The small aperture is moved to the right in panel B illustrating parallax changes, and finally in panel C the aperture is enlarged from location A to B to illustrate blur created by the superposition of all of the pinhole images generated by small pinholes filling the larger aperture. (See color figure in color plate section).

The pinhole-chambered eye of the Nautilus can sense the direction from which images are formed on its circular projection surface or retina. It has been demonstrated in laboratory tests that the Nautilus's visual system is capable of sensing the direction of motion [5, pp. 57+] of objects in its field of view. This behavioral response to motion is evidence for the evolution of a neural signal processing system capable of extracting affordances from the light field of even a simple eye. This behavior is valuable to the Nautilus in escaping predation. It is an example of image understanding in the Nautilus's visual system.

The individual rays that pass through the aperture can also be characterized by the time it takes for light to travel along each ray from a point on the surface of the object, to a location within the camera's entrance aperture, and then onto the projection surface. Most of these distances will be slightly different and, therefore, associated with slightly different times. This information is called phase. Biological systems do not respond fast enough to measure the time differences, but phase has spatial consequences on the projection surface.

As the size of the pinhole aperture increases, more light is imaged onto the retina, improving sensitivity – but spatial resolution is lost due to blur. As a result, a more complex chambered eye, with a lens in the entrance aperture, similar to most cameras today, evolved during the Cambrian Explosion. The common snail, Helix, is an example of a simple chambered eye with a lens. Eyes with lens are found in the fossil record of the eye, going as far back as the just a few million years after the Cambrian Explosion began. A complex chambered eye similar to the human eye evolved within the first 50 million years of the Cambrian period [5].

A lens placed in the aperture takes all the rays emanating from single points on the surface of objects that are at the focal distance of the lens passing through the entrance aperture, and brings these rays to focus at single locations on the projection. Image details are critical to affording critical behaviors (e.g., foraging) so, in biological systems, eyes have developed lens systems capable of changing focus to bring surface detail into sharp focus. The lens, then, is a solution for the resolution loss caused by blur.

The lens solution requires some means of controlling the lens focus. Surfaces that are not at the focal distance remain blurry. Points on out-of-focus surfaces come to focus either in front or behind the projection surface. For these bundles of rays, their projection on the projection surface of the camera or eye are dispersed as blur. For cameras and eyes where the aperture is circular, the blurred points form circles of blur on the projection surface. When a lens is added to the system, the size of the blur circle will depend upon the relative distance of the surfaces, either in front or behind the focal distance of the lens. Just as in the case of the pinhole camera blur not only limits spatial resolution, it also reduces image contrast by overlapping circles of blur.

A naïve view of biological vision is that chambered eyes are insensitive to phase. Blur information is the product of phase differences, as light at any moment in time reflecting from points on a surface arrives at the entrance aperture of the camera or eye at slightly different times. The focus mechanisms in biological vision use blur to close the control loop, bringing the lens to focus at distances determined by attentive mechanisms within the visual system.

The amount of blur in a lens system depends upon the aperture size and the relative distance to the in-focus surfaces in the camera's field of view. Blur has been shown in the laboratory to permit the human visual system to estimate depth when it is the only cue present in the image [11]. The visual system is, therefore, capable of extracting useful information generated by phase differences in the images projected onto the retina in this limited sense of using phase correlated information sampled from the light field.

As the location of the idealized pinhole changes within the aperture of a camera, so does the spatial information contained in the plenoptic function. These changes are due to parallax. Moving the camera or the eye to the left or right is often all that is required to see around an occluding object, as this action results in sampling a slightly different set of plenoptic functions in the light field. However, some of this occluded information is available at every camera or eye position, because the entrance aperture of the camera or eye is not a single point in space. Only the parallax information contained in rays for points on in-focus surfaces is lost, because the lens places all of these rays on top of each other within the image projected onto the projection surface of the camera or eye. Information from points on occluded surfaces remains available on the projection surface.

The top diagram in Figure 11.5, labeled A, shows a ball focused onto the projection surface of a single lens camera; the projection surface is shown as a vertical line in the figure. The bundle of rays show the collective paths of all of the rays emanating from a point on the ball as they are collected by the entrance aperture of the camera and focused onto a point on the projection surface. A triangle behind the ball is partially occluded. Nevertheless, rays from a point on the triangle's surface that are on the camera's optical axis and occluded by the ball are still imaged onto the projection surface. These rays enter the camera near the peripheral edge of the camera's entrance aperture and are blurred on the projection surface.

Focusing the camera at the focal distance of the green triangle would bring the rays from the triangle into sharp focus. This is shown in the bottom diagram, labeled B. Cameras change focus by changing the distance to the projection surface, as shown in Figure 11.5, diagram B. A chambered eye would change the focal length of the lens to bring a different focal distance within its field of view into focus on the eye's retinal surface. These are nearly equivalent ways of achieving the same goal. Now the point on the triangle is in sharp focus, despite being

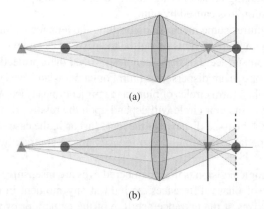

(a)

(b)

Figure 11.5 In the upper diagram in this figure, a ball with a triangle behind it is imaged onto the projection surface by a lens. In the bottom diagram, the triangle is brought to focus on the solid surface by changing the focal length of the lens. Despite the pyramid being occluded by the ball, it can still be brought to focus, because enough rays are collected through the periphery of the lens aperture in this illustration. The contrast of the triangle would be degraded (i.e., reduced) by the blurred image of the occluding ball. Changing focus would, in this limited way, be "looking through" the ball. (See color figure in color plate section).

occluded by the ball along the primary optical axis of the camera or eye. When this situation exists, changing the focal distance can make the occluded surface visible. The visibility of the occluded surface depends upon the size of the occluder, the separation between the occluded and occluding surfaces, image contrast, and the entrance aperture or pupil size. It is possible to see behind occluding surfaces by just changing focus.

Ordinary cameras do not capture the angular information contained in these rays and, once the image is captured, the information is lost. In biological vision systems, this information is sensed as blur and can be accessed by changing focus, or by translating the eye to a slightly different location. Because the center of rotation and the eye's optical nodal point are displaced relative to each other, a small eye movement produces a translational change and a new heading direction. Biological vision has evolved to use all of these techniques to access the parallax information in the light field.

Motion parallax is generated by object motion in the field of view of the camera or eye and by camera or eye motion. Motion parallax is captured in video recordings, and can give the viewer a strong sense of depth while watching a video sequence where objects in the field of view or the camera are moving. The Nautilus example mentioned earlier is evidence that the early evolution of biological vision systems created mechanisms capable extracting parallax information from the light field. Visual systems with two or more eyes with overlapping visual fields also extract parallax information from static imagery [5].

We are rarely aware of occasions when occlusions can be defeated by changing eye focus, but it is believed to be a common occurrence for small objects in the near field. Blur information in the image has been proven to be useful in perceiving depth [11]. Some of the parallax information in the light field that affords these behaviors is lost when imagery is gathered and archived by ordinary still and video cameras. Without this information, these behaviors are not possible. The ability to reconstruct this information in future imaging systems would, therefore, add useful information to current video imaging technology.

Cameras are not capable of understanding the image without an active process to analyze the image data. The only data captured by a traditional camera can be described as a two-dimensional array of intensities indexed by their location on the projection plane. In the case of the eye, the retina, where the image is formed, is populated with photoreceptors that encode the image as punctate neural signals. These signals are analyzed by a network of neurons within the visual pathways of the brain.

The distance to objects in the camera's field of view can be derived from the size of objects within the image, the occlusion of objects by other objects, from blur, from perspective, or from the loss of contrast due to scattering in the atmosphere. Parallax information projected as blur, even in static images, is a useful signal from the light field to derive object distances. In order for any of these signals to be used to extract distance information, the image processing must be able to segment the image into distinct objects. This is why biological vision and imaging processing for image understanding are both fundamentally object-oriented processes.

Our visual system has evolved neural mechanisms that use parallax to estimate the distance to objects and the closing rates of looming objects. These estimates are based upon data extracted from the light field sampled over a period of several milliseconds. Object recognition and identification, linear and aerial perspective, occlusion, familiarity, and other features of the imagery projected onto our retinas over time, are all used by the visual system to augment our image understanding of the location, size, range, surface relief, etc., of objects in the visual field.

11.3 Light Field Imaging

The information in the light field passing through the pupil becomes more or less accessible to visual cognition, depending upon the eye's focus. Focusing, as noted above, superimposes all of the rays passing through the pupil that originated at single points on the surfaces at the lens focal distance to single in-focus points on the retina. Rays originating on out-of-focus points are spread over the retina in circles of blur whose size depends upon focus and pupil diameter. Figure 11.6 illustrates focus for a schematic eye showing four bundles of rays that originated from each of the ends of two arrows. The arrow on far left is in focus while the other arrow is not.

Focusing the image on the projection surface aggregates information from all the rays passing through the pupil. When the information in these rays is correlated, (i.e., originated from the same point on an in-focus surface), the aggregate signal is strengthened. When the rays projected onto retinal locations come from several different surface points, and are therefore uncorrelated, information is mixed together and blurred, reducing the signal strength and image contrast. The information captured from the light field is not lost but to see it we must focus on different surfaces.

Image-capture and reconstruction in traditional 2D and s3D imaging systems do not support refocusing, and the only parallax information available in s3D is fixed by the location of the two cameras. When viewing natural scenes, we move our eyes and translate and rotate our heads to obtain more parallax information from the light field. We can refocus our eyes to rearrange and sort the bundles of rays coming from all the points that traverse our pupils.

The signal for focusing is missing with traditional image technology, as is a great deal of the parallax information, but our visual system has evolved to use this information. The result can be annoying and uncomfortable. For example, when viewing a large displayed image, the viewer might attempt to view an object on the screen that is out of focus or may move their head to see around something in the scene. No effort on the viewer's part can bring an out-of-focus object in a displayed image into focus, or allow them to see behind an occlusion.

A 2D video sequence, during which the camera moves, or in which objects are moving in the scene, can evoke the perception of depth in a scene. However, as soon as the motion stops,

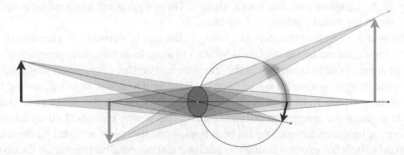

Figure 11.6 This diagram shows a schematic eye focused on the blue arrow. The transparent bundles are the rays from the end points of both arrows that form images of the end points on the retina. The leftmost arrow's image is sharp on the retina and the bundles converge to a point on the retina. The other arrow would come to focus on a projection plane behind the retina. Rays emanating from the this arrow's end points are spread or blurred over a wide area of the retina. When an out-of-focus object occludes an in-focus object, the blur reduces the image contrast of parts of the in-focus object. (See color figure in color plate section).

these motion-parallax-driven cues, along with the sense of depth, are lost. The inability to see around occlusions and the fixed image focus in standard 2-D and s3D imagery are the missing and/or inaccessible light field data.

Cameras that can retain the perceptually relevant information in the light field have been developed. The idea for this camera has many sources: Lippmann and Ives [12] and, more recently, Adelson and Wang [13] and Ng *et al.* [14]. Lytro recently introduced a light-field camera into consumer markets, and a commercial version of a light-field camera is available from Raytrix [15]. Understanding how these cameras capture the light field reveals what is required to build a true light-field display capable of reconstructing parallax information appropriate for any head position, and which supports attention-driven focusing.

The plenoptic camera design of Lytro and Raytrix is similar to an ordinary camera or to the eye in its basic design. To illustrate the principles of operation of this type of plenoptic camera, a spherical camera similar in geometry to the eye will be used. We will call the entrance aperture of the plenoptic camera the pupil, and the projection surface where images are formed will be called the retina. The eye's lens, called the primary lens, is located very close to the eye's pupil, and we will imply the same geometry for this plenoptic camera, although that is not a requirement [16].

A plenoptic camera has an array of tiny pinhole cameras located where the retina would be located in a real eye. These cameras can also have a lens, but that is not a requirement. The tiny cameras located on the projection surface capture images of the rays passing through the pupil. Every pinhole camera has its own unique entrance aperture, located at uniformly spaced positions within the array of tiny cameras. The uniform spacing is required to sample the light field uniformly, but this, too, is not a hard requirement. The human eye, for example, does not have a homogenous uniform array of photoreceptors, because the image we sense is a construction of the visual system and not simply a representation of the momentary image formed on the retina. Two of these pinhole cameras are illustrated, very much enlarged, in Figure 11.7; all of the other pinhole cameras are too small to be seen in this illustration.

Every tiny camera captures a slightly different image, depending upon its location in the array. Where rays from points in the 3D object space enter the plenoptic camera depends upon the location of these points in the visual field. Where rays traversing locations in the pupil of the plenoptic camera are projected onto the pinholes in the tiny camera array depends on the plenoptic camera's primary lens focus. A plenoptic camera does not require a primary lens that can change focus to make sharp images of any point in the visual field. A 2D image, reconstructed from the plenoptic camera's data and focused at a different focal distance than the primary lens, is made possible by rearranging the data captured by the array of tiny pinhole cameras.

Two rays from separate points in the 3-D object space entering the plenoptic camera at the same location in the pupil will generally be imaged to separate pinhole cameras. Two rays entering the plenoptic camera from the same point in 3-D object space will only be projected to the same pinhole in the tiny camera array if their point of origin is at the focal distance of the primary lens. This is illustrated in Figure 11.7, which traces four rays as they traverse the pupil and are projected onto the pinhole camera array. One ray from each of the two arrow tips passes through the center of the pupil; these rays are shown as dashed lines in the figure. A second ray from each arrow's tip passes through the same peripheral location in the pupil; these rays are represented as solid lines.

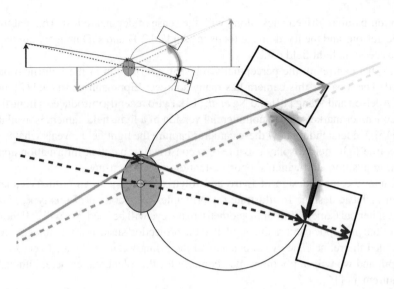

Figure 11.7 Ray tracings for two rays originating at the tip of the left arrow and two rays originating at the tip of the right arrow are followed as they come to focus at different depths in the upper small illustration. The left arrow is in focus on the retina, whereas the right arrow comes to focus behind the arrow. The enlarged illustration traces the rays as they are captured by the apertures, or miss them, in the case of the green dashed ray of two pinhole cameras located on the retina. (See color figure in color plate section).

The leftmost arrow is in focus; it is at the focal distance of the primary lens, which forms a sharp image of all points in the camera's field of view located at the lens' focal distance on the retina. The solid and dashed rays are both projected by the lens to the same pinhole camera location in the tiny camera array. The two rays are traced as they pass through the pinhole at this location and are projected onto the back surface of this tiny camera. The locations of the rays on the pinhole camera's back surface are correlated with the entrance pupil locations of the rays. These locations are correlated with phase differences in the rays traveling from the arrow's tip. This is the directional information that is lost in ordinary cameras. Most, or even all, of the rays projected onto this pinhole will have originated at the same point on the leftmost arrow's tip. The image projected on the back surface of this pinhole camera records each ray's location and phase in the plane of the pupil.

The dashed and green lines tracing the path of rays from the right arrow tip are projected to different locations on the pinhole array, because this arrow is not in focus. Different pinhole cameras will record the directional information contained in these two rays. The plenoptic camera nonetheless captures all of the directional information contained in these rays, so no directional information is lost. The plane of the pupil or camera aperture was characterized earlier as a collection of plenoptic functions each slightly displaced relative to the others within the aperture plane. The array of tiny cameras is sampling these plenoptic functions. Each unique location on the projection surface of this array of cameras corresponds to a single ray or heading direction from one of the plenoptic functions in the aperture plane. The plenoptic camera is effectively sampling all of these functions simultaneously.

Rearranging the data collected by the plenoptic camera allows the reconstruction of a 2D image that can be focused arbitrarily at a focal distance different from the primary lens focus of the plenoptic camera. The resolution of this reconstructed image is limited by the resolution of the micro camera array and the lens. A plenoptic imaging system capable of reconstructing a 2D image at an arbitrary depth cannot be a passive system; it requires image processing to work. The basic operating principles of this camera are similar to the system described by Lippmann [12].

In summary, rays originating at points in focus in the scene will be captured by a single pinhole camera in the array and rays originating from out-of-focus points will be captured by different pinhole cameras. Capturing the directional information is the key to capturing the light field in a small region in free space, i.e., where the plenoptic camera's entrance aperture is located.

Figure 11.8 illustrates what happens in a plenoptic camera when two points in 3D object space are located along the same direction from the eye, and one point is nearer to the eye than the other. This is illustrated with points on the base of the two arrows illustrated in Figure 11.7. These figures represent a top down view through the camera with the leftmost arrow pointing to the right from the camera's point of view and the right arrow pointing to the left. The latter arrow is out of focus and its base occludes some of the rays from a point on the base of the in-focus leftmost arrow.

The dashed lines in Figure 11.8 represent the rays from the base of the right arrow (ignoring for the moment the dashed lines continuing to the left from the base of the right arrow to the vertical line) that pass through the camera's entrance aperture and are projected onto the

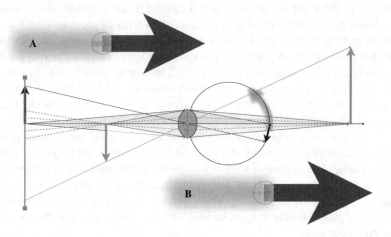

Figure 11.8   The middle of this diagram is a top down view of an eye that has focused on a green arrow pointing towards the observer's right. This arrow is partially occluded by an out of focus green arrow. In this diagram the purple line on the left represents a horizontal cross section of a light field display. The green dashed lines represent rays from the display surface that are discretely reconstructing the blur signal contained in the light field. Suppose the number of rays reconstructing the blur is sufficient to drive the eye's focus, which would mean that the observer viewing this display could focus at arbitrary depths within the viewing volume reconstructed by the display. Nonetheless the reconstructed blur might appear as a series of intense greenish spots as depicted on the retina in the insert labelled B in the lower right corner of the diagram or as overlapping small patches that merge smoothly into a blur as depicted on the left insert labelled A. The point here is that there might be a salient difference in the resolution required to drive the focus control mechanism of human vision and the number or rays required to produce good image quality in a light field display of the future. (See color figure in color plate section).

entrance apertures of five adjacent micro-cameras on the micro-camera array located on the camera's projection surface or retina. The micro-cameras are not shown in the diagram, but they would be located and centered at the locations where the dashed lines intersect the retina.

The spacing of the micro-cameras determines the spatial resolution of the captured image. In the insert in the upper left of this figure, labeled A, is an illustration of the image formed on the plenoptic camera's retina. The large circle within the base of the right arrow's out-of-focus image on the retinal surface is a large circle representing the blur from a single point at the base of this arrow. The five smaller circles within this circle represent the footprints of the pixels in the adjacent micro-camera array that are sampling the directional information from this point.

The micro-camera located on the optical axis of the primary lens will sample three rays, not shown in the diagram, from the base of the left arrow that is in-focus on the camera's retina. The left arrow occludes rays from the base of the right arrow that would pass through the left half of the primary lens. This data could be rearranged by image processing to render a 2D image reconstruction that changes the focus of the camera from the left arrow to the right one. In that case, the five samples from the base of the right arrow would be added together to form one pixel in the reconstructed image of this arrow that would now be in-focus. The three ray samples captured from the base of the left arrow by the micro-camera located on the optical axis of the plenoptic camera would be dispersed in this refocused and reconstructed 2D image as blur.

Now consider a different interpretation of the diagram in Figure 11.8. Suppose that the vertical line on the far left is the surface of a light field display viewed from above. In this interpretation of the diagram, the eye shaped camera is not a camera, it is a real eye. The display, represented by the vertical line, consists of an array of projectors, with each surface point reconstructing the plenoptic function at that location on the screen. The viewing volume, within which an observer can peer into a reconstructed light field, is located to the right of the display, where the eye is located in the diagram. The display must be able to reconstruct the plenoptic functions that would exist within this volume and within a finite volume behind the screen, so that when an eye is located at any point within the viewing volume, it can sample reconstructed plenoptic functions as if they were virtual.

There are at least two resolution requirements: first, for good image quality, the display element pitch much be good enough to reproduce the range of spatial frequencies that are considered acceptable. The window of visibility of the eye with regard to spatial details is bounded by several limits: brightness, contrast, and spatial frequency [17]. High-quality displays today are designed to produce approximately 30 line pairs of spatial information at the near viewing point. For a cell phone display that can be held very close to the eye, this means a dot pitch of over 200 pixels per inch.

In most viewing conditions, there is rarely sufficient contrast in the reconstructed signal to require more spatial information in the reconstructed imagery. Additional spatial resolution beyond this limit can nevertheless improve image quality when the spatial limit of the window of visibility has been exceeded. The added detail produces a spatial dither that can improve the tone scale rendering by effectively adding one or more bits of grayscale information to the reconstructed image. The ability to modulate the intensity of the picture elements over a range of at least three orders of magnitude, and the ability to produce a good black, are both critical determiners of subjective image quality.

The light field provides the signals that the nervous system uses to close the loop on focus, so these signals must be reconstructed, too. There is no reason to suppose that providing an adequate signal to drive focus will also provide an adequate signal to produce an acceptable level of image quality.

In this alternative interpretation of Figure 11.8, there is an object in front of the display – the right arrow – and one located in the plane of the display, the left arrow. The dashed lines represent rays projected from the light field display screen towards the eye viewing the display. In this example alone, in the horizontal extent of the display only, five light field picture elements project rays that pass through a point in space representing the base of the right arrow and additionally go through the eye's pupil, producing an image on the retina. If five rays must pass through the pupil on a diameter of the pupil to close the control loop for focus, then this would define the furthest distance from the screen that a viewer could view the display and focus within the viewing volume. A similar limit exists for the distance behind the screen.

When the eye is focused on the left arrow located on the light field display surface in this illustration, the five rays reconstructing the right arrow base must form a circle of blur on the observer's retina. Ideally, the rays within the circle of blur on the observer's retina would be overlapping and large, as in insert A in this figure. If, however, they are punctate non-overlapping and bright, as in insert B in Figure 11.8, then they might be sufficient to drive focus but insufficient to produce an adequate level of image quality. The optical requirements for these rays and the number of rays required to drive focus are unknown quantities today, because the appropriate experiments to determine them have not yet been performed.

Another aspect of the display that is illustrated in this diagram is the need for the display to sense the distance from the observer to the display screen. In this example, the right arrow is occluding the left arrow, so not all of the rays from the left arrow base are being rendered by the display. Suppose, however, that the observer were to move closer to the display. In this case, it might be possible to position the eye between the location of the virtual arrow reconstructed in front of the light field display screen and the screen. In a virtual reconstruction of the light field, it would be possible to physically pass through virtual objects reconstructed in front of the display. Once this has occurred, the occlusion created by the virtual object would vanish. The light field display would have to be able to sense the location of the observer relative to the display, and adjust the reconstruction imagery accordingly.

In s3D displays today, there is a clipping artifact that is called an edge violation in the entertainment video industry. In everyday visual experience, we perceive objects going behind occlusions, e.g., someone seen through a window walking past it inside a building. The disappearance of the person as they walk behind the wall is expected and ordinary. In stereo pair display reconstructions, especially in entertainment videos and cinema, it is commonplace to reconstruct objects and people in front of the display screen. When these objects and people move beyond the limits of the reconstructed image pairs on the right, left, top or bottom of the screen, the resulting appearance of an object or person being occluded without an obvious occluder is perceptually disruptive and odd.

The entertainment industry has adopted four workarounds and combinations of them to avoid these artifacts. Objects in front of the screen and near an edge can be rendered out of focus to reduce their salience in the scene and their physical contrast on the screen. In front of the screen, objects are rendered with a vignette or spatial apodization as they approach the screen edge. This also reduces the object's image contrast as it nears the edge. A floating window, or high contrast occluding edge, is created in one or both of the image pairs to produce a surface that, if it existed in the theatre, would occlude objects in front of the screen. A fourth workaround is to vary the flatness of the objects in various parts of the scene by changing the amount of image pair disparity – typically flattening objects in front of the screen and near edges, to reduce the perceptual salience of these artifacts.

Figure 11.9 illustrates the equivalent clipping artifact in a light field display reconstruction of a scene. In the top portion of this illustration, the upward-pointing and downward-pointing arrows are both fully rendered by rays produced by the light field display surface and that enter the obsever's pupil. The lower illustration shows the ray clipping that will occur if the imagery and observer are both translated to the left (recall that this is a top-down illustration). Now there is no display surface to reconstruct the tip of the downward-pointing arrow. It simply vanishes from view. This is essentially the more general case of the edge violation in today's s3D reconstruction. Whether or not similar workarounds to those used by the entertainment video industry today to mitigate this kind of artifact is another subject for future research and development.

In video communication systems, where imagery is captured or constructed by a computer graphics system, it is possible to manipulate the scale and perspective of objects rendered on the display screen. One can only speculate about how these parameters will impact the appearance of imagery reconstructed on a light field display. Another phenomenon unique to today's s3D imaging technology is the cardboard effect. Watching a soccer game on a s3D

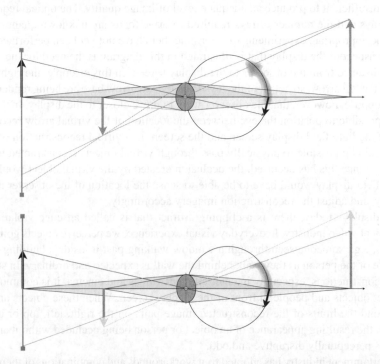

Figure 11.9 The top diagram in this illustration shows side view of a light field display (vertical lines on the extreme left) reconstructing an upward-pointing arrow that is in focus and a downward-pointing arrow partially occluding it and out of focus. Rays from the bottom (in the diagram) of the screen are necessary to render the blurred tip of the out-of-focus downward-pointing arrow. If the image and observer were to shift downward relative to the display screen, then a more general form of an edge violation will occur in this rendering of the light field and, just as in s3D displays, it is likely to appear very strange to an observer as the object that would be occluding the downward-pointing arrow's tip, which would logically be between the observer and the arrow is missing. (See color figure in color plate section).

display can make the players look like miniature people who are flat when viewed from a wide angle long shot from the top of a stadium. What is causing these perceptual phenomena is currently a research topic. It may be due to scaling issues created by exaggerating disparities in scenes with wide depths of field and sharp focus, but what exactly causes this perceptual artifact is yet to be determined. Nonetheless, one can expect similar issues to emerge as light field display and light field capture become part of the tool set for video communications.

The same spatial and temporal resolution requirements that apply to ordinary 2D displays will apply to a light field display. At half a meter viewing distance, a 100-dpi display produces about 15 line pairs per degree of visual angle, adequate for many display tasks at this viewing distance. For a handheld display that will be viewed closer to the eye, 200 dpi or more is appropriate. Temporal-resolution requirements are the same as in current displays. Avoiding or controlling temporal artifacts such as flicker, judder, motion blur, and a recently documented temporal artifact in s3D imaging [18], must be considered to determine the frame rates required for any specific task, especially if interlacing becomes part of the reconstruction architecture of these devices.

11.4 Towards "True" 3D Visual Displays

To summarize the discussion of the proceeding sections, the requirements for the optical system of a true 3D display can be understood by thinking of the display as a plane of plenoptic function generators. One way to visualize this is to consider the display system as a window. Consider a window to be divided into small area patches – so small that if we were to block off the entire window except a given patch, we would only see a color and intensity, as it is too small to see image detail. Considering the descriptions in the previous sections, the window can be considered to be an array of pinholes, where the light coming from each pinhole is half of the plenoptic function for the point located at the pinhole. In our depiction of the display system, each of these window patches will correspond to a pixel. For example, consider Figure 11.10, showing the light passing through a very small patch of a window near its center.

The color and intensity of light that comes through the small patch depends on its angle. From the angle shown in Figure 11.10 going to the viewer's eye, the viewer sees the color of

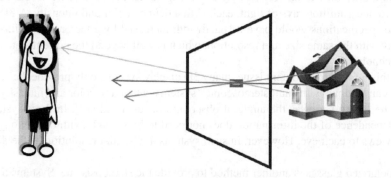

Figure 11.10 Depiction of the light rays going through one "pixel" of a window coming from different points of a toy house behind the window. To the viewer, in the position shown, the pixel will appear to be the color of the walls. (See color figure in color plate section).

the walls, but at other angles the viewer would see the color of the roof or the color of one of the windows. So, from each patch on the window, there is a bundle of rays emerging that is characterized by an angle, color, and intensity.

Thinking of the display as a window, we only consider displayed objects that are behind the display screen. As shown in the previous sections, it is possible to display objects in front of the display screen as well. However, in this case, the display cannot be considered to be a plane of points where the plenoptic function is uniquely defined. For objects in front of the screen, as pointed out earlier, the display would need to adjust its output, taking into account the location of the observer relative to the display.

The window analogy also points out that for some cases, the display needs to be physically large to provide a "true" life-like 3D image. For example, in Figure 11.10 the toy house is shown to be smaller than the display screen. If a real house were behind the display screen, a distance roughly equal to the size of the window, it would obviously only be possible to see a very small portion of the house.

From the window analogy for the display, it can be considered that the difference between a 3D and 2D display is the angular dependence of the pixel (plenoptic function) information. This angular dependent information conveys three aspects of the 3D scene:

1. Relative motion of objects (an eye sees a different view when it is moved).
2. Stereopsis (each eye sees a different view).
3. Focus (the angular spread of rays, intercepted by the pupil from a point in the scene, is determined by its distance from the viewer).

In adding these characteristics to a 2D display, it is also important that the resolution of the display remains high – near the limiting resolution of the eye – as the textural cues are important in perception of depth and "realism" of the image. If we consider the "ultimate" 3D display as one that can emulate a window as described above, we need to have a display with a high density of pixels that change color and intensity for different directions of view with very high angular resolution. If we say the eye collects a cone of rays spanning several tenths of a degree, and we would like have an adequate angular resolution to achieve a proper focus, it is likely that we would need a 0.1 degree angular resolution. If we then consider that we would like the window to be viewed over a 100 degree field of regard, we would need each pixel to provide 1 million rays of light, each with a defined color and intensity. A 3D display with these specifications would have its bandwidth increased by a factor of 1 million over a 2D display with the same size and resolution, which is well beyond the current state of liquid crystal technology.

Due to this information content issue, many 3D display systems only provide the stereopsis cue. This can be done with "autostereoscopic" systems, where the color and intensity of each pixel on the screen varies as the angle of observation is changed [19]. In these systems, the angular dependence of the information does not need to be so high, with only enough to get separate views to each eye. However, in such systems, the spatial resolution of the display is lowered.

Using polarized glasses is another method to provide the stereopsis cue. Systems using this approach can be divided into those that use "active" glasses and those that use "passive" glasses [20]. Passive glasses, however, have the advantages of light weight and being able to provide improved brightness, through the use of an active shutter as shown in Figure 11.11.

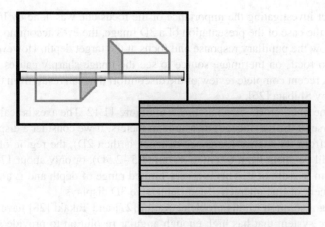

Figure 11.11 A display screen in the background and a polarization rotator panel in the foreground. The image is being rastered downwards, changing from the left eye view to the right eye view. The polarization rotator is designed to output light of one polarization state, that is transmitted by the left lens of the glasses (not shown) worn by the viewer, and another polarization state that is transmitted by the right lens.

The figure shows a display screen that field-sequentially shows the left and right eye view images, at the instant in time where the scanning of the display is halfway down the screen, erasing the previous right eye's view and writing the left eye view. In synchronism with this, the segmented active shutter, which controls the polarization state of light, changes its segments from passing the polarization sate that is transmitted by the right eye to that which is transmitted by the left eye [21].

In the case where the display screen is far from the viewer, the first and third cues listed above may be less important and ignorable (as in the case of a movie theater), and only the stereopsis is needed to provide a 3D image. However, this simplification may not work for scenes close to the viewer, such as would be portrayed on a desktop or mobile display. In this case, both the cues of relative motion and focus could be significant, especially the relative motion cue. The significance of the effect of relative motion is most easily understood by noting how the 3D information in a scene can seemingly jump out at the viewer if the viewer location is in motion. An excellent video that demonstrates this effect has been made by Lee [22].

Taking into account the relative motion cue, but not attempting to take into account the focus cue, reduces the requirement on the angular resolution to that which is required to produce stereopsis and smooth motion. The angular resolution, in this case, might be on the order of one degree, and can further be considered to be limited to only the horizontal direction. Here, the bandwidth of a 3D display is increased by a factor of "only" 100 over a 2D version. However, if a single viewer is considered and head tracking is used, systems with only two views are required.

Systems using passive glasses and head tracking have been demonstrated by zSpace [23], as well as autostereoscopic systems by SuperD [24]. Systems of this type that provide both the relative motion and stereopsis cues are quite effective. However for a "truer" 3D display, the focus cue should be added. While the need to include the stereopsis cue and relative motion cues are recognized as being essential to true three-dimensional displays, the importance of the focus cue may be less obvious.

An early paper investigating the importance of the focus cue was done by Inoue [25], who showed that, in the case of the presentation of a 3D image, the eyes accommodative response attempts to follow the pupillary response and focus at the target depth. However, the fact that the eyes need to focus on the image source to see the image sharply causes a conflict with this response. A recent complete review of the discomfort that can result from this conflict has been provided by Shibata [26].

Shibata's "zone of comfort" graph is shown in Figure 11.12. The axes is scaled in Diopters, that is, the inverse of the focal length measured in meters. If we consider a display screen that is half a meter from the viewer (vergence distance is then 2D), the region of comfort for a stereo image will be about from 67 cm to 40 cm (1.5–2.5D), or only about 17 cm behind the screen and 10 cm in front of it. This is a very limited range of depth and is a major issue for future development of true interactive and immersive 3D displays.

In considering a solution to this problem, Kajiki [27] and Takaki [28] have shown that an autostereoscopic system that has high enough angular resolution to provide several distinct views to the eye's pupil can cause the correct focus response. Figure 11.13 shows the idea that diverging rays from a 3D object are brought into focus on the retina by the appropriate focus response of the eye.

Takaki further showed a relation between the angular resolution of distinct rays intercepted by the pupil, and the depth range of the correct accommodative response of the eye [29]. One issue with the above approach is that the angular spread of rays is only along a horizontal direction and, therefore, may be considered to cause an astigmatism issue. One approach to alleviating this problem has been shown by Kim, who used slanted rays of light [30]. Figure 11.14 shows the idea where two rays enter the pupil along a slanted plane. The effect of having two or four rays enter the pupil is shown in Figure 11.15, where the focus is on one of three different objects.

A more complete solution may come from the use of integral imaging, where rays are emitted along all angles. A recent review of this approach has been completed by Xiao [31]. Figure 11.16 shows a focusing effect in these type of systems.

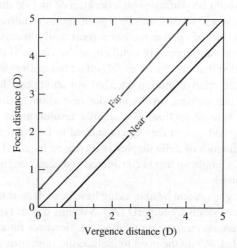

Figure 11.12   Zone of comfort is between the lines marked "Far" and "Near", where the convergence distance and focus distance are similar. Adapted from T. Shibata, J. Kim, D. Hoffman, M. Banks, 2011.

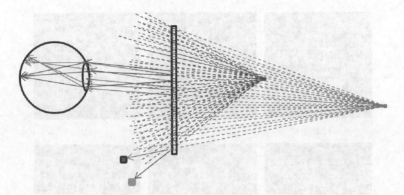

Figure 11.13 Accommodation with high-density of horizontal parallax images. Shown are two objects (dots) to be presented to the viewer by the display screen for a top view of the display screen that here has about 25 pixels. Each dashed line corresponds to the light ray from one of the dots through each pixel of the display. For the uppermost small detector, the bottom pixel on the display would appear the colour of the object further away, while for the lower detector it would appear the colour of the nearer object. For other detector positions, light from the dots would not be detected by those detectors looking at the bottom pixel. Rays from the screen related to the closer dot are more diverging than those from the more distant dot because of their relative distance behind the screen. The eye is shown focusing the rays from the closer dot on the retina. For this case the rays from the more distant dot would not be in focus. (See color figure in color plate section).

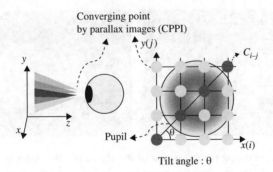

Figure 11.14 Intersection of rays of light from a common point with the pupil. Source: S-K Kim, S-H Kim, D-W Kim 2011. Reproduced with permission from SPIE. (See color figure in color plate section).

We have seen that including the focus cue with the above approaches requires that multiple rays enter the pupil of each eye. This added information can cause the bandwidth of the display to become very high, as mentioned earlier.

Therefore, to limit the bandwidth requirement of the display, our window needs to be smart enough only to direct light to the eyes of the viewer, instead of sending out rays to all directions. An eye tracking system could be used to determine the location of viewers' eyes, and each pixel of the display then only needs to be able to provide light of the colors and intensities corresponding to the angle of the rays leaving the pixel and headed for each eye (as shown in Figure 11.13). For example, if three rays are needed to be directed toward each pupil for it to focus properly, then a six-view system could be sufficient.

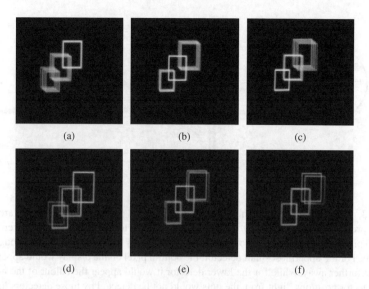

(a)　　　　　　　　(b)　　　　　　　　(c)

(d)　　　　　　　　(e)　　　　　　　　(f)

Figure 11.15 A comparison of the defocus effect in the case of using four rays from common points (top row), and with two rays from common points (bottom row), focusing on objects at: 0.25m (a) and (d); 0.6m (b) and (e); 1.8m (c) and (f). Source: S-K Kim, S-H Kim, D-W Kim 2011. Reproduced with permission from SPIE.

Figure 11.16 A computational reconstruction of the focusing effect for a type of integral imaging system. The picture at the left is where the closer car is in focus, and the picture on the right is where the further truck is in focus. Source: X. Xiao, B. Javidi, M. Martinex-Corral, A. Stern 2013. Reproduced with permission from the Optical Society of America.

An autostereoscopic system that uses head tracking to allow a high angular resolution of rays to be sent to the viewer, while maintaining a relatively high image resolution, has been shown by Nakamura *et al.* [32]. Another approach, where an array of lenses is placed on the display screen, has also been proposed [33]. Related to the idea of sending several light rays to each pupil is the idea of sending a non-planar wavefront to the pupil, as provided by a hologram. However, the bandwidth and technical problems associated with accomplishing this for a multiviewer display prevents high-resolution video rate devices that use this approach. A book chapter from Reichelt *et al.* makes clear these issues and demonstrates a solution that involves sending the holographic information only to areas in space where the viewers' eyes are located, as illustrated in Figure 11.17 [34].

While the above approaches are conceptually the most appealing, it is difficult to achieve both the desired image resolution and a sufficient number of rays to each eye to cause the focus cue. Another approach to the accommodation convergence problem is to consider an additional lens between the display and the viewer. Yanagisawa developed and analyzed a prototype system based on a display with an adjustable focus lens [35]. These concepts have been investigated in detail by Shibata [36].

These considerations lead towards solving the focus problem with a combination of the advantages of a volumetric display and a stereoscopic display. The advantage of the volumetric display is that it portrays focus naturally, but it has the disadvantage of not being able to handle occluded images well, and the bandwidth of the system is proportional to the number of depth planes. Love has proposed a system where a lens is placed in front of a flat panel stereoscopic display to provide field sequential focus planes of the displayed stereo images, as depicted in Figure 11.18 [37]. The field sequential approach requires the refresh rate to be multiplied by the number of focus depth planes. However, this is not as large a problem as for the case of a typical volumetric display, because the number of focus planes may be not be very large for an acceptable display. The number of depth planes where the eyes can converge is not limited as it is in a purely volumetric display.

Another approach to address the accommodation problem is a "fix" suggested by Bos to use multi-focal length lenses (like bifocal or progressive lenses), worn by the user [38]. The multi-focal length lens is used to allow the user's eye to have a focal length consistent with the target location of the 3D object, while the focal length of the user's eyes, combined with the corrective lens, would allow the image to be in focus on the retina. A practical implementation of this approach would be to use eye-tracking to allow the display system to measure the toe-in of the user's pupils to determine the depth of the user's gaze and, coupled with knowledge of the distance of the user from the display screen, to adjust the power of an electronic lens worn by the user.

As described in the chapter by Drewes (Chapter 8), advances are being made in eye gaze tracking technologies that are ushering in low-cost eye-tracker systems, which could be utilized for this application. The advantage of this approach over the others considered here is that no additional bandwidth or compromise of the image resolution is required from the display system over a conventional stereoscopic display. To be more specific about this approach, we

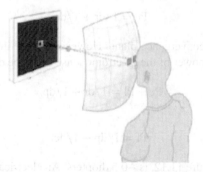

Figure 11.17 A method to decrease the information and diffraction angle requirements where holographic information is only directed toward the viewer's eyes. Source: S. Reichelt, R. Haussler, N. Leister, G. Futterer, H. Stolle, A. Schwerdtner 2010. Reproduced with permission of SeeReal Technologies.

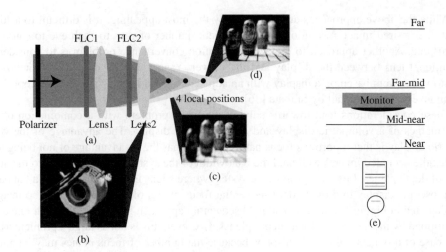

Figure 11.18 A multi-focal length lens on the left with an overview of the eye, lens, and display screen on the right. The four focal positions on the left correspond to the four depth planes on the right [37]. Source: G. Love, D. Hoffman, P. Hands, J. Gao, A. Kirby, M. Banks 2009. Reproduced with permission from the Optical Society of America.

can consider a "passive glasses" stereoscopic system with head tracking described above. The idea of the system is shown in Figure 11.19 [39].

If the corrective lenses are adjusted properly, we can have the focal power of the eyes to be consistent with focusing at the convergence plane, while at the same time have the image of the display panel to be in focus on the retina.

As mentioned above, detection of the "toe-in" of the viewer's pupils will allow the computer to know the convergence point of the eyes and, therefore, the distance to the object in the 3D scene being viewed. With the additional information of the distance of the viewer from the display screen, we can determine the power needed for the corrective lens. If we call the distances from the eye's lens to the panel "dp", to the convergence point "dc", and to the retina "dr", then we would need the power of the eye's lens to be:

$$Pe = 1/dr + 1/dc \qquad (11.1)$$

for it to be focused at the convergence plane. However, for the image to actually be focused on the retina, we need the power of the electronic lens "Pl" to be determined by:

$$Pe + Pl = 1/dr + 1/dp. \qquad (11.2)$$

This leads to:

$$Pl = 1/dp - 1/dc \qquad (11.3)$$

which, in the case of th Figure 11.12, is −0.5 diopters. An electrically controllable lens, applicable to this type of system, has been presented [40].

Figure 11.20 shows a picture of the lens that is approximately 1 cm in diameter. The ability of this electrically controllable lens to shift the focus point of an object is illustrated in

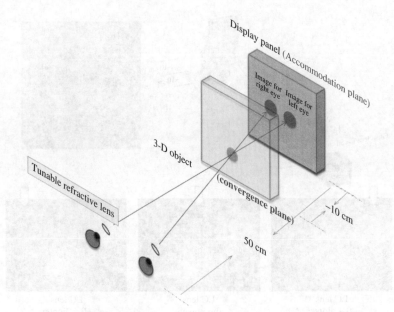

Figure 11.19 The use of tunable lenses near the eye to cause the eye to accommodate at the convergence plane, while the combination of the focusing power of the eye and the tunable lens causes the screen to be in focus on the retina. Source: P. J. Bos and A. K. Bhowmik 2010. Reproduced with permission from SID/Wiley.

Figure 11.20 A tunable liquid crystal based lens similar to that used in Figure 11.19.

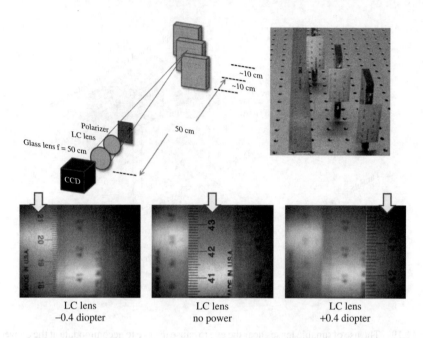

LC lens
−0.4 diopter

LC lens
no power

LC lens
+0.4 diopter

Figure 11.21 The lower three pictures are taken using the set up shown in the upper pictures, for three different sets of voltages applied to the tunable liquid crystal lens to provide the powers shown.

Figure 11.21, which shows that if we consider our eye to be accommodated to a 50 cm distance, the actual object can be between 40–60 cm from this lens. This demonstrates that if the display was at a fixed distance of 50 cm, the electronic lens could be used to cause the eye to refocus in the same way it would be required to if the display were between 40–60 cm.

This type of simple system could be useful in achieving 3D stereoscopic display systems that reduce the discrepancy between accommodation and convergence and, as a result, to reduce eye fatigue. However, it should be realized that this corrector-based approach is a "fix", as opposed to the more natural approaches of high resolution super-multi-view, integral imaging devices, holographic methods as presented by See-Real, or the volumetric display concept of Love. For example, one issue with this approach is that while the accommodation convergence ratio mismatch is addressed here, the method has the effect of unnaturally putting the entire display screen in focus at the depth of the particular aspect of the image being viewed. Artificially blurring images that are not at the depth of the object being viewed could alleviate this problem.

11.5 Interacting with Visual Content on a 3D Display

In the previous sections, we have discussed the fundamentals of 3D visual perception, 3D visual information capture, and the requirements of a "true" 3D display with an objective to deliver life-like visual experiences to the user. We then reviewed the progress in technologies towards achieving these goals. Here, we consider human inputs and interactions with the content displayed on such systems.

Historically, the development of 3D display systems focused on one primary application–presenting 3D photos and videos to the viewers to produce a depth perception. In recent years, displays have increasingly been adding touch sensing capabilities, especially on mobile communications and entertainment devices, thereby also becoming the primary human interaction interfaces. In addition, advances in real-time 3D imaging and computer vision techniques are starting to enable the implementation of user interactions in the 3D space in front of the display [41]. These developments promise to bring about a revolutionary change in human-computer interactions by making it possible to manipulate objects in the 3D environment directly and intuitively. Empirical studies have shown that when people are presented with 3D imagery in the 3D space in front of them, the natural tendency is to reach out and interact with them with their fingers [42]. This is not surprising, since we are used to doing so in the real world in our daily activities.

As we have discussed in previous sections in this chapter, a "true" 3D display must go beyond just presenting a stereo-pair of images for a scene being displayed (stereopsis cue). It must also provide seamlessly changing views corresponding to the movement of the head and the eyes of the viewer (motion parallax cue). In addition, the convergence of the eyes on the rendered object must be consistent with the focus of the lenses in the eyes (focus cue). While these are important requirements to produce a natural 3D viewing experience, they become even more critical for interactive applications where a user would reach out to manipulate a virtually rendered object in the 3D space with real body parts, such as the hands and the fingers.

Interacting with virtual objects in the 3D space with real body parts, e.g., "touching" with fingers or "grabbing" with hands, is limited to objects rendered with negative parallax so that they float in the space between the user and the display surface. Similar direct 3D interactions with the objects rendered with positive parallax, which float in the virtual space behind the plane of the display, are not possible, as the physical display surface would prevent reaching out through it. So, interactions with objects in these environments would require doing so with virtual body parts, such as hands rendered in the virtual environment and driven by the motions dynamically captured from the real hands of the users.

Let us consider the case of negative parallax, where an object is rendered between the user and the plane of the display. Although this allows physical interactions with the virtual objects rendered by providing the stereopsis cue, both motion parallax and focus cues become important because of the proximity of the user to the displayed content. As the user reaches out and grabs a rendered object floating in space, his or her visual system has to see both the virtual object and the real hand simultaneously. Absence of motion parallax cue in such cases, where the views of the displayed object on the retina remain fixed despite movements of the head and the eyes of the user, would cause confusion and discomfort, as the real hand obviously produces continuously varying images on the retina in coordination with the user's movements. Similarly, the absence of focus cue would create significant convergence-accommodation mismatches, so therefore either the object or the fingers will appear blurred, as both cannot be in focus at the same time. Since our perceptual system makes use of all these visual cues to understand the physical world and guide our interaction processes, the visual conflicts arising from the absence of these cues would impact our performance as we interface and interact with content displayed on a 3D display. A solution to the problem is to avoid bringing real and virtual objects in the view at the same time, and to use a virtually rendered equivalent of the hand for interactions with virtual objects.

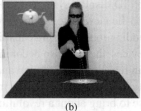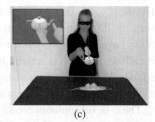

(a)   (b)   (c)

Figure 11.22 Illustration of visual conflicts during 3D selection of stereoscopically displayed objects: (a) The user is focused on her finger, with the virtual object appearing blurred. (b) Displaying a virtual offset cursor (white marker) at a fixed distance from the real fingertip reduces visual conflicts. (c) A virtual offset hand cursor provides familiar and additional size and distance cues for selection. Source: Bruder, Steinicke, Stuerzlinger 2013. Reproduced with permission from IEEE. (See color figure in color plate section).

The impact of these visual conflicts on the performance of the user, and the efficacy of the solution mentioned above, were evaluated with a Fitt's Law experiment by Bruder *et al.* [43]. Figure 11.22 illustrates the experiments and the key results from Bruder's work. Interestingly, as intuition would suggest, people were most efficient when using their real hands for interacting with the virtual objects rendered on a stereoscopic 3D display, but made more errors when compared to interacting with the same environments with virtual hands.

Using 3D visual displays for real-time interactive applications imposes additional requirements from the human factors perspective, besides the 3D visual cues described above. These include the need for fast response, such that there are no perceptible delays or lags between the human actions and the resulting system responses that lead to changes in the visual content.

The effects of inconsistencies between user actions and visual responses due to system lags have been studied and reported by La Viola [44]. Stuerzlinger *et al.* proposed limiting the degrees of freedom for user interactions within a virtual environment to reduce the failures and inconsistencies [45]. In addition, the fields of view for both the visual display and the real-time image capture device for human inputs are also a critical factor for ensuring good user experience. For example, it would be a frustrating experience for a user interacting with the content on a 3D display equipped with 3D gesture input capabilities, if the limited viewing angle of the 3D imaging device restricts user interactions to only a small active zone.

The user interface designs for interactive applications on systems and devices incorporating such 3D visual displays and 3D user input technologies would have to carefully consider and take into account human factors aspects of our interactions with the physical world, as well as the limitations with of the technologies and systems described above. For example, unlike a traditional 2D display with a touch-sensitive overlay, interacting with virtual objects in front of the display would provide no tactile feedback when touched. Therefore, the user interface designer needs to utilize other means of providing real-time feedback to the user, such as audio-visual cues that are appropriately designed in the context of the specific application and dependent on the extent of interactions. For example, squeezing a virtual ball would need to result in appropriate deformations of the ball, conveyed via changes in the shape and color that are consistent with the forces applied indicated by the motions of the fingers.

In addition, careful use of audio cues would also make the interaction "life-like", such as squeaking of the deforming ball in response and proportional to the squeezing action.

Similarly, carefully designed audio-visual cues can provide guidance to the user when the boundaries of the interaction zones are reached due to limited viewing angles.

In addition to 3D vision-based gestural interactions, multimodal user interfaces, including voice as an input as elaborated in the chapter by Breen *et al.* (Chapter 3) and eye gaze tracking technologies and interactions described in the chapter by Drewes (Chapter 8), can make the user interactions with content on a 3D display a more immersive and engaging experience. The considerations to make such interactions easy and intuitive for the user have been discussed in detail by LaViola *et al.* in the chapter on multimodal interfaces (Chapter 9).

11.6   Summary

While we use all our senses to perceive and interact with the physical world, vision plays the most important role. Visual displays have become an indispensable element of electronic devices of all form factors, ranging from the "small" displays on watches worn on the hand, to "medium" displays on smartphones, tablets, or laptop computers, to "large" displays on television sets or information kiosks. Visual displays have come a long way in recent years, with significantly improved visual qualities such as brightness, contrast, speed, and color performance. In addition, stereoscopic displays that can show visual content with stereopsis cues to the viewer have already entered the mainstream market. However, a "true" 3D display that would provide life-like immersive visual experience also needs to provide motion parallax and focus cues. In this chapter, we have considered the fundamentals of human vision and 3D visual perception to articulate the requirements of a "true" 3D display, and have reviewed the technologies required to realize such systems and their development status.

In recent years, displays have also been becoming highly interactive. Mobile displays have already become the primary human interface systems on mobile devices, by adding a touch-sensitive layer and associated user interface. As described in the chapter on touch technologies, touch-based user interfaces continue to be rapidly adopted in a wide range of devices and systems. The subsequent chapters presented the developments in voice and vision-based interfaces. The advances in speech recognition algorithms, 3D imaging and inference, and eye-gaze tracking technologies, are poised to bring about a revolutionary change to human interactions with the content on displays. These new human interface and interaction technologies, added to displays that will be capable of presenting "true" 3D visual content along with appropriately designed multimodal user interface schemes, promise to deliver life-like interaction experiences in the 3D space in front of the displays.

It is clear that the era of "interactive displays" is here. In the coming years and decades, we expect to see further proliferation of interactive displays into a broad range of devices and systems. This is a new area that requires further research and development, but it holds significant potential to bring about exciting new interactive applications and user experiences.

References

1. Watson, A.B., Ahumada, A.J., Farrell, J.E. (1986). Window of visibility: psychophysical theory of fidelity in time-sampled visual motion displays. *J. Opt. Soc. Am* **3**, 300–307.
2. Adams, A. (1948). *Camera and Lens: The Creative Approach*. ISBN 0-8212-0716-4.
   Adams, A. (1950). *The Print: Contact Printing and Enlarging*. ISBN 0-8212-0718-0. http://en.wikipedia.org /wiki/O_Brother,_Where_Art_Thou%3F#cite_note-CGS-7.

3. Shibata, T., Kim, J., Hoffman, D.M., Banks, M.S. (2011). The zone of comfort: Predicting visual discomfort with stereo displays. *J. Vis.* **11**(8), 11, 1–29.

    Banks, M.S., Akeley, K., Hoffman, D.M., Girshick, A.R. (2008). Consequences of incorrect focus cues in stereo displays. *Information Display* **7**/8, 10–14.

4. Needham, J. (1986). *Science and Civilization in China: Volume 4, Physics and Physical Technology, Part 1, Physics*. Caves Books, Ltd, Taipei.

    Richter, J.P. (ed.) (1970). *Aristotle, Problems, Book XV*. The notebooks of Leonardo da Vinci. Dover, New York.

5. Land, M.F., Nilsson, D.-E. (2001). *Animal Eyes*. Oxford University Press. ISBN 0-19-850968-5.

6. Faraday, M. (1846). Thoughts on Ray Vibrations. *Philosophical Magazine* S.3, Vol **XXVIII**, N188.

7. Gershun, A. (1936). *The Light Field*. Moscow. Translated by Moon, P and Timoshenko G. in *Journal of Mathematics and Physics* 1939 **XVIII**, 51–151.

8. Adelson, E.H., Bergen, J.R. (1991). The plenoptic function and the elements of early vision, In Landy, M., Movshon, J.A. (eds.) *Computation Models of Visual Processing*, 3–20. MIT Press, Cambridge.

9. Gibson, J.J. (1966). *The Senses Considered as Perceptual Systems*. Houghton Mifflin, Boston. ISBN 0-313-23961-4.

    Gibson, J.J. (1977). The Theory of Affordances (pp. 67-82). In Shaw, R., Bransford, J. (Eds.). *Perceiving, Acting, and Knowing: Toward an Ecological Psychology*. Lawrence Erlbaum, Hillsdale, NJ.

10. Dawkins, R. (2006). *The Selfish Gene: 30th Anniversary Edition*. Oxford University Press. ISBN 0-19-929114-4.

11. Held, R.T., Cooper, E.A., Banks, M.S. (2012). Blur and Disparity Are Complementary Cues to Depth. *Current Biology* **22**, 1–6.

    Held, R.T., Cooper, E.A., O'Brien, J.F., Banks, M.S. (2010). Using blur to affect perceived distance and size. *ACM Transactions on Graphics* **29**, 1–16.

12. Lippmann, G. (1908). Epreuves reversible donnant la sensation du relief. *J. de Physique* **7**, 821–825.

    Ives, H.E. (1930). Parallax panoramagrams made possible with a large diameter lens. *JOSA* **20**, 332–342.

13. Adelson, T., Wang, J.Y.A. (1992). Single lens stereo with a plenoptic camera. *IEEE Transactions on Pattern Analysis and Machine Intelligence* **14**(2), 99–106.

14. Ng, R., Levoy, M., Brédif, M., Duval, G., Horowitz, M., Hanrahan, P. (2005). *Light Field Photography with a Hand-held Plenoptic Camera*. Stanford Tech Report CTSR 2005-02.

15. www.lytro.com, www.raytrix.de

16. Xiao, X., Javidi, B., Martinez-Corral, M., Stern, A. (2013). Advances in three-dimensional integral imaging: sensing, display, and applications. *Appl. Optics* **52**(4), 546–560.

17. van Nes, F.L., Bouman, M.A. (1967). Spatial Modulation Transfer in the Human Eye. *JOSA* **57**(3), 401–406.

18. Hoffman, D.M., Darasev, V.I., Banks, M.S. (2011). Temporal presentation protocols in stereoscopic displays: Flicker visibility, perceived motion, and perceived depth. *JSID* **19**(3), 255–281.

19. Dodgson, N. (2005). Autostereoscopic 3D displays. *Computer* **31**(August).

20. Kim, J.H. (2010). Evolving Technologies for LCD Based 3-D Entertainment. *Information Display* **9**, 8.

21. Bos, P.J. (1993). *Stereo Computer Graphics and Other True 3D Technologies*, Chapter 6. McAllister D. (Ed.). Princeton University Press.

22. Lee, J. (2007). *Head Tracking for Desktop VR Displays using the Wii Remote*. http://www.youtube.com/watch?v=Jd3-eiid-Uw

23. http://zspace.com/

24. http://www.superd3d.com/

25. Inoue, T., Ohzu, H. (1997). Accomodative responses to stereoscopic three-dimentional display. *Applied Optics* **36**, 4509.

26. Shibata, T., Kim, J., Hoffman, D., Banks, M. (2011). The zone of comfort: Predicting visual discomfort with stereo displays. *Journal of Vision* **11**, 1.

27. Kajiki, T., Yoshikawa, H., Honda, T. (1996). Ocular Accommodation by Super Multi-View stereogram and 45-view Stereoscopic Display. *Proceedings for the third international display workshops* (*IDW'96*), **2**, 489.

28. Takaki, Y. (2002). Universal Stereoscopic Display using 64 LCDs. *Proc. 2nd International Meeting of Information Display*, 289, Daegu, Korea.

29. Takaki, Y., Kikuta, K. (2006). 3D Images with Enhanced DOF produced by 128-Directional Display. *Proc. IDW '06*, 1909.

30. Kim, S.-K., Kim, S.-H., Kim, D.-W. (2011). Full parallax multifocus three-dimensional display using a slanted light source array. *Optical Engineering* **50**, 114001.

31. Xiao, X., Javidi, B., Martinex-Corral, M., Stern, A. (2013). Advances in three dimensional integral imaging: sensing, display, and applications. *Applied Optics* **52**, 546.

32. Nakamura, J., Takahashi, T., Chen, C.-W., Huang, Y.-P., Takaki, Y. (2012). Analysis of longitudinal viewing freedom of reduced-view super multi-view display and increased longitudinal viewing freedom using eye-tracking technique. *Journal of the SID* **20**, 228.

33. Hong, Q., Wu, T., Lu, R., Wu, S.-T. (2007). Reduced Aberration Tunable Focus Liquid Crystal Lenses for 3D displays. *SID Symposium Digest* **38**, 496.

34. Reichelt, S., Haussler, R., Leister, N., Futterer, G., Stolle, H., Schwerdtner, A. (2010). Holographic 3D displays – Electro-holography with the Grasp of Commercialization. In Costa, N., Cartaxo, A. (eds). *Advances in Lasers and Electro Optics*, Chapter 29. INTECH.

35. Yanagisawa, N. *et al.* (1995). A focus distance controlled 3D television. *The journal of three dimensional images* **9**, 14.

36. Shibata, T., Kawai, T., Ohta, K., Otsuki, M., Miyake, N., Yoshihara, Y., Iwasaki, T. (2005). Stereoscopic 3D display with optical correction for the reduction of the discrepancy between accommodation and convergence. *JSID* **13**, 665.

37. Love, G., Hoffman, D., Hands, P., Gao, J., Kirby, A., Banks, M. (2009). High speed switchable lens enables the development of a volumetric stereoscopic display. *Optics Express* **17**, 15716.

38. Bos, K. (1998). Reducing the accommodation and convergence difference in stereoscopic three-dimensional displays by using correction lenses. *Optical Engineering* **37**, 1078.

39. Bos, P.J., Bhowmik, A.K. (2011). Liquid-Crystal Technology Advances toward Future True 3-D Flat-Panel Displays. *Inf. Display* **27**, 6.

40. Li, L., Bryant, D., van Heugten, T., Duston, D., Bos, P. (2013). Near-diffraction limited tunable liquid crystal lens with simplified design. *Optical Engineering* **52**, 035007-1.

    Li, L., Bryant, D., van Heugten, T., Bos, P. (2013). Physical limitations and fundamental factors affecting performance of liquid crystal tunable lenses with concentric electrode rings. *Applied Optics* **52**, 1978.

    Li, L., Bryant, D., van Heugten, T., Bos, P. (2013). Near Diffraction limited and low haze electrooptical tunable liquid crystal lens with floating electrodes. *Optics Express* **21**, 8371.

41. Bhowmik, A.K. (2013). Natural and Intuitive User Interfaces with Perceptual Computing Technologies. *Inf. Display* **29**, 6.

42. Grossman, T., Wigdor, D., Balakrishnan, R. (2004). Multi-finger gestural interaction with 3D volumetric displays. *Proceedings of the 17th annual ACM symposium on User interface software and technology*, 61–70.

43. Bruder, G., Steinicke, F., Stuerzlinger, W. (2013). Effects of Visual Conflicts on 3D Selection Task Performance in Stereoscopic Display Environments. *Proceedings of IEEE Symposium on 3D User Interfaces (3DUI)*. IEEE Press.

44. La Viola, J. (2000). A discussion of cybersickness in virtual environments. *SIGCHI Bulletin* **32**, 47–56.

45. Stuerzlinger, W., Wingrave, C.A. (2011). The Value of Constraints for 3D User Interfaces. In Brunnett, G., Coquillart, S., Welch, G. (eds). *Virtual Realities*, 203–223.

Index

Interactive Displays: Natural Human-Interface Technologies, First Edition. Edited by Achintya K. Bhowmik.
© 2015 John Wiley & Sons, Ltd. Published 2015 by John Wiley & Sons, Ltd.